Preparing for Sustainability Audits

This book is a practical guide to be used by people with little or no knowledge of sustainability as well as sustainability professionals and experts who intend to prepare software companies to undergo sustainability audits and assessments conducted by different organizations, successfully. With sustainability regulations and rules becoming stricter gradually, it will not be long until the mandatory requirements are needed to be fulfilled by IT companies and they will be expected to prepare for annual internal and external audits as they are for many other standard audits.

The book begins with explaining the main concepts of sustainability, continuing with presenting recommendations to turn those into practice. Upon completion, it will enable a reader to make an IT company ready for sustainability audits.

The first chapter makes the reader familiar with the sustainability dimensions from diverse perspectives. The second chapter explains the characteristics of the IT sector that make it unique in terms of the implementation of sustainability-related activities. Subsequently, the challenges and solutions on the way to prepare IT companies for sustainability audits and assessments are brought to attention, together with guiding the readers through this procedure. The third chapter focuses on documentation for sustainability audits, including but not restricted to statements, policies, procedures, and other supplementary documents. Chapter 4 contains the details regarding actions, from the very first stage of planning actions to revising and correcting them. Furthermore, it encompasses instructions on how to identify metrics for the activities planned. The fifth chapter guides the readers on how to measure the progress of actions, followed by discussions on how to analyze the data collected. The final chapter provides information on how to prepare and release sustainability reports.

Security, Audit and Leadership Series

Series Editor:
Dan Swanson, Dan Swanson and Associates, Ltd., Winnipeg, Manitoba, Canada.

The scope and mandate for internal audit continues to evolve each year, as does the complexity of the business environment and speed of the changing risk landscape in which it must operate. The fundamental goal of this exciting series is to produce leading-edge books on critical subjects facing security and audit executives and practitioners. Key topics addressed include Leadership, Cybersecurity, Security Leadership, Privacy, Strategic Risk Management, Auditing IT, Audit Management and Leadership, and Operational Auditing.

Philosophy.exe
The Techno-Philosophical Toolkit for Modern Minds
By Matthias Muhlert

Guardians of the Digital Age
Securing the Future
By Elçin Biren

Ironwill 360° Leadership
A Blueprint for Level 5 Unicorn Leadership
By Douglas P. Pflug

Behavioral Insights in Cybersecurity
A Guide to Digital Human Factors
By Dustin S. Sachs

Integrated Assurance
Unified Risk Strategy
By Patrick Hayes

Preparing for Sustainability Audits
A Practical Guide for the IT Sector
By Mahsa Fidanboy

For more information about this series, please visit: https://www.crcpress.com/Internal-Audit-and-IT-Audit/bookseries/CRCINTAUDITA

Preparing for Sustainability Audits
A Practical Guide for the IT Sector

Mahsa Fidanboy

CRC Press
Taylor & Francis Group
Boca Raton London New York

CRC Press is an imprint of the
Taylor & Francis Group, an **informa** business

Designed cover image: Shutterstock Image ID 2464128021

First edition published 2026
by CRC Press
2385 NW Executive Center Drive, Suite 320, Boca Raton FL 33431

and by CRC Press
4 Park Square, Milton Park, Abingdon, Oxon, OX14 4RN

CRC Press is an imprint of Taylor & Francis Group, LLC

ISBN: 978-1-032-95938-2 (hbk)
ISBN: 978-1-032-96015-9 (pbk)
ISBN: 978-1-003-58758-3 (ebk)

DOI: 10.1201/9781003587583

Typeset in Caslon
by codeMantra

To my all, my son, Vedat Ozan

and

my grandparents, Babaji and Mamaryam,

my mom, my sister, and my husband

Contents

Preface

It is always hard to start something new, whether it is a new habit or a new job position. My journey might not begin with a new job position, but it became more exciting and, at the same time, challenging!

Many of us change our jobs or evaluate the internal mobility opportunities to embrace challenges rather than continuing with the monotonous tasks, and it becomes more meaningful when your job is in line with the goal of saving our planet!

We, the sustainability experts, and in particular those of us working in the IT sector, are in charge of an invaluable task: contributing to the efforts of reducing the negative impacts of businesses on the society and environment. Additionally, sustainability experts in the IT sector are most likely equipped with the latest knowledge on technologies and the skills to leverage new applications to support sustainability initiatives.

You may have already known sustainability audits/assessments, or they may be new to you. However, sooner or later, many IT companies will adopt this approach to showcase their endeavor as regards sustainability.

This book aids the readers to make preparations for sustainability audits within IT companies and help them accomplish their desired audit results, while enhancing their sustainability initiatives.

Preface

Author Biography

Dr. Mahsa Fidanboy has been working in the IT sector for over ten years. Her areas of interest include agile methodology, software development standards and models, health information systems, IT service management, and sustainability. During her graduate studies, she conducted research on information systems and R&D performance. She is the author of the books *Organizations and Complex Adaptive Systems* and *Strategic Software Quality Management: How to Incorporate Sustainable Continuous Improvement Practices*. She has obtained international certifications in project management, agile methods, telecommunications, IT-related frameworks and models, and sustainability. Additionally, she has held various positions as a trainer, consultant, sustainability expert, and quality expert.

1
INTRODUCTION

Let me start this book with the question of what comes to your mind, when you hear the word sustainability. I ask my friends and colleagues to answer this question whenever they ask me about my job. The answers do not vary a lot as we have heard the word of sustainability many times and mostly recently, but we actually do not know the details and the scope of its usage until we are interested in the topic or our tasks are somehow intertwined with it.

The most common answer is the environment. It can be stated that the common sense in terms of sustainability is what benefits the environment. Similarly, a study which investigated the personal definitions for sustainability also found that sustainability is frequently perceived from the environmental viewpoint (Alexander, Jacovidis, & Sturm, 2022).

Among the other answers carbon footprint and the climate change take place. The growing popularity of these topics, especially due to the rising focus of media in the recent years, has played an important part in increasing the awareness of people. Similarly, the consequences of climate change that the world is dealing with constantly remind us of the unavoidable disastrous events that will become more intensified and excessive if serious actions are not taken promptly in order to reduce the negative impact.

It would be appropriate to start this chapter with the definitions of the frequently used words and terms related to sustainability, of which some of them are mistakenly used interchangeably or employed in the inappropriate context, regardless of the meaning they bear.

1.1 Definition of Sustainability

It would be hard to decide where to start from when trying to define sustainability, and many begins with "it depends on…". Reviewing the literature can provide us with some invaluable insights.

DOI: 10.1201/9781003587583-1

Sustainability was separated into two parts: normative and positive sustainability, where the first is about proposals and agreements, while the latter focuses on scientific analysis of sustainable development and sustainability (Ríos-Osorio, Cruz-Barreiro, & Welsh-Rodríguez, 2013).

In an attempt to suggest definition for sustainability, four uses together with relevant meanings were proposed as: a set of criteria, an approach, an object and finally, a goal, of which their common part is focusing on socio-ecological issues (Salas-Zapata & Ortiz-Muñoz, 2019). Thus, it can be implied that whatever the definition we choose, the association with society and environment is inevitable and cannot be overlooked.

The other inseparable part is that sustainability is not an operational level action per se but have to be accompanied by appropriate strategies at the highest level. It seems that at first the flow is top-to-bottom, nonetheless, beginning and continual of the operational level activities then triggers the bottom-to-top flow when decision makers and strategists of companies get feedback and recommendations from the lower level, and then, set the goals and strategies accordingly.

Sustainability can be defined in terms of its three well-known dimensions of social, environmental, and economics. Their focuses can be summarized as health and well-being; environmental aspects like water; and wages, prices, and local economies, respectively (Schiano & Drake, 2021). In practice, just like any other project, there is a need to define the scope of what is going to be included in sustainability practices and the audit. Sometimes, the scope of practices can be wider and more comprehensive than the scope of audits and this is what is recommended and expected actually for the reason that the goal of implementing sustainability practices should not be the audit, but making attempts to reduce negative impact on the environment, society, and economy and strengthening the positive impact briefly. Nonetheless, starting from studying the requirements and expectations of the audit and delimiting the scope of a company's practices is a practical approach and a quick-win solution. Preferring this approach does not wipe out the sustainability initiatives taken, in fact. To put it in a clearer way, what is expected to be performed is done with focusing on the least scope, subsequently, the least effort and time as well. This may result in the least value in terms of the impact, and the truth is that irrespective of bearing the most or the least value, it has a value, and it is of great significance for the new companies in their very first miles of their sustainability journey. Indeed, it is the choice between doing something and nothing.

One of the facts that is usually neglected when speaking of definitions is the misalignment between customers definitions and industry definitions of sustainability (Schiano & Drake, 2021). The definitions may also vary based on the individual differences, for instance, the academic or career background of people. This can easily be understood by asking people active in finance, legal affairs, and environmental engineering. Those in finance or with background in this sector may mainly think of green and sustainable finance. Similarly, the people associated with legal affairs view it from the point of ethics, environmental law, regulations, and compliance. When it comes to environmental engineering, the social and economic dimensions will take a back seat. Given all that has been mentioned, this is the case for sustainability audits as well. What is expected from a company in the food sector differs from a company in IT sector. In a similar vein, the initiatives that can be taken by these two companies are also dissimilar. Therefore, two important issues came to light: expectations and capabilities.

Weak and Strong Sustainability: Two views of sustainability have taken their place in the literature: weak and strong; the first is about leaving the same amount of human-made and natural resources we have received to the next generation, whereas the latter restricts it to the natural resources (Kuhlman & Farrington, 2010). Another point related to these two views is whether natural and human-made capital are substitutes to each other or they complement each other (Ayres, Van Den Berg, & Gowdy, 2001).

Corporate Sustainability: Corporate sustainability can be comprehended through six indicators suggested, which are the efficiency and effectiveness of ecological and social sustainability, sufficiency, and ecological equity (Dyllick & Hockerts, 2002). The key issues within this context are that the progress toward any goal, sustainability-oriented or not, and the quality of the progress, should be tracked in order to ensure the conduction of the right tasks in the right way. Creating strategies is just the first step and let us say the easiest one. The execution of sustainability strategies as seen within the IT companies is the hardest part since there is still a lot to learn, relearn, and alter. This is the reason why a remarkable part of sustainability experts in IT sector is doing research and searching for the latest information. To make it clearer, the research can encompass topics like energy efficient software, sustainable AI, increasing the energy efficiency of data centers, and so on. The rapid change and advance of technologies are just some drivers of the need for research. It is worth bearing in mind that any new technology may impose an impact on the environment, society, and economy and the movement toward a sustainable future should be accompanied by transparency regarding the sustainability characteristics of a product. Sustainability indicator is another related concept. Sustainability indicators can be applied to various levels, like sectoral, organizational, or even for groups of products, and are parameters for showing the improvable areas of performance, and they address where the changes were oriented toward (Clift, 2003).

A previous study narrowed its focus on explaining the misunderstandings with the concept of sustainability regarding seven dimensions (Vogt & Weber, 2019). The cultural dimension outlines that sustainability is more a new ethical and cultural orientation and not solely a social and technical program for resource conservation (Vogt & Weber, 2019). Organizational culture was delineated as an aspect of sustainability, addressing the relationship with all its facets between a company and its internal stakeholders (Aras & Crowther, 2008). This dimension was chosen among all since it plays a key role within the IT companies.

Organizational Culture: Delving into the concept of culture, we may notice the effects of the organizational culture on the perception of sustainability within a company, and consequently, on the actions the individuals perform.

A sustainability-responsible organizational culture does not emerge all at once on its own. It needs time and perseverance. This can be clarified through an example. Imagine a company that starts collecting batteries to be handed in to responsible recycling companies for their safe disposal. Emails, together with reminders, were sent, and boxes were placed in the offices. However, at the end of the first month, what is observed is that there are still batteries in trash bins, and when comparing the number of batteries used with the number of batteries disposed of and thrown into the box, they are not relevant. The organizational culture fails within this test. Even with the existence of the resistance to change, which hinders or/and delays the individuals to embrace new behavior and habits, the organization can still find solutions for building a sustainability-responsible culture. Incentives and rewards are common to encourage sustainability-responsible behavior, even though it will be hard to convince the top management to support these activities in many IT companies. It will be surprising, but at the same time shocking to see that even the employees of IT companies with sustainability units/departments do not have a clear idea about what sustainability means, and when they are asked to define it, they usually mention the environmental impact, and are not able to link it to the companies activities, since it is not among the manufacturing companies, in sectors like food or textile which according to common belief, sustainability is mostly connected with. In light of the points discussed, companies and, in particular, IT companies should make attempts to shape a common understanding of sustainability among their employees, revise and reshape their organizational culture accordingly.

Strategic Documents: Analyzing mission, vision, and value statements illustrates that they mostly reflect sustainability by addressing profit, people, and planet, respectively (Baral & Pokharel, 2017). Such strategic document is a great way to express companies' commitment to value creation for society (Baral & Pokharel, 2017). There are still IT companies that have not redefined their mission and vision statements. These companies are required to revise their strategic documents to be consistent with their sustainability goals and to highlight their sustainability responsibility.

1.1.1 Corporate Social Responsibility (CSR)

CSR is about corporate contributions to sustainable economic development, and CSR disclosure is performed for various purposes, such as providing insight into future financial performance, decreasing the asymmetry of information, and reducing capital cost (Tsang, Frost, & Cao, 2023). The growing interest of IT companies in CSR disclosure can be comprehended through the same reasons.

Looking closer at the IT sector highlights that CSR disclosure is perceived as a competitive advantage and most companies start their journey not to lose the competition with the belief that it will bring them financial benefits in long term, in line with the literature (Hastalona & Sadalia, 2021). It seems totally comprehensible since the investment in sustainability-related activities mostly becomes fruitful in long term, and this is the probable reason why many companies quit their journey partially or fully after a short time due to inability to earn any achievements and failure to obtain the desired results. What is meant by partially is that they may continue reporting their CSR activities from time to time, not on a regular basis.

CSR, as a term, is now perceived more comprehensively in comparison to the past and has turned from a company's internal operations into a sustainability-related societal discourse (Aslaksen, Hildebrandt, & Johnsen, 2021). The perception may differ from place to place or from time to time. It is better to give definitions for each term used within a company to develop a common understanding of the jargon. For instance, some prefer to remind the audience of the meaning they mean by a term at the beginning of their presentation, reports, or documents, and some take a further step, preparing a corporate sustainability glossary for their companies. This helps with more effective communication concerning sustainability. Having individuals and teams coming from different departments/units owing to the width of sustainability initiatives necessitates the existence of such a glossary within a company. It should be noted that some of the terms used in association with sustainability cannot be easily interpreted in the same manner, and this matter should not be underestimated. For instance, once in a meeting held in the company with attendees participating from finance, legal, and the supplier management department, there raised the need for clarification and explanation of two concepts of supplier procurement policy and supplier code of conduct. The other case took place when a participant in a meeting asked for the full form of DEI (diversity, equity, and inclusion), which then resulted in a debate on the difference between equity and equality. Such differences might seem minor; however, they hold completely diverse meanings in relation to sustainability.

CSR Initiatives: The answer to the question of what CSR attempts are, they can be summarized as all the policies and activities that evaluate, manage, and oversee the impact on, and a company's responsibility toward, society and environment (Christensen, Hail, & Leuz, 2021). There are several important issues that can be highlighted in this regard:

- CSR is not just about actions. The backbone of these actions is policies that were communicated and enacted. To put it differently, actions take their root from policies, and this does not mean the policies which have only been formed, but those which have been communicated with employees, and the engagement and commitment of them have been ensured. This is the reason why policies, despite being viewed as static and fixed, are required to be reviewed and updated from time to time, and the employees should be able to share their opinions and feedback on this matter.

  Another issue is to remind the employees about the policies from time to time and maintain policies that can be accessed at any moment by employees. Nowadays, we do not suffer from the problems paper-based archives and non-digitalized company used to have where retaining a document from archives needed much effort and time. The corporate websites are now a good place to share the policies with our stakeholders.

- There exists a myth that when policies are established and actions are taken, then the goals are met, and it is considered completed, no need to do more. Unfortunately, it is a common belief between managers and team leaders of many companies, and this gets worse when they deliberately overlook it in order to avoid putting effort and time into it. Yes, deliberately! In some cases, following and tracking the actions are not purposefully done because they believe that it would be a waste of time, and it would be wise to save their time for other tasks. What can be said with great emphasis is that following up, monitoring, controlling, tracking – no matter what term is used – are part of the completion of the task. Getting a task done is realized when this step has been performed otherwise companies will be trapped in the honeymoon situation, if we may call it this way when during audits and a short-time after audits everything is going fine in accordance with the standard, or model and after some time, company will go back to its previous state where individuals begin to do things in the old manner. When the time for the recertification comes, then all should start from scratch! This is the reason why the significance of conduction of surveillance audits and recertification audits should not be underestimated. You might say that sustainability audits are not mandatory for IT sector. It is right. However, the key to success is to use the way that works for the same or similar cases, and if this means to plan for surveillance audits, or periodic internal audits, it should be done. Most sustainability audit/assessment entities expect IT companies to conduct annual controls or/and internal audit reports, and present any kind of related evidence.

Experience and observation show that biannual controls will work better since a big portion of sustainability departments' tasks comprises research, and these audits should leave a space for it as well.

Quarterly audits will be the most ideal, nonetheless, the problem with them is that preparation and reporting tasks call for resources and budget, and for many reasons which mentioned within this book, neither much resources, nor much budget are made available to most sustainability departments/units within IT companies for the time being. Therefore, sticking with the minimum number of audits/internal controls will be the cost-effective option. You might raise the question of whether increasing the number of audits could help improve your performance in the audits. The answer is that in almost all cases, it is not true. Fulfilling the minimum number is what is expected, and that literally means if it is met, then the requirement is fulfilled and you are given the score. In this scope, the critical issue is performing these internal audits/controls properly and comprehensively with no missing points, missing documents, and evidence. Being quantitatively above the minimum will do no good if the minimum level of qualitative satisfaction is not met. This is one of the many mistakes for failure in sustainability audits: To sacrifice quality for the sake of quantity!

Investment in Sustainability: It is worth taking a look at investments in sustainability since we talked about budgets in the last paragraph. Some studies examined the reasons why companies make investments in sustainability. One of these studies drew attention to multinational enterprises from developing countries and indicated their reasons behind investments. Some of these reasons include strengthening and supporting institutional insufficiencies in their country and demonstrating their commitment and reliability to their stakeholders (Doh, Littell, & Quigley, 2015). Observing the IT sector points to findings similar to those in the literature. The current investments of IT sector in sustainability-related initiatives are much more aiming at the attempts to be viewed as responsible and credible in the stakeholders' eyes. This is certain that in the near future, as the negative impact of human on the earth and the consequences of past actions are becoming more and more visible and perceptible, the need for being considered responsible will turn into the must to be responsible, or to say it in a different way, the choice will no longer stay as a choice, and become a mandate soon. The advance of AI and technology contributes to this situation actively for the reason that energy is at the center of all the upcoming technologies. Today, the talk revolves around AI and what changes it can bring to our lives. Nonetheless, after this phase, the focus shifts to the challenges regarding AI usage, such as how to make AI technologies more energy-efficient; how to cut the related costs; and how to decrease their dependence on natural resources.

Commitment: The question that comes to mind is how companies show their sustainability commitment to their stakeholders. The most probable answer is by means of their corporate websites. This issue was investigated for the case of India comprehensively in the literature where the paper suggests not confining the action to just publishing the CSR/sustainability information on the corporate websites, but to view it as a tool and chance to enrich the stakeholder engagement and dialogic communication, for example, through adding links to social media (Jain & Winner, 2016). This is actually the problem that we can pinpoint after taking a look at companies in the IT sector. Apart from a few large, world-renowned companies that are at the forefront of CSR/sustainability reporting and come up with innovative ideas in their report each year, many of the remaining companies just follow their lead and structure their reports accordingly. Given all the innovations in marketing and the myriad ways and prescriptions on how to attract the attention of customers, the question of why adequate effort is not being made in this regard is still unanswered.

In a similar vein to what was stated earlier, CSR/sustainability reporting is mainly directed toward being noticed by stakeholders, and creating the perception of a sustainability-responsible company among them; nonetheless, all these endeavors should also be analyzed from the marketing standpoint. It should not be confused with greenwashing or any other fraudulent behavior of companies. Let me shed light on this with an example. Imagine you have been working on a report for several months, and you are supposed to present it to the top management in a meeting. You have chosen the old school days method of presenting by using lengthy textual slides. The second option is to prepare visuals, printing the main visuals and distributing them among the attendees, together with adding some scientific humor to your presentation, interacting with your audience throughout the presentation. You may be right to defend the first option. You may say that the goal is reporting, and that you have achieved the goal. But reporting is not about pronouncing the data at the meeting, but making the data meaningful to the audience and keeping them engaged. The influence of marketing comes forward to help the report catch the attention of stakeholders. Much more should be done on the communication side of CSR/sustainability reporting, and leveraging innovative solutions like AI can be an ideal alternative.

Stakeholder Engagement: Stakeholder engagement was briefly pointed out in the previous paragraph. Regarding materiality, engaging the stakeholders either indirectly or directly assists the companies in getting to know their stakeholders' expectations and needs better, which can be useful in forming their reports (Torelli, Balluchi, & Furlotti, 2020), and disclosing the specified information accordingly.

Stakeholders: It is worth reviewing the literature for the definition and details of stakeholders. Stakeholders were categorized as primary and secondary, where without involvement of the first group, the company cannot survive, consisting of customers, employees, shareholders, suppliers, investors, and the public group, whereas the latter group is impacted by or impacts the company (Clarkson, 1995). Identification of the details of the relationships with each of these groups is critical. These include details of:

- How often are they communicated with?
- Who is in charge of communication with each group?
- What kind of communication channel do they prefer or is it suitable for their case?
- What is the level of their influence on the sustainability-related decisions?
- What is their priority for the company?
- What is their priority for each dimension of sustainability? For instance, employees are much more affected by the decisions relevant to the social dimension.
- Are there any cross-influences between the groups? If yes, determine which influences the other and to what extent?
- Concerning communication related to different dimensions of sustainability, which group should be communicated first for the purpose of disseminating the information? Or should the groups be targeted simultaneously? Sometimes, it is better to communicate with a group and let them spread the word!

Relationships between the organizations and stakeholders were investigated in the available literature, and six types of stakeholders were proposed considering factors such as the direction and extent of the influence, and their predominancy within the relationship (Mainardes, Alves, & Raposo, 2012). When preparing the stakeholders' list, these factors can be added in the format of columns, and each stakeholder can be rated according to them. Such a document should be available to those in charge of the communication, and whenever communication takes place with the stakeholders. This listing can be part of a bigger practice named "stakeholders mapping", which was described as grouping stakeholders into more particular classifications, turning a generic map into a stakeholder-specific map (Hutt, 2010).

A relevant, and at the same time interesting study classified stakeholders in sustainability marketing into four groups of: economic, environmental, social, and regulatory based on influencing or being affected during the accomplishment of goals related to each dimension, except for the last which involves the overall sustainability-related goals (Kumar, Rahman, & Kazmi, 2016).

Identifying the stakeholder can be facilitated by using six categories of conditions from a managerial point of view which are: type of claim, nature of relationship, legitimacy basis, and natures of risk, power and obligation, and additionally, not all conditions should be met for being considered as a stakeholder (Miles, 2017). As you can see, the issues associated with stakeholders cover a wide range. For sustainability audits, after identification of the stakeholders, the issues related to them should be well managed. These issues can range from the safety of internal stakeholders to encompassing privacy and security of the information of both internal and external stakeholders.

Relationship between Company and Stakeholders: Stakeholders and companies' relationships can be recognized through restrictions, exchanges, and value creation (Tokoro, 2007). Managing the relationships requires to know the type of relationship the company has with each stakeholder group. The relationships might even differ within the groups. For example, a company's relationships with all its customers are not the same. As a result, a possible solution can be adding the type of relationship to the shareholders' map or it can be more detailed as needed. This helps the employees in charge of the stakeholders' communication to pay an extra attention on how to communicate, while keeping in mind the type of relationship. Given all mentioned, the stakeholder dialogue should be emphasized within companies by heeding the demands and ideas of diverse stakeholder groups and infusing them into business management processes (Tokoro, 2007).

It is also worth noting that the role of business managers in managing the relationship with stakeholders should not be overlooked, and according to the literature, business managers' approach should be less firm-focused, and they should observe the relationship with and between stakeholder groups (Arenas, Lozano, & Albareda, 2009). Similar to the notion covered in the questions presented before, we may detect cross-influence between stakeholder groups, and they can be affected by or affect each other. The stakeholder group that is the key influencer or the groups with high influence are better to be determined, and their relationship with the other stakeholder groups should be checked out closely. For the positive issues, it can be regarded as an advantage to the company; nonetheless, its critical role comes to light in connection with the negative issues, by worsening the situation and amplifying the negative effects which should be carefully taken care of beforehand.

Regarding CSR reporting, in addition to the stakeholders already mentioned, which were labeled as the traditional stakeholders, an extension consisting of external stakeholders was introduced, including NGOs, insurance companies, territorial communities, citizens, external R&D, various associations, and other companies (O'Connor & Spangenberg, 2008). NGOs and other secondary stakeholders were proved to play key role in CSR; however, their role and their legitimacy are disputed (Arenas, Lozano, & Albareda, 2009).

A study considering stakeholder groups as managers' motivators toward CSR concluded that owners took the first, customers the second and employees the fourth places, while in a what-should-be scenario, customers became first, succeeded by employees, and owners at the third (Ditlev-Simonsen & Wenstøp, 2013).

Relationship between four types of CSR and financial performance was proved to vary among diverse sectors, and for IT sector, positive relationships were found between financial performance and CSR categories of employees and environment, and not for society and market (Feng, Wang, & Kreuze, 2017). This is just one of the reasons for considering sustainability according to the sector.

A main challenge which IT companies encounter is retaining their employees and decreasing the turnover rate. This originates from the large number of job positions available online; the simplicity to apply and start the new jobs; the tendency of individuals to work at different companies in order to learn working with various technologies and tools; and the high competition between companies to employ the quantified employees which compel them to offer more benefits to the human resource. These all can explain the key role of employees in terms of CSR, and their impact on companies to consider executing more and more actions as regards the well-being and health of the employees that, at last, result in more contributions toward the third SDG.

The Role of Sector: The practice of sustainability in each sector bears differences from the other, even if not in all areas. Sustainability experts, regardless of being called sustainability experts, irrespective of the sector they are active in, should first learn about the sector their company is working in in order to be able to plan and execute the strategies, goals, and actions in an efficient and effective manner. Like the relationship between CSR and financial performance, the importance of each CSR type for each sector within the sustainability audits differs, and that is the reason why the audit companies query details about your company, and its business activities. The audit questions and the whole assessment and evaluation performed during audit is heavily dependent on the sector of your company. Therefore, taking the sector into consideration while researching for guides, best practices, or real-world cases, the role of the sector should not be neglected.

The Role of Context: Another interesting subject is the difference in the CSR-related perceptions, which are context-dependent (Barkemeyer, 2011). What is observed in practice is that this situation becomes an issue in multinational companies and also in those with diverse people on the executive board. Since the significant decisions are made at this level, it is better to first reach a consistency between the members' mindset, and then disseminate the common mindset throughout the whole company and cultivate the company's culture. Not being on the same page is a barrier on the way to CSR initiatives. The simplest example is that sustainability might be a priority for some board members, while it is ranked comparatively lower for others. Even in such a case, the initiatives and plans will vary, and so do the goals, objectives, and final destination. Never forget that this is the road that takes us to the destination(goal), and selecting the right one is as important as setting the goals!

CSR Assessment: When it comes to the assessment of CSR, theoretical models consider one, some, or all of the dimensions of social, economic, and environmental; nevertheless, they are not engaged with the inherent conflicts or synergies originating from pursuing all dimensions simultaneously (Abdelhalim & Eldin, 2019). Assessment of multi-dimensional concepts has always needed more care and effort compared to the uni-dimensional concepts. This comes from the probable relationships between dimensions, and before everything else is done, these questions should be thought about:

- Do the dimensions have effects on each other?
- How do they affect each other? Positively or negatively? To what extent?
- Is the total point the mathematical sum of the dimensions' points or is there a synergy present, causing the total to be bigger than the mathematical sum?

These are the theoretical parts of the debate. In companies, theories are less respected or let us smooth the statement by saying that theories are less taken into consideration in practice. There are many top managers and managers who believe that experience is of more weight compared to theory. Due to this mindset unfortunately, much cannot be done in terms of sustainability except imitating what other companies do for the reason that as sustainability is gaining more attention day by day, the new standards, regulations, practices, and theories are highly likely to emerge, and the negligence of these will cause great loss to the company by misleading it on its sustainability journey. Given all mentioned, for obtaining quick wins, it is recommended to stick with the previously implemented solutions and methods, for instance, by looking at what others in the same sector or with the most similar context have done so far. Hence, in long term, it is better to review all the aspects of a problem deeply and widely. As for the assessment, current assessment models can be employed in first steps, and then, in the further steps, the details stated above within questions can be inspected, and the right methods and tools can be chosen.

CSR Models: Two models for CSR were introduced: the constituency model and the sustainability model of which first views non-shareholders' interests as an extra expense and the second as a vital success factor with its focus on long-term benefits (Millon, 2011). As you can guess, today, most shareholders and investors of IT companies seem to put more stress on current earnings than future gains, consistent with what was stated in the conclusion section of a relevant article (Millon, 2011). This can be due to the high level of uncertainty within the IT sector that makes the current survival much more crucial than the unseen yet-to-come future. If you ask the shareholders of some IT companies, their answer is most likely that they are unsure whether they can survive the next wave of technological advancements or not, let alone make investments in CSR in the hope of future pay-offs!

CSR Pyramid: Pyramid of CSR (Carroll, 1991) is helpful in understanding the responsibilities related to each dimension. Philanthropic responsibilities at the top involve contributing to society and life quality, followed by ethical responsibilities concerning being ethical, while the third is legal responsibilities regarding complying with laws, and economic responsibilities are at the bottom, indicating profitability (Carroll, 1991). Among these dimensions, the one that is harder to be practiced, and there is much to be investigated with regard to its role in sustainability audit, and if we may say, any kind of audits, is the ethical domain. Sometimes, it seems really hard to differentiate what is legal from what is ethical. That might be due to the paradoxical-seeming situations of being legal, but not ethical, and being illegal, but ethical, although in rare cases. All these highlight the importance of having the knowledge of the context of the company you are working for as a sustainability professional, together with being familiar with the national laws, regulations, and the cultural norms and beliefs, which are also of equal significance.

1.1.2 Environmental, Social, and Governance (ESG)

ESG indicators help companies to report on their activities of protecting the environment, being socially responsible and showing governance integrity (Jiang, Feng, Wang, & Chang, 2024). This paper also underlines the fact that ESG performance at microlevel have been studied more comparing to the macroeconomic level (Jiang, Feng, Wang, & Chang, 2024).

ESG was mentioned as both a strategy and a standard for investors to assess corporate behavior together with the future financial performance (Li, Wang, Sueyoshi, & Wang, 2021). Two databases for ESG scores, KLD and Thomson Reuters ASSET4, were also explained in the same article (Li, Wang, Sueyoshi, & Wang, 2021). ESG scores are becoming more and more popular for the sustainability evaluation of companies. The main issue with the ESG score is that, albeit regarding the three dimensions in the rating, it is indeed a measure for the sustainability performance of an organization, not engaging with sustainability nor the impact of the company on the world (Clément, Robinot, & Trespeuch, 2022).

Companies responsible for ESG can benefit from several competitive advantages, which result in higher corporate sustainable performance, while social performance is highly related to ESG disclosure positively and environmental and economic performance come next, according to results for Asian firms (Alsayegh, Rahman, & Homayoun, 2020).

ESG Disclosure: Looking at the IT sector, the ESG disclosure is currently about demonstrating the company's care for their environment and society, which helps with building trust with the customers whose purchasing behavior is becoming more and more responsible in terms of sustainability. The responsible production and consumption, as implied by its name, has two sides: companies and customers. The rising awareness of companies goes hand in hand with more customers preferring sustainable products, and what is noteworthy here is that the key role is played by customers in this situation. More ESG responsibly purchases are what can shift the market toward responsible production, and this means there is much work awaiting the customers. The reason is that the impact of IT companies, particularly on the environment, is easily overlooked because some or all of their products are not physical, and it will be hard to link waste, water, or air pollution to them at first attempts.

The other issue with ESG disclosure is the lack of unification in reporting. This is not limited to the IT sector, actually. What companies do in practice is review several reports and ESG disclosures from the leads in their sector or in a closely related sector, tailoring them based on their context and needs, and then prepare their reports. These attempts, indeed, help with harmonizing the ESG disclosure although indirectly.

ESG Dimensions: The findings of another study show that the social dimension of ESG is the dimension with the largest effect on economic sustainability (Simone, Petracci, & Piva, 2022). This result plays a pivotal role in the strategy making of the IT companies since it can influence their choice of strategies, and subsequently, their tasks and activities. The good news is that enhancing the activities associated with the social pillar of ESG is the simplest compared to the other two in practice. The dimension that mostly puts IT companies in a tough spot is the environment. This is for the reason that environmental best practices and sample applications are harder to find in IT sector in comparison with manufacturing. Therefore, tailoring those initiatives requires much time and effort, together with a sufficient level of knowledge and expertise to identify what requirements should be met and what can be done to fulfill those for the audits.

In contrast, the social initiatives can be found widely on social media, reports, corporate websites, and so on. Even simple research on the topic can enable companies to integrate the same or similar practices into their plans and reinforce their social pillar of sustainability.

1.1.3 ESG versus CSR

ESG is about incorporating the ESG factors into their business models by investors and companies, whereas CSR mainly focuses on social responsibility and improving corporate citizenship (Gillan, Koch, & Starks, 2021).

Another clarification of these two concepts describes CSR as a general framework for sustainability primarily used by firms, while ESG is employed by investors, considered as a measurable assessment, and the same study stated that they are complementary to each other (Kaźmierczak, 2022).

When it comes to the IT sector, these two terms are mostly used interchangeably in many academic works. This may be due to the reason that the details of their distinction do not matter to them, and they view the big picture. What is of importance to them is defining the relevant sustainability actions and strategies, and then, ensuring a reasonable disclosure and reporting of them, ensuring their sustainably-responsible image creation in both the customers and investors.

In a similar vein, another paper addresses the evolution of ESG 2.0 from CSR, where the key feature of ESG 2.0 is incorporating sustainability into core business strategies (Passas, 2024). Agreeing with what was mentioned in the same study, this integration requires a change in the mindset (Passas, 2024). This change should be on the way to stop viewing sustainability practices as a burden, specifically a financial burden to the company, and begin taking it into account as an enabler of innovation and competitive advantage, as pointed out in the literature (Passas, 2024). With the intense competition in the market, what should be done is to leverage the innovative potentials, and recognizing the aspects of such potentials would not happen till a company gets involved in the environmental and social issues (Daub, 2007).

Unfortunately, what can be seen within IT companies is that this transformation has not been commenced yet or has not taken place completely, and still, many companies are stuck in the first mindset. A big cause of this situation is that sustainability practices have not been implemented widely, and there still remains a lot of unknown in this regard. Many companies have no or little idea of where to begin with particularly those who provide services to their customers.

This is the case for IT sector as well, where we have digital products intertwining with services.

Lack of sufficient knowledge on the implementation of sustainability initiatives results in two situations: Companies may perceive the practices as much costly than they really are, or they will refer to consultancy companies for benefiting from their services. Both these two cases bring about an extra expenditure for companies. In the first case, they will spend more money because of not knowing a better economic solution, and in the second case, they will pay for consultancy services.

A good distinction between concepts related to sustainability was performed by depicting policy scope on the x-axis and policy focus on the y-axis, where the y-axis ranges from environment to ESG at the top (Sheehy & Farneti, 2021). According to this paper, SDG and CSR are common in encapsulating ESG notions, though they differ in their scope of policies, that is, organizational for CSR and international for SDG (Sheehy & Farneti, 2021).

1.1.4 Sustainable Development Goals (SDGs)

You might have heard the term SDGs in meetings, reports, or on the news. This stands for Sustainability Development Goals introduced by the United Nations (UN), which provides the list of goals, targets, and descriptions within a human-righted-oriented approach toward CSR, including totally 17 goals (Meseguer-Sánchez, Gálvez-Sánchez, López-Martínez, & Molina-Moreno, 2021). The SDGs reports are published annually and demonstrate the global progress on the 2030 Agenda for Sustainable Development (United Nations Department of Economic and Social Affairs, 2024).

Companies have been using these SDGs as a guide to build their strategies and initiate their activities. Similarly, their progress has been communicating by relating it to the SDGs. This global harmonization helps a lot with speaking the same language concerning sustainable development.

You may also be familiar with the term "SDG index". Its goal is to estimate the success of diverse countries in terms of the execution of SDGs, and the interesting facts are that it turned out to be negatively related to unemployment and positively to life expectancy (Tu, Kubatko, Piven, Kovalov, & Kharchenko, 2023).

SDGs at Different Levels: SDGs can be pursued at different levels, like national and organizational. The number of IT companies that have aligned their initiatives with these goals and have disclosed their progress toward these goals is remarkable. Turning to a higher level, countries can provide advantages to their population through the execution of SDG-oriented actions, which enable them to enrich equal opportunity, encourage economic empowerment, and foster sustainable development efforts within the country (Filho et al., 2019).

For the case of IT companies, upholding the SDGs can be advantageous to the employee, company, environment, and society. Some of the SDG-oriented practices of IT companies are mentioned below, along with indicating the SDG or SDGs they can be associated with.

- A company can plan for well-being programs for its employees or it can offer them health insurance or take an active role in practices to support its employees physically and mentally (addressing SDG 3).
- A company can provide training and development opportunities to its employees or organize workshops where instructors can be external or even internal to the company (addressing SDG 4).
- A company can uphold gender equality within all its human resources procedures, like retention, recruitment, and so on (addressing SDG 5).
- A company can engage in innovative ideas for saving water or energy consumption, even partnering with other companies (addressing SDGs 6, 7, 9, 17).
- A company can invest in research and development activities with the aim of producing energy-efficient products (addressing SDG 7, 12).
- A company can support the infrastructure of smart cities by volunteering to supply part of its IT structure (addressing SDG 11).

1.1.5 Sustainability Management

A sustainability management framework consisting of five building blocks of knowledge and commitment of management, stakeholders, and workforce, and strategic and operational planning was introduced and explained in the literature, which can help companies establish sustainability management systems (Maas & Reniers, 2014).

Corporate sustainability management is expected to take into account some prominent factors, including innovation, stakeholders' requirements, effectiveness, and efficiency of the operations of the business consistent with the Framework for Strategic Sustainable Development (FSSD) (Baumgartner, 2014). Some features of FSSD are presenting a five-level model entailing subjects, such as system, guidelines, actions, and methods, and providing an operational procedure for organizations to facilitate their strategic movement toward sustainability (Broman & Robert, 2017).

After doing a sufficient amount of research and studying the best practices of sustainability management systems in companies, it can be well realized that such a system goes beyond what is usually perceived as management systems. You may have heard of quality management systems, information security management systems, and privacy management systems, which are founded on the periphery of a certain subject, quality, information system, and privacy, respectively. You may justify that it is also the case for a sustainability management system. That might be partially true, but not exactly and thoroughly.

Sustainability management is actually an umbrella term that encapsulates diverse notions like energy management, water management, waste management, occupational health and safety management, and so on. An integrated sustainability management system can be recommended to companies that wish to cover all aspects of sustainability. Indeed, the company's strengths regarding each of these aspects may vary. In the next paragraphs, the details are shared.

To implement the environmental side of a sustainability management system, the first step can be searching for the standards and models related to an environmental management system. The further steps can be investigating the model and standards related to the energy management system, together with checking for the requirements of water and waste management system. Since the scope of this book is IT sector, waste and water management rank lower in regard to priority comparing with the energy notion.

When it comes to the social side of a sustainability managements system, guides and instructions on social responsibility can be referred to. Social dimension is also about the health, safety, and well-being of the employees, and thus, standards and models related to workforce health and safety, such as occupational health and safety management systems, can be placed under this category.

Risk Management: Risk management is present in every aspect of systems since risks are an inseparable, incorporated part of each management system. Managing a system cannot be fully realized without actively managing the risks, including but not limited to actions like risk identification, evaluation, prioritization, and then, determining the strategy like strategies to mitigate the risk, or escalate it to a higher level in an organization's hierarchical structure, or to avoid it. In this respect, standards, models, and guides can be benefited from on how to handle different types of risks.

Risk management, despite being of paramount significance in management, is often underrated, and risks are preferred to be ignored till they occur and become an issue. The negligence is somehow due to the belief in the stability of risks throughout time because some risks are identified repeatedly during risk assessments each year. This does not mean all risks are fixed, but as you know, each risk is assigned a probability and an impact. After the prioritization, the highly ranked risks will be acted on, while the others might be left for some time in the future, and this is the real reason why some risks stay in the risk log for a longer time.

Performance Management System: The findings of previous research have shed light on the fact that to execute strategies related to sustainable development, its different dimensions should be infused into the performance management systems of companies, and a framework is introduced in the paper (Hristov & Chirico, The Cultural Dimension as a Key Value Driver of Sustainable Development at a Strategic Level: An Integrated Five-Dimensional Approach, 2023). Managers are well aware that setting the goals should be accompanied by other practices. These practices may include but are not limited to:

- Defining targets,
- Operationalizing the concepts,

- Determining performance indicators,
- Establishing a system for tracking the progress toward goals, with the ability to display real-time data, compare the data across time periods, and filter the data based on desired variables.
- And establishing a performance management system.

AI and Sustainability: The other issue that is gaining popularity in the sustainability area, just like many other areas, is AI. Leveraging AI technologies was stated to bring about three primary advantages, which are automation, analysis of data and presenting insights, and the ability to combine many resources for solving problems of high complexity (Nishant, Kennedy, & Corbett, 2020).

Getting the right solutions from AI depends highly on feeding it with reliable and valid data. Therefore, before taking any steps forward, we should ascertain that the right data are collected in a proper manner. The AI's ability to analyze the data empowers us to inspect various scenarios and models and refine our decision-making process You might have watched movies where a meteorologist employs AI tools to model the weather patterns, or to pinpoint the changes within the climate patterns due to an abrupt change in a factor.

As for IT companies, they are probably among the forerunners of AI users. They can capitalize on this situation by:

- Utilizing AI in improving their reports,
- Aiding them to come up with the right choice of visuals,
- Helping with the analyses and the interpretation of data,
- Predicting the future financial gain from sustainability initiatives,
- Suggesting new ways to implement SDGs,
- Bringing about new ideas to foster sustainable development,
- Calculating the values for different metrics and indicators based on a specified guideline or standard,
- And preparing videos, slides, and images to support the sustainability practices.

It is worth mentioning that the AI wave has already started to come and has found its place in the design of the sustainability reports.

1.2 Sustainability Dimensions

The dimensions proposed for sustainability vary between different studies and sometimes between different sectors, as is the situation for the definition. Today, the lack of a common language is one of the serious problems with sustainability. Sustainability has been defined and measured differently, and this makes it hard to establish new policies and procedures, as the current situation is vague. A company that is perceived to be green from one point of view can be considered "not yet sustainable" from another viewpoint. Let alone the concept of greenwashing and the ethics that lie not just behind the sustainability audits but also all kinds of audits. Setting and having a common definition for sustainability may help the companies and, at a higher scope, may help the world in many ways, including:

1. Being able to benchmark against the best and create future strategies accordingly.
2. Being able to compare the companies within and between different sectors.
3. Being able to put the impact at the center of our talks instead of being distracted by trivial details of the activities, sector, location, and many other factors.
4. Being able to track the progress clearly to set the roadmap, policies, and determine future actions.

The *first advantage* is of notable significance to the strategists of the IT companies. As we know, strategies are not created considering only the factors internal to the organizations; nonetheless, the external factors, among which competition and competitors take place, are taken into account as well. The very simple example is the use of SWOT analysis. Thus, in order for the right strategies to be formed righteously, we need to be able to compare our companies in terms of sustainability against the best practices and high performers.

The *second advantage* points at the fact that the rise in the awareness of customers, and investors pushes them toward making comparisons between companies in terms of sustainability. Customers are becoming more and more prone to purchase responsibly. Therefore, the companies that are capable of showing their sustainability capabilities are the ones that can survive the battlefield. The extent of the severity of the competition may change from sector to sector or even product to product.

The *third advantage* helps us with removing the excuses that companies use for their insufficient sustainability activities. The key issue here is to focus on the maturity of sustainability practices, not the barriers and obstacles. Unfortunately, sometimes, companies focus on the difficulties and what they cannot do regarding sustainability. Instead of this approach, we can put our efforts into what we can do. At first stages, our practices might not be perfect and to the desired level, however, we should not forget that any steps toward sustainability are really valuable and make a difference, even by showing awareness and trying to make more sustainability-supportive decisions. Briefly, IT companies need to do their best, even if it is not the best, for the time being regarding sustainability, and throughout time, the sustainability initiatives will become more and more comprehensive and efficient.

After forming the strategies, we need to pave the way for their implementation at different levels. The implementation does not occur without close watching and tracking the progress, and this addresses the *fourth advantage*. You may have heard about the different methods companies employ for monitoring their progress toward their goals, such as objectives and key results (OKR) methodology. This is not only limited to the sustainability-related strategies, but inclusive of all strategies. Therefore, the IT companies should design measurement and reporting systems for sustainability, which is very advantageous to them when it comes to sustainability audits as well. Auditors are also well aware of the fact that actions are only actions on paper if they are not measured and reported, and when the results are not analyzed. Hence, the general in-a-nutshell to-do list for preparing for sustainability audits contains the following steps:

1. Form the sustainability-related strategies.
2. Define the tasks at different levels including tactical and operational levels.
3. Define how the tasks completion is measured and what is meant by done (completed).
4. Measure the progress.
5. Report the progress.

6. Analyze the progress to elicit best practices, improvement areas, and lessons learnt.

The above process bears characteristics of agile practices, and as we are aware that agile methodology is not restricted to IT projects, it can be taken advantage from in sustainability management as well.

Five dimensions were suggested for sustainability, including three dimensions of place, along with the other two dimensions of permanence and persons (Seghezzo, 2009). In the article, the shortcomings and limitations of the previous definition were artfully stated, and a triangle of sustainability was introduced to depict the image to the reader more vividly. This paper reminds of the fact that when we talk of environment and society, we should not take these two concepts as just two words, but the width of the notions they embrace should be carefully paid attention to. To put it in other words, the environment and society could mean different things when used in various places and at different points in time. It can be clarified through an example. Imagine company A is a textile company located in Asia and company B is a software developer in America. The environment and society to these companies may bear commonalities, but they can actually differ. As a textile company, the part of the environment that is mostly affected by company A's activity is the region it is located, the nearby town or city. At the same time, the emissions, due to the greenhouse effect, have impact on the global climate. You can view the problem from the view of the butterfly effect, all the activities and the impact they leave on the world make more sense! Back to our example, the society dimension of sustainability for company A can be the local society at first place. Although, we now know that the internet has made it possible for a small size company in Asia, for example, to donate to the movement of saving the polar bear in the Antarctica. Company B may make donations to support the restoration of rain forests and protect them. What sustainability shows us is that we, as either individuals or companies, leave impact on the planet we all share, and this impact varies due to several factors of which the most crucial ones are:

- Proximity to the source of change/impact.
- Direct or indirect relationship with the source of change/impact.

- Extent of dependency on the source of change/impact.
- Frequency of transaction with the source of change/impact.

The critical issue is that companies are systems in a complex environment and they should generate value for two main purposes: to survive and to provide advantages to ecological, social, and other systems (Bansal & Song, 2017). What is meant by system is a whole containing numerous parts which are in relationship with each other either directly or indirectly, and all work to achieve the common goal of the system (Fidanboy, 2022). Viewing the companies as systems, although it seems to be difficult and detailed at first sight, can always be of help to understand the internal dynamics and their relationship with the environment. IT companies are not alone in this environment, and they are engaged in different relationships with other companies, like a buyer–supplier relationship.

Supply Chain: As you know, sustainability reporting does not end with disclosing the information of a company, yet disclosing information of the company's downstream and upstream value chain is also required for two purposes: evaluating compliance of the value chain, and exhibiting contributions to users (Wagenhofer, 2024). This is one of the key points where IT companies encounter a hurdle. As an IT company, the upstream value chain contains companies from which our software and hardware are supplied. These are mostly among the well-known companies, for instance, laptop manufacturers. During the preparations for sustainability audits, companies can be asked to monitor their value chain activities concerning environmental and social issues or even conduct on-site audits to assess them. This raises an issue here on how to monitor and perform audit on these globally recognized, and nudge them toward completing the corrective actions. This can be true for the downstream value chain as well, after the product becomes available, for example, evaluating the activities of the company responsible for the marketing of your products. When looking at the whole notion, it can be realized that the main goal is to be responsible as a company, and push the others over whom you may have some kind of power or with whom you have business relationships, to take their responsibilities. To clarify it, as a customer, you can impact the supplier by demanding specific features in the products, like asking for energy-efficient products, and vice versa, as a supplier, you have the ability to shape the behavior of customers to prefer and purchase more sustainable products. Given all that, assessing our supply chain is what is expected by auditing entities, and if there is a chance to fulfil it, then doing so is the first option. The other option can be researching the sustainability-related activities of the companies we are working with on the internet, looking for news and posts, and examining their social media, and scoring them based on the sustainability criteria we have previously defined for the evaluation of our supply chain. These criteria may include subjects such as disclosing a sustainability report, holding sustainability-related certifications, taking part in socially responsible projects, or building networks and partnerships for the betterment of society and/or environment.

Another solution that is widely employed is using a self-assessment questionnaire on sustainability for managing the supplier chain. Within a study, two types of supplier sustainability questionnaires, internally developed and industry-wide developed, were compared; the topics covered embody business conduct and compliance, occupational health and safety, social standards, sub-supplier management, environment, and company management (Fraser, Müller, & Schwarzkopf, 2020). Your company may have received supplier self-assessment questionnaires before. The subjects included in the questionnaires are almost similar with minor differences. Some IT companies, due to their tendency toward digitalizing the processes, have already developed platforms for managing the supplier chain. With the growing global emphasis on extending the sustainability initiatives beyond the border of companies, to their value chain, and also being expected to report carbon emissions for different scope of carbon encompassing upstream and downstream emissions, sooner or later, the rest of the companies are so probable to commence developing similar platforms, and adopt the practice. It may not only entail the supplier assessment questionnaires but also through expanding its features, the details and visual reports can be brought to the user by just a click, accessing invaluable information such as: when a specific supplier filled out the questionnaires; what were uploaded as evidence and supporting materials to the questions; what percentage of our suppliers have filled out the questionnaires which is a key value asked for during sustainability audits. The supplier management platform can also be enhanced to contain features such as on-site audits, supplier monitoring and controlling modules, supplier ESG-related practices enabled with file uploading feature, supplier ESG certificates and external endorsement, and an agreement management module.

During sustainability audits, companies might be required to report the percentage of suppliers who have signed the ESG-related articles or appendix, addition to their agreement. As you know, companies may use different types of agreement, and it is essential to know which supplier signed which agreement, not by searching a local computer, or looking through physical files, but it would be much convenient to be able to view it and the related visual reports on such an online platform. Thus, an online agreement/contract management system integrated within the platform would be an ideal solution.

Certificates and external endorsement are not of less importance. Similar to the quality practices, having the information on which supplier hold ESG-related certificates and endorsements help the companies through their decision-making process, and besides, companies can disclose that, for instance, 65% of our suppliers have environment management systems, which is an indicator of managing the supply chain, and a bonus during the sustainability audits.

Auditing and monitoring of suppliers have gained the attention of many practitioners and the media. The number of news articles on the companies that lack a monitoring mechanism for their suppliers is not few. Day by day, customers become more aware of the notion of responsible consumption, and they are more conscious of the sustainability impact of their purchases. For this reason, failing to notice the importance of monitoring and controlling your suppliers seems risky, in addition to not being able to meet the requirements of sustainability audits. Given all the above, planning for the audits/checks, entering the audit results into the platform, creating corrective actions, and assigning them to the relevant individuals on the supplier side, and retaining such information via a single platform will be practical.

The other values which should be presented during sustainability audits are percentages of suppliers who have accepted and signed various codes of conduct, policies, statements, and documents other than sales agreements. These approvals are to demonstrate that the suppliers are aware of their costumers' approaches and are informed about what approach is expected from them. An example can be supplier code of conduct.

Leveraging the available tools and applications within IT companies, and making them suitable for use within such a context (platform for managing suppliers), is another widely implemented solution. IT companies have access to numerous applications for their production. These tools can be redesigned to be utilized in this context. This roots from the standpoint of IT companies regarding the sustainability budget. They are reluctant to assign extra budget to suitability initiatives when they can find a workaround and keep going, saving the budget for production pipeline.

Looking at the other study depicts another image for sustainability consisting of four dimensions of environmental, social, ethical, and economical where it also validates the independence of these four dimensions from the sector (John & Narayanamurthy, 2015). The same paper then elaborates on their model and maps to the company's planning at three levels of strategic, tactical, and operational. The planning notion will be discussed later.

In line with the first item of the to-do list mentioned before, dragging the sustainability into the strategies of the companies is almost inevitable since the trend and the vital need for such actions have long been started, and many of strategies have already been realized and communicated with their related stakeholders within companies. This does not mean that when the sustainability-oriented strategies have been implemented, there is no more a need for other mature strategies. On the contrary, it asks for more strategies toward sustainability. The reason why this notion was emphasized here is that unfortunately, many companies that have already reached a remarkable or let us say acceptable level of sustainable maturity will no longer want to exceed on this path, but to maintain their current situation. We can see many of these companies in IT sector specifically. This is mainly due to the fact that they are at a point much further from the other companies, among which are the ones that have not made any sustainable strategies yet or even have no plans. This makes them more reluctant or, let us say, careful to take further steps as fast as before. By careful, it is meant that the aforementioned companies prefer to be more cautious with the resources they allocate to sustainability-related activities since each task getting done needs budget, time, and effort regardless of the benefit that it may bring back.

The other reason is that though sustainability reporting and action taking are mandatory for some companies and sectors, it is completely or partially voluntarily for the others. The voluntary nature of such practices causes the related decisions and the following actions to take place more slowly, over much longer period of time.

In another study, first, seven dimensions for sustainability was illustrated in a table, including environmental, social, legal, political, cultural, ethical, and economic, which after analysis, political, social, and ethical dimensions, were found to play the key roles in the dynamics of sustainability processes (González, Martín, Vaca-Tapia, & Rivas, 2021).

Another study (Lozano, 2018) proposed an organizational sustainability framework that can be referred to by companies as a guide for their sustainability practices. The framework contains inputs, system elements, outputs, and purpose fulfillment components, and the paper defines organizational sustainability as an organization's contributions to sustainability equilibria consisting of economic, environmental, and social dimensions, considering the factor of time and their interrelationships as well (Lozano, 2018). What makes me excited a lot about this paper is taking into account the factor of time. The problem that we face in implementing sustainability in companies is overlooking time and what happens throughout the time. The definition here reminds us of systems thinking where an organization is viewed as a system with its elements in exchange of information and energy with its environment, and the change over time has an impact on the system. From this viewpoint, boundaries, the factors that affect and are affected by the system or its elements, should be well investigated.

Almost all literature agrees on the fact that numerous dimensions and definitions for sustainability, sustainable development, and the related concepts exist. For instance, the difference in terminology employed for sustainable development was stated in a paper where the terms "approaches" or "tactics" were used for representing groups of principles associated with the same topics, introduced under categories of environmental, societal, and economic (Glavic & Lukman, 2007). What is detected in practice is that the companies and individuals coming from the same background, sector, or industry usually do not bear any hardships in understanding each other, and they perceive the terms similarly. This varies when it comes to people from different industries. For this reason, trying to choose the words and terms more cautiously and with reference to the literature can be a solution. There is a hope that as sustainability is becoming more and more popular and the day to day rise in the awareness of people is continuing, we will reach a point where all the terms will be used and perceived as they are meant to be, with no conflict. The ambiguity and vagueness are mostly the case when something is getting popular at the early stages. Remember when AI introduced and everything smart was somehow supposed associated with it. However, after some time, AI, machine learning, business intelligence, etc., were began to be used not interchangeably, but exactly where they are correctly fitted.

A very significant truth about the dimensions of sustainability is that they are not separated and isolated concepts. On the contrary, they are intertwined. A paper examines this issue, pointing out that there will be "tensions" between different dimensions when they are practiced simultaneously as per the integrated approach where the role of governance, business models and measurement, reporting, and management systems becomes intensely critical (Giovannoni & Fabietti, 2013). The application of sustainability in the firms shows the same issues and problems. Firms struggle with balancing their strategies and goals in different dimensions of sustainability. The term I used for such a case is "labyrinth maze game with marbles". Imagine the board is the company, and the holes are the goals the company wants to achieve. The marbles are the strategies related to the dimensions of sustainability, regardless of their number. You, as the decision maker or strategist, should keep the board in your hands in a balanced way while trying to direct the balls to the holes. Focusing on a ball drifts you away from concentrating on the other balls, and the balls may go off the board and fall or, let us say, the strategies fail in meeting the expectations. The solution is looking at the big picture, making efforts to take a holistic approach, in our example, keeping an eye on all the marbles while taking careful, thoughtful steps toward directing them toward the holes. The good news is that findings of a study indicated that even companies of high sustainability are not successful in combining all parts of sustainability into a balanced model (Danciu, 2013). The integrated approach is actually the one which I found absolutely compliant for the IT companies. Let me elaborate more on the related situations in IT sector. The tension between the dimensions is excessively felt between financial and the other two dimensions in IT sector. This may be due to several reasons which are summarized below:

Sustainability seems rather as a voluntary initiative for IT companies rather than a mandatory one, and this leads to the difficulty in justifying the budget required to be allocated to. You may say that with the rise in the number of responsible purchases and the switch of customers' behavior toward companies with sustainability activities, they will feel the pressure and be bound by the mandate. The answer is yes, although it may take some time for this pressure to become tense and irresistible. Besides, the dynamic environment of IT sector makes the problem of survival, which is mainly and mostly about the financial matter, a company's first priority. For this reason, IT companies, before taking actions toward environment and society, first intend to guarantee their stable financial situation, consequently, taking tiny actions, creeping through their sustainability journey, which is one of the solutions the companies apply to. The second is dispensing the practices over time, maintaining a steady pace, the one I labeled as "penguin walk". In other words, the general approach by the stakeholders of a company is that they will allocate the money when it becomes mandatory, but for now, just keep going with the minimum money that is possible and delay the request for more budget as long as it is possible.

There exists a second myth or, let us say, a common misunderstanding that "digital is already sustainable". Most people compare the situation with that of the past and strongly believe that digitalization does not result in any environmental or social impact. That might be the case when looking at the benefits digitalization has brought to our lives such as decrease or elimination of paper use, a reduction in the number of carbon emissions due to transportation, and so on. Just remember when we had to write and prepare documents in hard copies, taking miles to handle them to the offices or sign the papers. They have all been facilitated owing to digital libraries, archives, e-signature software, and many similar programs. With all that in mind, there are still some points we may overlook: the energy. All the devices we use consume energy, and this energy, if not coming from renewable energy sources, causes carbon emissions. The very simple example is the case we have all heard about; even a search on an internet search engine contributes to carbon emissions. It also becomes more and more important as AI is becoming more widely used, for the reason that all the training of AI is energy-costly.

The other comparison, which makes the impact of the IT companies on the environment less visible, is the huge impact of manufacturers. For instance, companies in chemistry or textile industries leave a more observable impact on the environment. Therefore, the effects of the activities of IT companies can be simply underestimated. Moreover, when talking of sustainability, the social practices should not be neglected, and the effects on the employees and the society are required to be paid attention to.

You may say that just like many other industries, the IT industry has been turned into the red ocean. That is right. Though the case is a bit different and severe for IT sector, for the reason that startups and new entrants can be established easily and quickly when compared with other sectors. For example, they may start from releasing a computer game and then, expand their products. Furthermore, IT solutions can be imitated and copied if preventive actions are not taken properly at the right time. All that is mentioned makes the financial dimension to appeal the most attention and the matter of survival. Thus, the other two dimensions come next, and they are considered less advantageous from the views of the decision makers.

The other issue, which is noteworthy to be mentioned, though it would be better if not to be seen at all, is greenwashing. It is one of the terms begun to be widely used lately due to the number of companies involved in it. It delineates the false and misleading claims of the companies regarding their sustainable attempts. Such companies make endeavor to demonstrate that their products and services are green and ignite the perception that they care for the environment, society, and economy. You might say that such practices can be pretended to a limited extent and that a company cannot claim what does not exist at all. You are right. The vast number of greenwashing cases is those that exaggerate sustainability-related activities and aim at magnifying the tiniest task, picturing it as a huge contribution. It is not paradoxical to the fact that even a small pace toward sustainability counts. To clarify it, greenwashing is done with the purpose of deceiving the customers, stakeholders, and investors, and it is not performed in good faith.

The last reason is somehow related to the previous reason. Numerous measures and measuring methods provide companies with options that can be seen as favorable at first sight for the reason that they enable companies to make a choice and choose the right method and metric. Although by looking at what is happening in practice, the fact will come to light that companies may abuse the existence of a variety of methods and metrics. They may report the ones in which they can be shown much greener! Even when reporting sustainability in subsequent years, some companies have been noticed to bounce between different methods to showcase their contributions in a deceptive manner.

References

Abdelhalim, K., & Eldin, A. G. (2019). Can CSR Help Achieve Sustainable Development? Applying a New Assessment Model to CSR Cases in Egypt. *International Journal of Sociology and Social Policy*, Vol. 39, No. 9/10, pp. 773–795.

Alexander, R., Jacovidis, J., & Sturm, D. (2022). Exploring Personal Definitions of Sustainability and Their Impact on Perceptions of Sustainability Culture. *International Journal of Sustainability in Higher Education*, Vol. 23, No. 3, pp. 686–702.

Alsayegh, M. F., Rahman, R. A., & Homayoun, S. (2020). Corporate Economic, Environmental, and Social Sustainability Performance Transformation through ESG Disclosure. *Sustainability*, Vol. 12, No. 3910, pp. 1–20.

Aras, G., & Crowther, D. (2008). Governance and Sustainability; An Investigation into the Relationship between Corporate Governance and Corporate Sustainability. *Management Decision*, Vol. 46, No. 3, pp. 433–448.

Arenas, D., Lozano, J. M., & Albareda, L. (2009). The Role of NGOs in CSR: Mutual Perceptions among Stakeholders. *Journal of Business Ethics*, Vol. 88, pp. 175–197.

Aslaksen, H. M., Hildebrandt, C., & Johnsen, H. C. (2021). The Long-Term Transformation of the Concept of CSR: Towards a More Comprehensive Emphasis on Sustainability. *International Journal of Corporate Social Responsibility*, Vol. 6, No. 11, pp. 1–14.

Ayres, R. U., Van Den Berg, J. C., & Gowdy, J. M. (2001). Strong versus Weak Sustainability: Economics, Natural Sciences, and "Consilience". *Environmental Ethics*, Vol. 23, No. 2, pp. 155–168.

Bansal, P., & Song, H.-C. (2017). Similar But Not the Same: Differentiating Corporate Sustainability from Corporate Responsibility. *Academy of Management Annals*, Vol. 11, No. 1, pp. 105–149.

Baral, N., & Pokharel, M. P. (2017). How Sustainability Is Reflected in the S& P 500 Companies' Strategic Documents. *Organization & Environment*, Vol. 30, No. 2, pp. 122–141.

Barkemeyer, R. (2011). Corporate Perceptions of Sustainability Challenges in Developed and Developing Countries: Constituting a CSR Divide? *Social Responsibility Journal*, Vol. 7, No. 2, pp. 257–281.

Baumgartner, R. J. (2014). Managing Corporate Sustainability and CSR: A Conceptual Framework Combining Values, Strategies and Instruments Contributing to Sustainable Development. *Corporate Social Responsibility and Environmental Management*, Vol. 21, pp. 258–271.

Broman, G. I., & Robert, K.-H. (2017). A Framework for Strategic Sustainable Development. *Journal of Cleaner Production*, Vol. 140, pp. 17–31.

Carroll, A. B. (1991). The Pyramid of Corporate Social Responsibility: Toward the Moral Management of Organizational Stakeholders. *Business Horizons*, Vol. 34, No. 4, pp. 39–48.

Christensen, H. B., Hail, L., & Leuz, C. (2021). Mandatory CSR and Sustainability Reporting: Economic Analysis and Literature Review. *Review of Accounting Studies*, Vol. 26, pp. 1176–1248.

Clarkson, M. B. (1995). A Stakeholder Framework for Analyzing and Evaluating Corporate Social Performance. *The Academy of Management Review*, Vol. 20, No. 1, pp. 95–117.

Clément, A., Robinot, É., & Trespeuch, L. (2022). Improving ESG Scores with Sustainability Concepts. *Sustainability*, Vol. 14, No. 13154, pp. 1–15.

Clift, R. (2003). Metrics for Supply Chain Sustainability. *Clean Technologies and Environmental Policy*, Vol. 5, pp. 240–247.

Danciu, V. (2013). The Sustainable Company: New Challenges and Strategies for More Sustainability. *Theoretical and Applied Economics*, Vol. XX, No. 9(586), pp. 7–26.

Daub, C.-H. (2007). Assessing the Quality of Sustainability Reporting: An Alternative Methodological Approach. *Journal of Cleaner Production*, Vol. 15, pp. 75–85.

Ditlev-Simonsen, C. D., & Wenstøp, F. (2013). How Stakeholders View Stakeholders as CSR Motivators. *Social Responsibility Journal*, Vol. 9, No. 1, pp. 137–147.

Doh, J. P., Littell, B. B., & Quigley, N. R. (2015). CSR and Sustainability in Emerging Markets: Societal, Institutional, and Organizational Influences. *Organizational Dynamics*, Vol. 44, pp. 112–120.

Dyllick, T., & Hockerts, K. (2002). Beyond the Business Case for Corporate Sustainability. *Business Strategy and the Environment*, Vol. 11, pp. 130–141.

Feng, M., Wang, X., & Kreuze, J. G. (2017). Corporate Social Responsibility and Firm Financial Performance; Comparison Analyses across Industries and CSR Categories. *American Journal of Business*, Vol. 32, No. 3/4, pp. 106–133.

Fidanboy, M. (2022). *Organizations and Complex Adaptive Systems*. New York: Routledge.

Filho, W. L., Tripathi, S. K., Guerra, J. B., Giné-Garriga, R., Lovren, V. O., & Willats, J. (2019). Using the Sustainable Development Goals Toward a Better Understanding of Sustainability Challenges. *International Journal of Sustainable Development & World Ecology*, Vol. 26, No. 2, pp. 179–190.

Fraser, I. J., Müller, M., & Schwarzkopf, J. (2020). Dear Supplier, How Sustainable Are You?; A Multiple-Case Study Analysis of a Widespread Tool for Sustainable Supply Chain Management. *Nachhaltigkeits Management Forum*, Vol. 28, pp. 127–149.

Gillan, S. L., Koch, A., & Starks, L. T. (2021). Firms and Social Responsibility: A Review of ESG and CSR Research in Corporate Finance. *Journal of Corporate Finance*, Vol. 66, No. 101889, pp. 1–16.

Giovannoni, E., & Fabietti, G. (2013). What Is Sustainability? A Review of the Concept and Its Applications. In C. Busco, M. L. Frigo, A. Riccaboni, & P. Quattrone (eds.), *Integrated Reporting; Concepts and Cases That Redefine Corporate Accountability* (pp. 21–40). Cham: Springer International Publishing.

Glavic, P., & Lukman, R. (2007). Review of Sustainability Terms and Their Definitions. *Journal of Cleaner Production*, Vol. 15, pp. 1875–1885.

González, A. L., Martín, J. Á., Vaca-Tapia, A. C., & Rivas, F. (2021). How Sustainability Is Defined: An Analysis of 100 Theoretical Approximations. *Mathematics*, Vol. 9, No. 1308, pp. 1–20.

Hastalona, D., & Sadalia, I. (2021). Literature Review ESG and Sustainability Finance. *Budapest International Research and Critics Institute-Journal (BIRCI-Journal)*, Vol. 4, No 3, pp. 3548–3557.

Hristov, I., & Chirico, A. (2023). The Cultural Dimension as a Key Value Driver of Sustainable Development at a Strategic Level: An Integrated Five-Dimensional Approach. *Environment, Development and Sustainability*, Vol. 25, pp. 7011–7028.

Hutt, R. W. (2010). Identifying and Mapping Stakeholders: An Industry Case Study. *Corporate Communications: An International Journal*, Vol. 15, No. 2, pp. 181–191.

Jain, R., & Winner, L. H. (2016). CSR and Sustainability Reporting Practices of Top Companies in India. *Corporate Communications: An International Journal*, Vol. 21 No. 1, pp. 36–55.

Jiang, P.-C., Feng, G.-F., Wang, H.-J., & Chang, C.-P. (2024). CSR from Different Perspectives: The Global ESG Indexes Updated. *Corporate Social Responsibility and Environmental Management*, Vol. 31, pp. 4694–4714.

John, L., & Narayanamurthy, G. (2015). Converging Sustainability Definitions: Industry Independent dimensions. *World Journal of Science, Technology and Sustainable Development*, Vol. 12, No. 3, pp. 206–232.

Kaźmierczak, M. (2022). A Literature Review on the Difference between CSR and ESG. *Scientific Papers of Silesian University of Technology, Organization and Management Series*, Vol. 162, pp. 275–289.

Kuhlman, T., & Farrington, J. (2010). What Is Sustainability? *Sustainability*, Vol. 2, pp. 3436–3448.

Kumar, V., Rahman, Z., & Kazmi, A. (2016). Stakeholder Identification and Classification: A Sustainability Marketing Perspective. *Management Research Review*, Vol. 39, No. 1, pp. 35–61.

Li, T.-T., Wang, K., Sueyoshi, T., & Wang, D. D. (2021). ESG: Research Progress and Future Prospects. *Sustainability*, Vol. 13, No. 11663, pp. 1–28.

Lozano, R. (2018). Proposing a Definition and a Framework of Organisational Sustainability: A Review of Efforts and a Survey of Approaches to Change. *Sustainability*, Vol. 10, No. 1157, pp. 1–21.

Maas, S., & Reniers, G. (2014). Development of a CSR Model for Practice: Connecting Five Inherent Areas of Sustainable Business. *Journal of Cleaner Production*, Vol. 64, pp. 104–114.

Mainardes, E. W., Alves, H., & Raposo, M. (2012). A Model for Stakeholder Classification and Stakeholder Relationships. *Management Decision*, Vol. 50, No. 10, pp. 1861–1879.

Meseguer-Sánchez, V., Gálvez-Sánchez, F. J., López-Martínez, G., & Molina-Moreno, V. (2021). Corporate Social Responsibility and Sustainability. A Bibliometric Analysis of Their Interrelations. *Sustainability*, Vol. 13, No. 1636, pp. 1–18.

Miles, S. (2017). Stakeholder Theory Classification: A Theoretical and Empirical Evaluation of Definitions. *Journal of Business Ethics*, Vol. 142, pp. 437–459.

Millon, D. K. (2011). Two Models of Corporate Social Responsibility. *Wake Forest Law Review*, Vol. 46, pp. 523–540.

Nishant, R., Kennedy, M., & Corbett, J. (2020). Artificial Intelligence for Sustainability: Challenges, Opportunities, and a Research Agenda. *International Journal of Information Management*, Vol. 53, No. 102104, pp. 1–13.

O'Connor, M., & Spangenberg, J. H. (2008). Methodology for CSR Reporting: Assuring a Representative Diversity of Indicators across Stakeholders, Scales, Sites, and Performance Issues. *Journal of Cleaner Production*, Vol. 16, pp. 1399–1415.

Passas, I. (2024). The Evolution of ESG: From CSR to ESG 2.0. *Encyclopedia 2024*, Vol. 4, pp. 1711–1720.

Ríos-Osorio, L., Cruz-Barreiro, I., & Welsh-Rodríguez, C. (2013). The Concept of Sustainable Development from an Ecosystem Perspective: History, Evolution, and Epistemology. In A. Yáñez-Arancibia, R. Dávalos-Sotelo, J. W. Day, & E. Reyes (eds.), *Ecological Dimensions for Sustainable Socio Economic Development* (Vol. 64, Chapter 2, pp. 29–46). Boston, MA: WIT Press.

Salas-Zapata, W. A., & Ortiz-Muñoz, S. M. (2019). Analysis of Meanings of the Concept of *Sustainability. Sustainable* Development, Vol. 27, pp. 153–161.

Schiano, A. N., & Drake, M. A. (2021). Invited Review: Sustainability: Different Perspectives, Inherent Conflict. *American Dairy Science Association*, Vol. 104, pp. 11386–11400.

Seghezzo, L. (2009). The Five Dimensions of Sustainability. *Environmental Politics*, Vol. 18, No. 4, pp. 539–556.

Sheehy, B., & Farneti, F. (2021). Corporate Social Responsibility, Sustainability, Sustainable Development and Corporate Sustainability: What Is the Difference, and Does It Matter? *Sustainability*, Vol. 13, No. 5965, pp. 1–17.

Simone, L. D., Petracci, B., & Piva, M. (2022). Economic Sustainability, Innovation, and the ESG Factors: An Empirical Investigation. *Sustainability*, Vol. 14, No. 2270, pp. 1–16.

Tokoro, N. (2007). Stakeholders and Corporate Social Responsibility (CSR): A New Perspective on the Structure of Relationships. *Asian Business & Management*, Vol. 6, pp. 143–162.

Torelli, R., Balluchi, F., & Furlotti, K. (2020). A Materiality Assessment and Stakeholder Engagement: A Content Analysis of Sustainability Reports. *Corporate Social Responsibility and Environmental Management*, Vol. 27, pp. 470–484.

Tsang, A., Frost, T., & Cao, H. (2023). Environmental, Social, and Governance (ESG) Disclosure: A. The *British Accounting Review*, Vol. 55, No. 101149, pp. 1–21.

Tu, Y.-X., Kubatko, O., Piven, V., Kovalov, B., & Kharchenko, M. (2023). Promotion of Sustainable Development in the EU: Social and Economic Drivers. Sustainability, Vol. 15, No. 7503, pp. 1–15.

United Nations Department of Economic and Social Affairs. (2024). *The Sustainable Development Goals Report 2024.* New York: United Nations Publ.

Vogt, M., & Weber, C. (2019). Current Challenges to the Concept of Sustainability. *Global Sustainability*, Vol. 2, No. e4, pp. 1–6.

Wagenhofer, A. (2024). Sustainability Reporting: A Financial Reporting Perspective. *Accounting in Europe*, Vol. 21, No. 1, pp. 1–13.

2
Sustainability in IT Sector

Sustainability journey of different companies, their start points, check-points, and their finish point may differ from each other. Exploring the literature portrays some critical points of this journey:

- Implementing environmental and social management systems and reporting their progress.
- Getting certified on environmental and social issues.
- Utilizing market mechanisms, like using eco-badges.
- Reporting and disclosing financial statements, followed by social and environmental information (Chabrak, 2015).

Some sectors might have already reached all the points, and they may have added some new destinations to their journey as well. However, when looking at the IT sector, some points have not been arrived at; some are missed intentionally; and the remaining points are where they are currently present. Let us go through each of the points mentioned above in the next paragraphs:

Firstly, sustainability management system, particularly integrated sustainability management systems, are gaining popularity, and there is a need for more experts and consultants in the related fields.

Secondly, certifications are of great value among IT companies, specifically those related to employees' and customers' privacy and information security, which can be grouped under the social dimension of sustainability. Certifications associated with the workforce welfare, health and safety, and environment are becoming appealing to IT companies as well, and in the very near future, competition will grow fiercer in this regard.

DOI: 10.1201/9781003587583-2

Thirdly, what seems to be a necessity is an eco-badge for software products, including digital and AI-based solutions. With the increasing adoption of AI solutions, the need for entities to control, monitor, and audit them concerning their sustainability will be amplified, and the eco-badge is a way for companies to communicate their sustainability commitment and responsibility to their stakeholders, specifically their customers. Some companies have begun adding a carbon footprint calculator to their corporate websites, to inform visitors of the carbon emissions they contribute to, aiding a lot with raising awareness.

Fourthly, sustainability reports of many IT companies can be accessed online, and some have been releasing their reports for many years. The IT companies are becoming more and more interested in disclosing their sustainability reports owing to the fact that it can be considered as a great way to draw the attention of investors, and investment in their new ideas and innovations is what matters a lot to IT companies. You might have come across posts and news about IT companies that have been successful in receiving investments, and as you can guess, it is neither easy nor uncompetitive.

As a concluding point, it is noteworthy that there is a hope for integrating sustainability into the products of IT companies, like many other practices like privacy and security. This can be a notion similar to or the same as the concept of "design for sustainability", which is systematically taking sustainability into account over the whole life-cycle of product and process (Bakshi & Fiksel, 2003).

2.1 IT Sector Characteristics

Dynamic and Turbulent: The changes occurring within the IT sector make it dynamic. These changes do not only encompass the advances of new technologies. The changes can be associated with producers, suppliers, solutions, customers, methodology, frameworks, models, and so on.

Emergence of New Tools and Technologies: Many may defend that new tools are being introduced into many sectors and not just IT. Thus, it cannot be considered a unique characteristic. The truth is that we experienced a similar situation when AI began to become part of our daily lives. Examples are not few. The main difference is the impact. The impact of IT technologies on our lives, on other sectors, and on other fields of study is huge. This can be easily perceived by looking at leveraging AI in multi-disciplinary areas. For example, models are now helping medical doctors to predict the possibility of a disease occurrence in individuals before the symptoms even start. The food industry is utilizing AI to understand when the substance of a product begins to change during time, or they even use it for enriching the taste of their products. Thus, at the first place, this is the impact of IT tools and technologies which make IT sector distinguishable from others.

Secondly, the other sectors rely heavily on the IT sector. For example, medical records are recorded on health information systems. Designs are mostly prepared by using computers in different sectors. Monitoring and controlling of processes in manufacturing is being carried out via various software programs. There are myriads of examples that can highlight this dependency.

Rapid Formation and Dissolution of Companies: Many IT companies have started their business from a small room, or a computer, and large premises are not needed in the early stages. This leads to rapid changes in the dynamics of the companies entering the sector and those leaving it. Not all start-ups fail, even though the percentage is not low. There are many start-ups that have been successful in making their way to become unicorns. The ease of entering the sector tempts those with new ideas to establish companies, and the high competition, together with the quickness of imitating the products, in particular, the digital products, are influential factors in the formation and dissolution of companies.

Newly Introduced Products: A wide range of new products and services are introduced every day. This can be a computer game or an integrated sustainability management system solution. In most cases, they are introduced quickly so that the company can exceed the rivals and sell the products till the other companies come up with an alternative. For this reason, their impact on society, environment, and economy cannot be well inspected. This is like the case for the solutions that consumed enormous amount of energy.

2.2 Challenges

1. The main challenge in IT companies is creating a sustainability department/unit or if this department already exists, taking sufficient budget for the initiatives.
2. The other difficulty is to internalize the values of sustainability-oriented behavior. For instance, employees who turn off the lights of empty meeting rooms, even if they were not the ones who used the room before, are those who have already internalized these values.
3. Owing to the wide range of sustainability activities, collaboration is strongly required between the departments. Sometimes, departments other than sustainability see the related work as an extra workload, and the endeavor which is not counted anywhere. This then results in performing the tasks incompletely or delaying them as far as they can.
4. Lack of the sufficient knowledge is another problem. Those who know sustainability, do not have much information about the IT sector and vice versa.
5. Legal aspect of sustainability covers a huge range of topics. For example, which documents should be approved and by whom, or which documents should be signed, or how to practice due diligence.
6. Within IT companies, we are often impatient to see the results because we have gotten used to viewing them in real time.
7. Due to the dynamism of the sector, identification of the stakeholders is difficult if the changes are associated with them. The changes with the products and solutions also compel the company to focus on how to react to the changes. Therefore, their priorities are altered quickly.
8. There are expectations from IT companies to contribute to sustainability by using their latest tools and technologies. This can become like a pressure from the public and the shareholders.
9. The instability of the market in terms of entrants and leavers makes the survival the first locus of focus.
10. The new products either digital or physical may not be inspected for their impact on sustainability, and raise doubt in shareholders, including customers concerning their purchase or for investors, regarding their investment options.

2.3 Solutions

Numbering of the solutions is consistent with the number of corresponding challenges.

1. The solution to the first challenge seems to be only possible through emphasizing the financial gain of these initiatives, and embellishing it with cases, examples, and studies that have indicated the role of sustainability activities on improving the financial status of companies. Sustainability by design or incorporating the sustainability into the production of IT products is the ideal one, however, it has a long way to go.
2. By investing in cultivating the organizational culture, and trying to strengthen the organizational values, this situation can be improved.
3. There are two possible ways to motivate employees for collaboration. First is to justify that what is done within the scope of sustainability, and in particular in terms of the social dimension, is reflected in their working conditions and health at the workplace. Since the social dimension begins with concentration on the employees, examples can be brought to help them understand the significance of actions. Secondly, sustainability-related goals can be integrated into their objectives to assure them that what they do is counted and is of value.
4. It is an easy obstacle to overcome if the employees are open to learning. The repetitive tasks of sustainability in IT sector are search, research, and action. How repetitive can it be when it is spoken of research!?
5. Several actions can be considered for removing the hurdle. Building cross-functional sustainability team with members from legal department is one way. Offering related trainings to the sustainability experts within the company can be the other option. Benefiting from consultancy services can be another solution.
6. Patience is the fundamental value for sustainability activities. You, as the sustainability experts, should be patient with other teams and departments, like when explaining the concepts to

them or when asking for their collaboration. Moreover, you should be patient to see the gains of sustainability actions, and so should the top management and the other stakeholders. In particular, the financial gain from sustainability initiatives is not accomplished in a short time.

7. Having a sustainability department with sufficient budget can enable the company to engage with other changes, while the sustainability department continues its daily activities.
8. Partnerships with other companies who are running sustainability projects is a great choice as the resources and the tasks required are distributed between different parties. Besides, the workload would not be as much as carrying out a project on their own. In addition, being equipped with knowledge and skill from different areas of expertise facilitates creating more innovative and practical solutions.
9. Sustainability-initiatives can be considered as competitive advantage which can turn them into an influential factor in the survival of companies. Therefore, this topic should be discussed in meetings, and its importance should be underlined.
10. Products can be mapped in accordance with their contribution to different sustainable goals, like SDGs. Many IT companies have already mapped their products; nonetheless, there is a need for much more initiatives to be taken.

References

Bakshi, B. R., & Fiksel, J. (2003). The Quest for Sustainability: Challenges for Process Systems Engineering. *AIChE Journal*, Vol. 49, No. 6, pp. 1350–1358.

Chabrak, N. (2015). Promoting Corporate Social Responsibility and Sustainability: A Model of Integrity. *Society and Business Review*, Vol. 10, No. 3, pp. 280–305.

3
Documentation

On the way to make an IT company prepared for a sustainability audit, you may find that some documents should be revised or new documents should be created. The objectives of these documents are beyond the audit and based on their types, they have a goal to accomplish. The samples of such documents can be easily found on the internet. However, the format of all the documents within a company should be consistent and should be created based on the company's corporate format. Being sure to keep versions' history and let time for the policy to be implemented within a company after its creation or revision are the points that need extra care. A document that was written a day or a week before the audit does not fulfil its main purpose, but it is perceived to be developed just to be there, not more.

3.1 Types of Documents and Their Structures

3.1.1 Statement

The first notions that come to mind when talking of a statement in the context of businesses are the vision and mission statement, which show where a company goes and what it does, respectively. The examples of these statements are placed on the corporate websites, either under the same headings, or as where we go and what we do. The vision and mission statements of the companies would be more appropriate to entail sustainability-related material for the reason that these statements are among the basic information investors and other stakeholders want to learn about a company at the first step.

3.1.2 Policy

Policies were defined as the principles that aid with decision making in terms of practices, procedures, and actions (Randolph, 2006). Policies are

DOI: 10.1201/9781003587583-3

required to be structured according to a corporate template within a company, like other documents. The template should be better to be selected with care, preferring the one that is detailed and comprehensive. This will guarantee that the policies meet the requirements of most auditors in terms of the format. The policy may encompass the following details:

- The definition of the terms used.
- The main body of policy consisting of principles.
- Laws, regulations, and rules the policy is related to and/or based on.
- The scope regarding time, place, and entities (individuals, units, or departments), and the purpose of the policy.
- The consequences of violating the policy.
- Contact information if further communication is needed.
- Details related to who prepares the documents, together with approvals, changes, and revisions.
- Duties and responsibilities of different entities with regard to the policy.
- Main quantitative goals within the context of policy. For instance, achieving zero customer grievance cases.

3.1.3 Procedure

Procedures are steps written in order that should be taken for policy execution (Randolph, 2006). In IT companies, procedures are usually illustrated alongside visuals since employees are used to creating procedures for many production-related subjects as well. Procedures are like tools assisting the sustainability experts, picturing the steps to reach the goals for other departments. To make it clearer, if data related to ethics are required, the sustainability unit/department can visualize the procedure to facilitate the understanding and collaboration of the other department. Suitability experts should not forget that they are the ones who know what is required to be done in terms of sustainability and guiding the other departments is among their responsibilities.

3.1.4 Code of Conduct

A code of conduct is a document including the principles a business complies with, together with expressing the expectations of the company from different stakeholders, such as employees, suppliers, and business partners. An example is the supplier code of conduct when the principles according to which the company builds relationships with its suppliers and treats them, along with the expected behavior and actions it expects from suppliers take place. In most cases, codes of conduct should be both communicated and accepted by the party it is targeted at.

3.1.5 Report

Reports are documents used to transfer information to the other parties other than the reporters. Based on the context and the purpose of reporting, they may or may not include visuals. Regarding sustainability, two types of reports are discussed within this book: sustainability reports, which are comprehensive reports illustrating the contributions of a company to sustainability, like their initiatives to decrease the negative environmental or social impact. The second is the internal report that might also be used by external parties like auditors, exhibiting the progress of the actions taken to accomplish sustainability goals, sustainability KPIs, or any other sustainability indicators. The example can be the DEI statistics report, which can be created annually and helps managers to track the values and assess the efficiency and effectiveness of their actions. Such a report can be used by human resource managers, sustainability managers, or top managers to review their strategies. Sustainability auditing organizations will also accept these reports if they are meeting the requirements, such as being in the desired format, containing information regarding the time the data correspond to, having a document number just like any other document in the company, and having evidence that they are accessible by the relevant stakeholders. These reports are of great importance because they ensure the auditing entity that everything is under control and is being tracked continuously, and as it is obvious, evidence for performing, monitoring, and controlling processes is among the main practices sought by auditors during audits. These provide answers to three main questions:

- Do you do practice A?
- Do you track the progress of practice A periodically?
- Do you control whether the desired results were gained or not?

When looking at reports in general, the first questions that come to mind are:

- What is expected to be reported?
- What can be reported?
- How can it be reported, in terms of the tools and methods which are employed?

You may have seen different types of report in your career journey or even on the news in daily life. Some reports may contain more words and qualitative information while the others include numbers and quantitative data. Those who are familiar with research methods know the difference between these two approaches well. While the qualitative approach tries to describe phenomena and examine it with the use of textual data, quantitative methods are employed when numerical data is required. It encapsulates works like finding trends in the data, positive or negative relationships, causal relationships, patterns, and so on. Benefiting from these notions in sustainability can equip us with a good knowledge of a wide range of methods and tools, enable us to collect appropriate data, followed by preparing suitable reports. Concerning sustainability audits, there are several goals for quantitative reporting:

- Depicting the current situation or to be clearer, being able to show the values for current time and place.
- Maintaining the old data correctly and being able to retrieve it.
- Containing understandable and meaningful visuals (graphs, charts, etc.) illustrating patterns and trends in data

3.1.5.1 Reporting Standards

Giving a clear definition for sustainability reporting and comprehending its role are difficult in light of three factors: the existence of numerous and diverse notions, the presence of several organizations associated with reporting, and different regulations concerning reporting, whether voluntary or mandatory (Stolowy & Paugam, 2023). The notable issues for IT companies are:

- Determining the audience of your company's sustainability reporting, whether it is national or international, or both. This helps with the selection of the right reporting standard.

- Identifying your customers, or more specifically, to which country your products, even digital ones, are planned to be sold to. Some countries require the imported goods to comply with specific regulations and disclose specific information.
- Staying updated and informed about the regulations and reporting standards. Since sustainability is being paid more and more attention to gradually, new standards may emerge, the existing ones may be modified, or multiple reporting standards can be integrated. Hence, following the trends will benefit the company in capturing the right expectations and, consequently, planning for the appropriate response.

Sustainability reporting standards provide guidance on what to report and how to report it. Sustainability assessors and/or auditors might require companies to conform to these standards. These standards generally include the metrics of diverse dimensions of sustainability, and companies should be able to report the values for those metrics.

The reporting standards can be viewed under two categories of industry-focused standards and industry approach standards, where the examples of the first is SASB and for the second is GRI, as briefly mentioned in a study (Torres, Ripa, Jain, Herrero, & Leka, 2023). By looking at the examples, it can be realized that industry-focused standards are those primarily prepared for specific industries while the industry approach is the one taking the industry into consideration within their content; however, they were created as a general standard at the very first sight.

One of the serious problems with the sustainability reporting standard is that they were formed in response to the meeting the expectations of some key actors, and the expectations of a large group of stakeholders are not taken into account, for instance, focusing on the investors' and market participants' expectations solely (Ali, Fukofuka, & Narayan, 2023). Thus, some kind of tailoring or more precisely, adding issues coming from the materiality analysis to the report can be fruitful in minimizing this problem. When looking at the sustainability reports, it can be realized that most companies are reporting in the same manner, incorporating information and topics they have found suitable, and then specify and address the parts according to the reporting standards of their choice, for instance, using the GRI content index.

It would be useful to have a look at the difference between sustainability reporting among different organizations. In a study, engagement with the UN SDGs and reporting were indicated to vary between companies and educational organizations in their defined sample, where findings showed that for-profit companies tackle more UN SDGs, and release their sustainability reports more (Hamilton & Waters, 2022). Examining the sustainability report of the IT companies also implies that a remarkable number of these reports have incorporated UN SDGs as the main body of their reports or have acknowledged them numerous times within their content. This can be due to the fact that UN SDGs are among the best-known goals and targets by stakeholders, and associating with them makes the efforts and initiatives of the companies in terms of sustainability easily and simply more visible and comprehensible.

3.1.5.1.1 Global Reporting Initiative (GRI) GRI is a leading global authority that has created a standard framework for sustainability reporting (Dilling, 2010). The same study found that being based in Europe, or being part of energy or producing sectors, or having a greater profit margin, increases the probability of generating sustainability reports of high quality (Dilling, 2010). There are many IT companies that publish their report compliant with the GRI standards. Some companies include their GRI content index at the end of their annual or periodic sustainability reports, and some add it as a separate document to their corporate website, most probably with hyperlinks to the report. The importance of an interactive report that make the stakeholders engaged with it should neither be ignored nor underestimated. Digital reports are of paramount significance for two reasons: the first is that when we are talking of sustainability, digital reports are mostly desired, and preferred over physical and printed reports; the second reason is that where stakeholders find information about companies and gain knowledge about them is their corporate websites, not the physical archives, and what can be added to these websites are digital reports, of course. Thus, both the content and the design of the reports should be prepared taking the following main factors into account: the reporting standard requirements, stakeholders' expectations, and the company.

3.1.5.1.2 Sustainability Accounting Standards Board (SASB) SASB has made attempts to merge financial materiality with ESG by gaining benefit from materiality for financial reporting purposes, contributing to dragging sustainability into the mainstream of corporate businesses (Jebe, 2019). The importance of this integration rests in the fact that corporate performance measurement is no longer based only on financial values, and environmental, social, and economic contributions play key roles in the success of companies (Hastalona & Sadalia, 2021).

Exploring the literature helps with understanding which companies select GRI over SASB and which favor SASB for their reporting. It was indicated that factors linked to the companies' backgrounds in finance resulted in the selection of SASB, while the factors relevant to sustainable development led to the preference of GRI (Pizzi, Principale, & De Nuccio, 2023).

3.1.5.1.3 ISSB (International Sustainability Standards Board) The International Financial Reporting Standards (IFRS) Foundation has gathered entities providing investor-oriented sustainability standards under the umbrella of its ISSB, addressing how opportunities and risks associated with social and environmental issues affect the disclosing companies, contrasting with stakeholder-oriented standards (De Villiers, La Torre, & Molinari, 2022).

3.1.5.1.4 Task Force on Climate-related Financial Disclosures (TCFD) It was first issued in 2017, and the guidelines contains recommendations for accounting when evaluating the impact of climate-related issues on the performance of the company, and when this information will be embodied in annual report or along with it (Baumüller & Sopp, Double Materiality and the Shift from Non-Financial to European Sustainability Reporting: Review, Outlook, and Implications, 2022).

3.1.5.1.5 EU Non-Financial Reporting Directive (NFRD) Non-financial reporting is considered an important initiative toward the integration of sustainability-related issues into the companies' strategies (Aluchna, Roszkowska-Menkes, & Kaminski, 2023). The objective of NFRD is to enhance the disclosure of non-financial information by public interest entities having more than 500 employees, which should publish reports on their policies like due diligence process, business models, risk management, KPIs, and outcomes (European Parliamentary Research Service (EPRS), 2025). NFRD was mentioned to make room for companies to disclose information as they deem appropriate (Aluchna, Roszkowska-Menkes, & Kaminski, 2023).

3.1.5.1.6 Directive on Corporate Sustainability Reporting (CSRD) CSRD was aimed to replace reporting requirements proposed for EU accounting law by NFRD in 2014, and through scope extension, like including small and medium-sized companies, and introducing new topics for reporting, it redesigns the European reporting framework regarding sustainability (Baumüller & Grbenic, Moving from Non-Financial to Sustainability Reporting: Analyzing the EU Commission's Proposal for a Corporate Sustainability Reporting Directive (CSRD), 2021). The companies impacted by CSRD, publishing reports in 2025 covering the financial year of 2024, have to adopt the new rules (European Commission, 2025).

3.1.5.1.7 UN Global Compact's COP (Communication on Progress) Its mission was mentioned as fostering businesses to enact policies concerning sustainability and social responsibility, and reporting on them which is based on ten key principles, and its scope embodies labor (four principles), human rights (two principles), environment (three principles), and anti-corruption (one principle) (Jose, 2017; United Nations Global Compact, 2025).

3.1.5.1.8 CDP (Carbon Disclosure Project) CDP is a non-profit organization, and a database for disclosure of strategies related to GHG, water usage, and climate change, globally, and factors accounted for the assessments entail risks and opportunities due to climate change, and GHG supporting data management (Siew, 2015).

3.1.6 Materiality Assessment

Companies are required to evaluate and compare the materiality of ESG topics for their business in order to determine which non-financial data to include in their report (Garst, Maas, & Suijs, 2022). This article suggests a six-step guide for conducting materiality assessment (Garst, Maas, & Suijs, 2022).

There are materiality assessment tools and templates available online, and companies can take advantage of these tools. The key point here is to review the tool and its guide in order to become more familiar with how to conduct materiality assessment. At the end of the assessment, the tools usually provide a materiality matrix, which depicts the importance of issues to stakeholders and their impact on business; the first variable is posited on the y-axis and the second variable is shown on the x-axis. To clearly communicate the strategies of a company, this matrix, a related table, or just the descriptions of the matrix can be added to the sustainability report of the company.

Materiality assessment helps companies with:

- Deciding on which issues the company should focus on and to what degree.
- Evaluating the company's impact on each issue.
- Orchestrating sustainability management with sustainability strategies.
- Determining the components of the report (Taubken & Feld, 2018).

The other subject that needs to be paid enough attention to is the identification of the stakeholders. This is of much significance because the inability to incorporate the stakeholders' views in the materiality analysis well and failing to conduct a good materiality analysis may cause conflicts with the stakeholders (Beske, Haustein, & Lorson, 2020). Thus, likewise in the literature, capturing all the aspects of importance from different viewpoints of stakeholders, internal analysis, and strategy is a critical issue (Calabrese, Costa, Ghiron, & Menichini, 2017).

Not including the process through which the materiality assessment is performed is another issue, leading to a situation where the way the reporter identifies material issues is vague to the readers, and the findings obtained from the sample of a study points to a decrease in the disclosure of such information, implying that managerial capture is present (Farooq, Zaman, Sarraj, & Khalid, 2021).

The other problem identified with the materiality analysis is the lack of a structured materiality approach (Calabrese, Costa, Ghiron, & Menichini, 2017). This can be resolved by using scientific methods of decision-making support tools and methods like the Multi-Criteria

Decision Making (MCDM) method, which, according to literature (Calabrese, Costa, Ghiron, & Menichini, 2017), was stated to be used earlier for this purpose.

Another relevant concept is double and triple materiality. By double materiality, it is indicated that risks and opportunities are significant from financial and non-financial viewpoints, while triple materiality additionally brings the sustainability context into attention (Beyne & Moratis, 2025).

The other recommendation suggested by a study is to combine financial materiality in evaluating ESG performance for the following reasons:

- Behind investing socially responsibly lies maximizing the profit,
- And internal governance like financial materiality is advocated by investors (Madison & Schiehll, 2021).

You may be among those who defend their standpoint regarding taking financial and non-financial (ESG) reporting apart from each other. It is a debatable topic. Nonetheless, we should not forget that one of the reasons behind sustainability reporting is attracting investors, and it cannot be denied that investment is about money and profit. Thus, trying to balance reporting to meet all the above purposes will be beneficial to the company.

The other factors that can influence a company to disclose sustainability reports are the structure of the ownership, the size of the company, and media visibility, whereas corporate governance has an impact on the presence of audits or a sustainability committee within a company (Dienes, Sassen, & Fischer, 2016).

The other study, within its sample, found that utilizing sustainability reporting as a public relations strategy incentivizes companies to publish their sustainability reporting, and sustainability reporting has turned into a part of a business strategy for obtaining a competitive advantage (Amran & Haniffa, 2011). IT companies have already begun to use their reports for attracting the attention of potential customers.

Going through the sustainability reports of IT companies published so far signifies that some financial data are included in the reports, for instance, the amount donated to charity organizations.

or the monetary value corresponding to voluntary work. It seems that companies are already aware of the truth that reporting some financial values, like the contributions, can help with the understanding of the effort, time, and resources spent more clearly.

What should not be ignored during all the sustainability initiatives is transparency. Every detail that takes place in any kind of documents, including reports, should be clearly described. During audits, it even becomes much more significant and serious, and it can be said that the number of cases the companies fail to meet the requirements of sustainability audits due to the ambiguity and vagueness of the documents provided is not low. You don't want to lose points and be assigned corrective actions for the lack of transparency while you have done everything perfectly and timely.

3.2 Common Mistakes and Tips

- Sometimes, it is mistakenly believed that the strategies are basis for materiality analysis and the analysis should be performed with regard to the sustainably strategies of the company. This occurs when materiality analysis is run just for the purpose of being done, and being posited in the report, not for engaging the stakeholders and involving them in the sustainability-related strategy making. That might seem to be a short-term solution; however, it would not be useful in the long term, and the consequent strategy-making process will become problematic and challenging. It would be more suitable to conduct the materiality assessment and then formulate, revise, or alter the sustainability-related strategies. Based on the literature, materiality assessment will assist in understanding how a company's sustainability-related strategies are set (Beyne & Moratis, 2025).
- As indicated in the previous section, transparency should be placed at the top of our communication, either written or spoken. Deciding on the content of various documents is affected by the guides, and requirements expected to be met; nonetheless, the final decision is still ours, even though partially. In this case, make an endeavor to keep consistency between the reports, and report the same values for different time periods.

For example, if the company could not achieve 100% of its ESG goals but achieved 70%, there is still a percentage to be reported. Keeping in mind that every step counts in sustainability initiatives, 70% means efforts were made, time was spent, and resources were allocated.

The other truth is that as you are continuing on the path of sustainability, the expectations become more difficult to fulfil and the goals become more arduous to achieve. Hence, what has been really implemented within a company is supposed to be reported during audits and within the sustainability reports clearly. Sustainability auditors are well aware of all the facts mentioned, and they can almost predict how well a company can progress within a specific amount of time if the influencing factors remain almost the same.

To sum up, transparency and consistency are required to be paid attention to for both the sustainability audits and sustainability reporting.

- Before preparing any type of documentation for the sustainability audits, reviewing all the materials, guides, and instructions will save time and effort. There are companies that present their documentation in the formats not accepted and approved by the auditors. Their documents might not embody all the sections and details required, and unfortunately, this is the reason for many low scores, and subsequent corrective actions as a result of audits. What should not be misunderstood here is that we are not supposed to prepare documents for the purpose of audit, but to prepare documents for the use and implementation within the company, which at the same time, meet the requirements of audits. Documentation requirements of the sustainability audits are mostly consistent with widely known standards, frameworks, and guides. Thus, taking these internationally known standards as the basis, taking a glance at what sector leaders, competitors have done till now, and evaluating the best practices will assist in deciding the right content and format. The recommendation here is to conform to the most comprehensive format, and if needed, integrate several formats into one all-encompassing one to avoid further rework or failure. It is noteworthy that this cannot be interpreted as leaving the documents there in the archives, but they should be reviewed and modified periodically to be conformed with the latest standards, regulations, and frameworks, and a critical issue is to keep the history of versions and changes.

– Another issue worth noting is that documents should not be necessarily created from scratch. In most cases, documents are available, but lack some sections. Therefore, they should be enhanced and modified; some sections may be added, some sections may be replaced, and some should be expanded. Going with enriching the available documentation is the first and easiest choice. In case a specific document does not exist, a new one can be created after examining the requirements, guides, best practices, rivals' and sector leaders' documents available online, and so on.

References

Ali, I., Fukofuka, P. T., & Narayan, A. K. (2023). Critical Reflections on Sustainability Reporting Standard Setting. *Sustainability Accounting, Management and Policy Journal*, Vol. 14, No. 4, pp. 776–791.

Aluchna, M., Roszkowska-Menkes, M., & Kaminski, B. (2023). From Talk to Action: The Effects of the Non-Financial Reporting Directive on ESG Performance. *Meditari Accountancy Research*, Vol. 31, No. 7, pp. 1–25.

Amran, A., & Haniffa, R. (2011). Evidence in Development of Sustainability Reporting: A Case of a Developing Country. *Business Strategy and the Environment*, Vol. 20, pp. 141–156.

Baumüller, J., & Grbenic, S. O. (2021). Moving from Non-Financial to Sustainability Reporting: Analyzing the EU Commission's Proposal for a Corporate Sustainability Reporting Directive (CSRD). *Economics and Organization*, Vol. 18, No. 4, Special Issue, pp. 369–381.

Baumüller, J., & Sopp, K. (2022). Double Materiality and the Shift from Non-Financial to European Sustainability Reporting: Review, Outlook, and Implications. *Journal of Applied Accounting Research*, Vol. 23, No. 1, pp. 8–28.

Beske, F., Haustein, E., & Lorson, P. C. (2020). Materiality Analysis in Sustainability and Integrated Reports. *Sustainability Accounting Management and Policy Journal*, Vol. 11, No. 1, pp. 162–186.

Beyne, J., & Moratis, L. (2025). A Material World: How Can Materiality Assessments Be Used to Define Organizational Sustainability Priorities, While Taking into Account the United Nations' SDGs? *Corporate Governance: The International Journal of Business in Society*, Vol. 25, No. 1, pp. 47–70.

Calabrese, A., Costa, R., Ghiron, N. L., & Menichini, T. (2017). Materiality Analysis in Sustainability Reporting: A Method for Making It Work in Practice. *European Journal of Sustainable Development*, Vol. 6, No. 3 pp. 439–447.

De Villiers, C., La Torre, M., & Molinari, M. (2022). The Global Reporting Initiative's (GRI) Past, Present and Future: Critical Reflections and a Research Agenda on Sustainability Reporting (Standard-Setting). *Pacific Accounting Review*, Vol. 34, No. 5, pp. 728–747.

Dienes, D., Sassen, R., & Fischer, J. (2016). What are the Drivers of Sustainability Reporting?; A Systematic Review. *Sustainability Accounting, Management and Policy Journal*, Vol. 7, No. 2, pp. 154–189.

Dilling, P. F. (2010). Sustainability Reporting in a Global Context: What Are the Characteristics of Corporations That Provide High Quality Sustainability Reports – An Empirical Analysis. *International Business & Economics Research Journal*, Vol. 9, No. 1, pp. 19–30.

European Commission. (2025, April 13). *Corporate Sustainability Reporting.* Retrieved from European Commission; The Official Website of the European Union: https://finance.ec.europa.eu/capital-markets-union-and-financial-markets/company-reporting-and-auditing/company-reporting/corporate-sustainability-reporting_en

European Parliamentary Reasearch Service (EPRS). (2025, April 13). *Non-financial Reporting Directive.* Retrieved from European Parliament: https://www.europarl.europa.eu/RegData/etudes/BRIE/2021/654213/EPRS_BRI(2021)654213_EN.pdf

Farooq, M. B., Zaman, R., Sarraj, D., & Khalid, F. (2021). Examining the Extent and Drivers of Materiality Assessment Disclosures in Sustainability Reports. *Sustainability Accounting, Management and Policy Journal*, Vol. 12, No. 5, pp. 965–1002.

Garst, J., Maas, K., & Suijs, J. (2022). Materiality Assessment Is an Art, Not a Science: Selecting ESG Topics for Sustainability Reporting. *California Management Review*, Vol. 65(1), pp. 64–90.

Hamilton, S. N., & Waters, R. D. (2022). Mainstreaming Standardized Sustainability Reporting: Comparing Fortune 50 Corporations' and U.S. News &World Report's Top 50 Global Universities' Sustainability Reports. *Sustainability*, Vol. 14, No. 3442, pp. 1–17.

Hastalona, D., & Sadalia, I. (2021). Literature Review ESG and Sustainability Finance. *Budapest International Research and Critics Institute-Journal (BIRCI-Journal)*, Vol. 4, No. 3, pp. 3548–3557.

Jebe, R. (2019). The Convergence of Financial and ESG Materiality: Taking Sustainability Mainstream. *American Business Law Journal*, Vol. 56, No. 3, pp. 645–702.

Jose, T. (2017). Need for Harmonisation of Sustainability Reporting. *Journal of Finance and Economics*, Vol. 5, No. 6, pp. 253–258.

Madison, N., & Schiehll, E. (2021). The Effect of Financial Materiality on ESG Performance Assessment. *Sustainability*, Vol. 13, No. 3652, pp. 1–21.

Pizzi, S., Principale, S., & De Nuccio, E. (2023). Material Sustainability Information and Reporting Standards: Exploring the Differences between GRI and SASB. *Meditari Accountancy Research*, Vol. 31, No. 6, pp. 1654–1674.

Randolph, S. A. (2006). Developing Policies and Procedures. *AAOHN Journal*, Vol. 54, No. 11, pp. 501–504.

Siew, R. Y. (2015). A Review of Corporate Sustainability Reporting Tools (SRTs). *Journal of Environmental Management* 1, Vol. 164, pp. 180–195.

Stolowy, H., & Paugam, L. (2023). Sustainability Reporting: Is Convergence Possible? *Accounting in Europe*, Vol. 20, No. 2, pp. 139–165.

Taubken, N., & Feld, T. Y. (2018). Impact Measurement and the Concept of Materiality – New Requirements and Approaches for Materiality Assessments. *NachhaltigkeitsManagementForum*, Vol. 26, pp. 87–100.

Torres, L., Ripa, D., Jain, A., Herrero, J., & Leka, S. (2023). The Potential of Responsible Business to Promote Sustainable Work – An Analysis of CSR/ESG Instruments. *Safety Science*, Vol. 164, No. 106151, pp. 1–10.

United Nations Global Compact. (2025, April 04). *The Power of Principles; The Ten Principles of the UN* Global Compact. Retrieved from United Nations Global Compact: https://unglobalcompact.org/what-is-gc/mission/principles

4

Actions

Sustainability-related issues and actions are illustrated as a hierarchy of four levels; levels 1 and 2, located at the bottom, address humans' survival and health; level 3 indicates extinction of species and human rights; and level 4 encompasses controversial subjects mainly (Marshall & Toffel, 2005). Coming to the practice, companies can benefit from such hierarchies and methods to determine and prioritize their sustainability-related actions. Since this book's locus of focus is sustainability audits, we can take a deeper look into this from a sustainability audit's perspective. The auditing entities inform the auditees (the companies that will be audited) about the scope of the audit, and for the case of sustainability, they may provide supporting documents and materials, or they may direct you to check some references and standards before the audit. They may tell you about the relative importance or even give you the weights corresponding to ESG dimensions, clarifying which is of more significance to your company, taking into account many factors, such as the sector, the size of the company, the annual revenue of the company, and the location of the company. In accordance with the weighted dimensions or the referred documents, the prioritization of the actions can be conducted. Discussing the usual practices in IT sector, the important issues in various dimensions of sustainability can be summarized as follows:

Concerning the *environmental* dimension:

- Reducing the energy consumption through controlling heating, air conditioning, and cooling (HAC) control systems.
- Increasing the usage of clean and renewable energy.
- Utilizing and producing energy-efficient products.
- Identifying the actions that lead to carbon emissions, creating a greenhouse gas (GHG) inventory, and measuring and reporting emissions across different scopes.

DOI: 10.1201/9781003587583-4

- Controlling for the water consumption, and searching for new solutions to optimize it.
- Monitoring the waste, particularly the e-waste, which is already a great challenge for the IT sector.

Regarding the *social* dimension:

- Incorporating gender equality into human resources practices, including recruitment, retention, training and development, promotion, and so on.
- Helping with the economic growth at diverse levels. For instance, by training people or encouraging your employees to develop their skills, leading to the improvement of the household economy.
- At a higher level, the contributions can be made by growing as a company in size or/and revenue, becoming a global company, sending your products beyond the borders of the country, and providing wider employment opportunities.
- Leveraging their skills, knowledge, and resources to back sustainability-related innovations and bring the creative ideas to life.
- Innovation is a closely tied concept with the IT companies since they are at the frontier of new technologies. For instance, by partnering for developing technologies for irrigation, water management, and agriculture.
- Developing and enriching practices to maintain and upgrade well-being, health, safety of employees, and their families, and to expand these practices to the society the company is in contact with.
- Upholding the privacy and security of the information of our employees, customers, stakeholders, and whom we keep their information.
- Contributing and making donations to reduce hunger, poverty, and support marine and terrestrial life.
- Devising a disaster plan for the company, and taking an active role in case of soothing the pain and alleviating the loss caused by disasters, especially natural disasters, within the society.
- Following the labor and human rights regulations and standards, and update the policies accordingly.

- Integrating the newly introduced or newly emphasized topics into the company's practices. For example, paying extra attention to the diversity of workforce and implementing the promotion, retention, and other human resources practices accordingly. Another example can be bringing the concept of forced labor or the right to disconnect into actions executed within the company.

In relation to the *governance* dimension:

- Planning for internal controls, both financial and non-financial controls, and communicating them with the relevant employees. These controls are better to be performed on a regular basis, so that diversion from the plans and desired states would become less likely to occur, and be corrected quickly.
- Integrating sustainability strategies into daily business operations and internalizing the sustainability-related responsibilities within the company.
- Controlling for the progress toward sustainability goals, and reflecting it in the company's performance.
- Maintaining an active and regular internal and external communication in terms of sustainability. Internal communication can be sending emails and organizing events on sustainability-related days like Earth Day.
- For external communication, the support of the marketing department and the public relations side of the communication should not be forgotten.
- Ensuring the collaboration of relevant departments. In accordance with what is stated within the book, sustainability initiatives should be governed by different departments within a company. As a solution, some companies have followed the principles used for quality and have delegated process owners for sustainability processes.

  The other solution is empowering the sustainability departments/units to carry out the processes and getting support from the other departments wherever required.

  Another solution is forming a sustainability committee within companies, comprising members coming from various departments, resulting in the team taking advantage of different insights and multiple perspectives. Furthermore, having a sustainability committee facilitates the decision-making, leading to much wiser decisions because of considering numerous viewpoints. It can be elaborated upon with an example. Human resources, legal, finance, and sustainability departments will have different perceptions when talking about diversity. The legal department thinks of its legal regulations, and the consequences of discriminating between employees, while human resources may consider it as the number of females in a sector like IT, and the finance department might regard it as a cause for more diverse expectations toward compensation and benefits. For the sustainability department, diversity is a favorable aspect, resulting in a better score and performance in audits, apart from all the advantages diversity brings to the company. Such a holistic view of a subject, as depicted in the paragraph, is of great value in making decisions.

- Managing the sustainability risks. Similar to other practices within an IT company, like security audits, quality audits, or privacy audits, risks are not overlooked in relation to sustainability. Auditors expect companies to provide risk logs for the sustainability risks, like the risks related to not renewing the old electricity outlets, which leads to higher electricity consumption, and indirectly influence the carbon emission if the electricity is not generated from clean and renewable energy. The challenge here is that due to the broad scope of sustainability, these risks may overlap with the risks originating from privacy management and information security management. Thus, the importance of communication between actors should not be underrated.
- Categorization of risks is another challenge. Like what was previously mentioned, sustainability covers many subjects, and so do the risks. Governance risks took the first place among the categories of risks in terms of being reported by companies, and social risks came second, compared to other groups indicated in the study, like reputational and environmental (Karwowski & Raulinajtys-Grzybek, 2021).

Looking at the field of risks, risk management rests upon the same principles, for whatever the context. To put it differently, the steps and actions taken in managing risks are similar to those followed in privacy or quality. The basics are identifying the risks, assessing them, prioritizing them, selecting the risk strategy, creating and executing the risk response plan, communicating the risks and actions taken, and monitoring them.

A previous study identified four groups of strategies associated with risk management chosen by supply managers: avoiding risks, mitigating risks through monitoring or through collaboration, and accepting risks (Hajmohammad & Vachon, 2016). Supply chain risks, including social and environmental risks, are those expected to be handled by companies as well. Related reports and documentation are asked to be presented during the sustainability audits. This includes the risk strategies chosen. If your company uses an integrated risk management system, be sure that sustainability risks are retained there. If there is no integrated management system, then try to implement a sustainability risk management consistent with the other risk management systems inside the company and, more favorably, conform with widely recognized risk management principles and/or standards.

- Identifying the sustainability opportunities and managing them. Opportunity is a word that usually accompanies the word "risk". When we talk of risk, we are supposed to talk about opportunities as well. You may have heard about turning the risks into opportunities for growth and innovation. Sustainability opportunities should be watched carefully and searched actively. For instance, the company may have the chance to partner with another company to develop a software solution for agricultural drones. This project, besides being environmentally beneficial, can enhance the image of the company, and within the audits, such sustainability-oriented projects are welcomed and valued. It worth reminding that all the initiatives should be recorded with evidence such as photos, news, and other documents. These, can be then referred to during audits, or in the course of sustainability report preparation. This is of high value for two reasons: first, evidence constitutes a key part of audits, and whatever is claimed during audits, either verbal or written, should be backed by evidence. You don't want your company to be falsely misunderstood, or being accused of greenwashing, or deceiving the stakeholders. Thus, evidence should be recorded systematically, in order to be accessible at the moment of need. The second reason is transparency. As what emphasized throughout this book, transparency should be upheld during audits and within reports. The evidence contributes to the transparency, too.

- Managing the relationships with the stakeholders. Within the governance dimension, stakeholder management and the initiatives for the purpose of attracting investors should be put forward among the actions. Stakeholder engagement comprises many practices, some of which are communicating with them, learning about their interests, valuing their interests and incorporating them into goals and strategies, ensuring them of transparency, and so on. Therefore, planning is required to manage them effectively and efficiently.
- Reflecting stakeholders' interests in reports. Sustainability reporting and upholding the stakeholders' interests during its preparation should be focused carefully on for the reason that what a company has executed so far regarding sustainability is mainly exhibited in the report, and undermining its role will result in a failure to illuminate all the efforts and endeavors made. As you know, sustainability reports make up an important part of the sustainability audits since their content can be considered evidence. Hence, a well-structured, appropriate, and transparent content plays a critical role in the sustainability audits.
- Additionally, there are competitions that award certifications, being held worldwide or at the national level. The juries of these competitions evaluate the sustainability reports of the participants of the competition based on numerous criteria, such as design, quality of the components, and relevancy of the content. Annually, many companies upload their sustainability reports for their corresponding year, and compete with rivals for winning the title of best sustainability report. This offers them a great opportunity for:
    - Showcasing their reporting capabilities,
    - Highlighting their sustainability initiatives,
    - Bearing a competitive advantage over rivals,
    - Drawing the attention of the potential investors,

- Motivating their employees specifically the sustainability department/unit/team,
- Building a good corporate image and gain reputation,
- Leveraging the win for marketing purposes,
- And, last but not least, importantly, obtaining an extra score during the sustainability audits and improving their sustainability audit performance. If none of the above is satisfactory, it can be an award and/or a trophy on the dusty shelves of your company!

4.1 Planning the Actions

Planning is recommended to be done from the higher levels and then broken down to the lower levels. Levels of the formal corporate strategic planning are: corporate, business, and functional, at 12 steps briefly embodying identifying vision, posture, and mission, formulating business and functional strategies, combining them at the corporate level, evaluation of action programs, resource allocation, and budgeting (Hax & Majluf, 1984).

The planning process can be formulated in four different steps, typically according to the literature:

1. Determine and define objectives.
2. Create strategies.
3. Evaluate strategies.
4. Tracking the outcomes (Boyd & Reuning-Elliott, 1998).

The logic and the backbone of the planning process in IT companies are similar to the above model. The steps an IT company usually takes for their *planning* consist of:

- Studying the market, company, and factors influencing the goals
- Setting the goals.
- Communicating the goals and objectives with the employees.
- Ensuring the commitment and ownership of the goals by employees.
- Having the goals and strategies at different levels of the hierarchical structure finalized by the related employees. At this stage, there is a need for collaboration between the employees

and the company's strategists. It can be done and facilitated through delegating certain employees to help the others with understanding the goals of the higher levels correctly, and as a result of the thorough comprehension, the goals and plans for the lower levels are highly likely to be fitted with the others.

- Deciding on the time, budget, and resources for tasks.
- Defining progress indicators for each goal; how to measure and control for the progress toward the goals, and how often to perform the monitoring.
- Reminding the employees of the goals and progress checks.
- Evaluating the individual and overall progress at the end of the period, goals were generated for.
- Organize closing meetings, where the flow of the meeting conclusions is from the lower levels to the highest level, comprising the best practices, lessons learnt, and the probable improvement areas which will be benefited from for the next period's goal setting.

When planning for sustainability, the *corporate level of sustainability* should be the focal point. This brings several considerations to light:

- Sustainability experts of companies should work hand in hand with companies' strategists and decision makers. This roots from the fact that the sustainability department or unit cannot succeed without the righteous relevant strategies. This is not a one-level process, but different hierarchical levels are involved and are required to be harmonized for achieving the corporate sustainability goals.
- Another perspective, which should not be ignored and underrated, is that sustainability tasks and activities are neither restricted to a team, unit, or department, nor assigned to and performed solely by one. It is where collaboration should be stressed and called for. For instance, for issues associated with diversity, inclusion, equity (DEI), sustainability unit or department is not able to complete all the tasks or decide on all the aspects. There is a need for an orchestration between sustainability and human resources units/departments. The lack of sufficient and efficient collaboration between various

business units is seen in many real-world cases, which end with the inability or insufficiency in fulfilling the expected requirements. Such situations do not end here and go on with the departments blaming each other for the deficiencies that have occurred, which will have a negative impact on the organizational culture and overall organizational performance.

To stress the relationship between sustainability practices and human resources, there has emerged a concept as "sustainable HRM" which is in association with the side of sustainability relevant to managing people (Ehnert, Parsa, Roper, Wagner, & Muller-Camen, 2016). A really interesting finding of the same paper, which is totally consistent with what is observed in the IT sector as well, is that companies disclose information connected to their internal labor force more than their external human resource (Ehnert, Parsa, Roper, Wagner, & Muller-Camen, 2016). Taking a glance at what is happening at IT companies' sustainability reports calls attention to the fact that most of the metrics and values contained in their reports are actually those coming from the internal workforce. To look deeper into the problem, what is really meant by the external workforce seems to be vague for them; who it includes and who is excluded. This turns into a critical issue in the IT sector for the reason that the outsourcing of part of the software development lifecycle (SDLC) is very common. For example, the testing of the use cases can be delegated to external testers. To make it clearer, the first group that comes to mind when mentioning external workforce in IT companies is subcontractors. The external workforce and the relevant concepts are not only significant in the reporting, but also during the audits/assessments. Many auditing entities expect the IT companies to demonstrate evidence regarding their policies, practices, and the results with regard to the external force. The well-being, labor, and human rights of the workforce are expected to be valued and treated like those of the internal employees, and as for audit, saying or writing will not suffice, and there needs to be evidence of actions taken, especially quantitative reports.

- Similar to individual responsibility toward sustainability in a society, the responsibility in a company should be felt by all the employees, and they should take ownership of the sustainability tasks.
- One of the other notable problems which is widely seen in the companies is the planning and assigning of the tasks. These activities are determined by the sustainability unit/department and assigned to relevant individuals or teams from diverse departments. These assignments that are conducted without consultation with the assignees and in the absence of their consent lead to the reluctance of the assignees to accomplish tasks thoroughly or at the desired time. What is recommended is to plan for the tasks with the related stakeholders and assignees, not to plan for them. Sustainability, as outlined earlier, covers a wide range of concepts. Thus, the best choice is to take advice from the experts and professionals of various departments, and prepare the schedule and action list only after discussion with them. For instance, due diligence, which is a popular sustainability topic these days, encompasses notions from finance and law. The actions to be carried out should be determined and decided on with their agreement and commitment.
- Support of top management is essential for achieving sustainability objectives, just like many other applications, for example, implementing a standard within an organization. This support, apart from the financial aspect, is beneficial to the organizational culture, employee motivation, their commitment to sustainability, and their ownership of tasks.

Before taking any steps toward the preparation of a sustainability report, the answers to the following questions are better to be sought, which helps with the planning process with regard to the sustainability reporting:

- Who are the stakeholders?
- What is of importance to the company and its stakeholders?
- How important are diverse sustainability-related concepts to the stakeholders? (materiality analysis)

- How can these concepts be defined and measured? (Definition is needed to be documented in order to keep consistency within the company).
- What kind of data can be collected for each notion?
- How can the data be collected? By whom? From whom (individuals or departments)? When (specifying the intervals)?
- How can the data be analyzed? By whom?
- How can the report be presented? By whom?
- How long does it take to analyze the data?
- To whom is the report presented?
- What is the most appropriate reporting (taking the audience and the context into consideration)?
- What is the best time to disclose the report? (For example, before/on a specific day or holiday, during an event, etc.)
- What are the best channels to disclose the report? (On corporate website, on social media accounts, etc.) In which order? (For instance, first, creating a post, next, sharing a video on social media, and then publishing the report on the corporate website or all simultaneously).

4.2 Identifying the Metrics and Measures

In consistency with the application of this book which is IT sector, the standpoint is in a similar vein with the literature, stressing the importance of developing sector-specific standards for materiality by identifying the important sustainability-related issues in a sector, and key performance indicators (KPIs) to report them which leads to increase in the ability of companies in reporting their ESG performance (Eccles, Krzus, Rogers, & Serafeim, 2012).

Seeking the right KPI might take time since they should be experienced and proved to indicate sustainability correctly, and the paper came up with two KPIs for sustainability: value-added distribution and profitability stability for integrated reporting (Oshika & Saka, 2017). Like what was stated, finding the appropriate KPIs, as it needs time in research, does not happen quickly in practice either. Many IT companies that have been defining sustainability KPIs for years, do struggle with KPI determination. It is like the process of trying and learn. If it doesn't work, then modify it or replace it completely. In addition to this, the rapid development of the theory, introducing new KPIs, is also a key player in reviewing the sustainability KPIs regularly.

In an attempt to define KPIs, a paper created KPIs for key dimensions of sustainable development with their concentration on environment, economy, stakeholders, organization, culture, and the last, which addresses all dimensions, sustainable development (Hristov, Appolloni, & Chirico, 2022). Similarly, another study grouped KPIs under social, environmental, and economic indicators, portraying generic KPIs that, irrespective of the size and the production type of the company, can be made use of across all industrial contexts (Contini & Peruzzini, 2022). Dividing the KPIs based on ESG dimensions is a common and practical way. Complementary dimensions can be added if there are any KPIs that should be included, particularly those in close association with the company and its context. Some of the KPIs heavily depends on the sector, and for this reason, it is vital to bring the sector in as a key factor into the process of research for KPIs.

Another aspect of KPIs which should be taken into consideration is that KPIs should be selected in a way to exhibit future financial performance of the company associated with risks, specifically downside risk, because investors are in need of such information (Jørgensen, Mjøs, & Pedersen, 2022)

Furthermore, KPIs have strong correlations with corporate goals, and some of the criteria for their selection are being able to: describe value creation process effectively, and outline the perspective and situation of the company (Hristov & Chirico, The Role of Sustainability KPIs in Implementing Sustainable Strategies, 2019). The same paper (Hristov & Chirico, The Role of Sustainability KPIs in Implementing Sustainable Strategies, 2019) encompasses KPIs that can be of great help in determining the KPIs for your company. To elucidate the concepts, goals are the ends, and KPIs are the means by which the progress toward the goal is tracked.

Data collection systems within companies have developed and it roots from the tendency toward disclosing the data (Adams & Frost, 2008). This tendency also resulted in having social and environmental data incorporated into decision-making, risk, and performance management practices (Adams & Frost, 2008).

4.3 Executing the Actions

To execute the planned actions, the following issues should be given attention:

- Ensure employees' commitment.
- Ensure ownership of the tasks.
- Build a collaborative culture.
- Internalize responsible consumption and production.

A remarkable number of sustainability-related actions are performed by different departments within a company. If the actions should be done by other departments, the correct term is that the sustainability expert is accountable for the actions, and the employees of other departments are those responsible for doing them. For instance, conducting financial internal controls is not part of the responsibilities of the sustainability department, but this department should ensure that controls are carried out as planned, and the results required to be presented to the auditors are recorded.

To make it clear, apart from the tasks which should be conducted by the sustainability department directly, the other tasks are those assigned to other departments, and the responsibilities of the sustainability department regarding such tasks are:

- communicating clearly with other departments about the expectations,
- convincing the departments to collaborate,
- tracking the progress but do not command, and avoid ordering them.
- Collecting the results and evidence from them for the sustainability audits.

4.4 Monitoring and Controlling the Actions

The activities needed to be performed for monitoring and controlling the actions are:

- Specifying the time monitoring and controlling take place.
- Determining the additional monitoring and controlling situations.

- Determining where to keep the data and results.
- Deciding on when to report the results, how, and to whom.

Monitoring and controlling the activities internal to a department are among the department's responsibilities. In some cases, the sustainability department can guide and support such initiatives. For instance, the sustainability department can assist the legal departments in conducting anti-corruption audits or controls. Various sustainability auditing entities ask for different evidence concerning monitoring and controlling activities. Thus, communication of these requirements to the corresponding department and giving them guidance should be done by the sustainability department/unit.

4.5 Revising and Correcting the Actions

Within the scope of sustainability, many revisions and modifications might be required, specifically for companies that are new to sustainability initiatives. Sustainability audits can be considered as a way to learn about the strengths and weaknesses of a company, and what needs to be improved, excluded, or continued. The comments and results of the audit can provide companies with some clues on what to do next. In some cases, the company cannot do more in a specific area, due to many reasons, of which the most seen one is the lack of resources. In this regard, the company can concentrate on the other activities and enhance its other practices. In reality, the practices of a company are not balanced in terms of sustainability dimensions, and the good news is that the weight of these dimensions is not usually the same in the sustainability audits as well.

References

Adams, C. A., & Frost, G. R. (2008). Integrating Sustainability Reporting into Management Practices. *Accounting Forum*, Vol. 32, pp. 288–302.

Boyd, B. K., & Reuning-Elliott, E. (1998). A Measurement Model of Strategic Planning. *Strategic Management Journal*, Vol. 19, pp. 181–192.

Contini, G., & Peruzzini, M. (2022). Sustainability and Industry 4.0: Definition of a Set of Key Performance Indicators for Manufacturing Companies. *Sustainability*, Vol. 14, No. 11004, pp. 1–37.

Eccles, R. G., Krzus, M. P., Rogers, J., & Serafeim, G. (2012). The Need for Sector-Specific Materiality and Sustainability Reporting Standards. *Journal of Applied Corporate Finance*, Vol. 24, No. 2, pp. 65–71.

Ehnert, I., Parsa, S., Roper, I., Wagner, M., & Muller-Camen, M. (2016). Reporting on Sustainability and HRM: A Comparative Study of Sustainability Reporting Practices by the World's Largest Companies. *The International Journal of Human Resource Management*, Vol. 27, No. 1, pp. 88–108.

Hajmohammad, S., & Vachon, S. (2016). Mitigation, Avoidance, or Acceptance? Managing Supplier Sustainability Risk. *Journal of Supply Chain Management,* Vol. 52, No. 2, pp. 48–65.

Hax, A. C., & Majluf, N. S. (1984). The Corporate Strategic Planning Process. *Interfaces*, Vol. 14, No. 1, pp. 47–60.

Hristov, I., Appolloni, A., & Chirico, A. (2022). The Adoption of Key Performance Indicators to Integrate Sustainability in the Business Strategy: A Novel Five-Dimensional Framework. *Business Strategy and the Environment*, Vol. 31, pp. 3216–3230.

Hristov, I., & Chirico, A. (2019). The Role of Sustainability Key Performance Indicators (KPIs) in Implementing Sustainable Strategies. *Sustainability*, Vol. 11, No. 5742, pp. 1–19.

Jørgensen, S., Mjøs, A., & Pedersen, L. J. (2022). Sustainability Reporting and Approaches to Materiality: Tensions and Potential Resolutions. *Sustainability Accounting, Management and Policy Journal*, Vol. 13, No. 2, pp. 341–361.

Karwowski, M., & Raulinajtys-Grzybek, M. (2021). The Application of Corporate Social Responsibility (CSR) Actions for the Mitigation of Environmental, Social, Corporate Governance (ESG), and Reputational Risk in Integrated Reports. *Corporate Social Responsibility and Environmental Management*, Vol. 28, pp.1270–1284.

Marshall, J. D., & Toffel, M. W. (2005). Framing the Elusive Concept of Sustainability: A Sustainability Hierarchy. *Environmental Science and Technology*, Vol. 39, No. 3, pp. 673–682.

Oshika, T., & Saka, C. (2017). Sustainability KPIs for Integrated Reporting. *Social Responsibility Journal,* Vol. 13, No. 3, pp. 625–642.

5 Results

5.1 Preparing the Data and Metrics

Before any steps toward data preparation, there is a need to review what data we own and how they are different from the expectations. At the planning stage, which was fully explained in Chapter 4, we have already planned for our actions to fulfil most of the requirements. "Most" is preferred here because in practice, sometimes, it is not feasible to meet all the expectations, and some of them need to be delayed or tailored majorly, concentrated on, a further point in time. Therefore, the main notions here are:

1. Reviewing the requirements.
2. Identifying the requirements we can fulfil as a company.
3. Make decisions for those that haven't been taken into the current plan.
4. Selecting the right metrics and indicators.
5. Listing the appropriate data collection methods for each indicator/metric.

CSR performance indicators vary at diverse levels, such as the Global Competitiveness Index, the Environmental Performance Index, the UN Human Development Index for different CSR dimensions at country level, and the Dow Jones Sustainability Index (DJSI), and the Global 100 Index at the company level (Strand, Freeman, & Hockerts, 2015). It is noteworthy that as a sustainability expert of a company, one should be totally aware of the national and international policies and regulations together with the sustainability status of the country or countries the company has branches in for the reason that they are not separated, but connected, being affected and having effect on each other, particularly and more perceptibly at the macro-to-micro flow.

DOI: 10.1201/9781003587583-5

For instance, companies located in Europe are supposed to be compliant with laws and regulations concerning sustainability, or there are countries that have already started standardizing the sustainability reporting, have determined their reporting expectations publicly and precisely.

A topic which has gained the attention of sustainability academics and practitioners recently is introducing maturity levels within the context of sustainability. Maturity levels were defined as a way to demonstrate the current status through dividing the progress into classes within a certain dimension (Eisner et al., 2022). This is a familiar concept for IT companies because they are already familiar with the various maturity models. Different aspects of corporate sustainability dimensions were framed based on four maturity levels: level 1 addresses respecting compulsory laws and rule; level 2 points at elementary efforts; level 3 indicates a satisfactory position; and level 4 represents remarkable endeavor (Baumgartner & Ebner, Corporate Sustainability Strategies: Sustainability Profiles and Maturity Levels, 2010). Even if the companies do not intend to publish their sustainability maturity levels, such papers and works can be invaluable in assisting them in interpreting their current situation from a sustainability standpoint. Simply, if we don't know where we are on the sustainability path, we cannot make a proper estimate of how much way is left to the goal, how much time and resources are essential for achieving the goal, and toward what direction our attempts should be directed. The consequence is ending in a "hold on state", which many IT companies are actually found in. Hold on state happens when companies do not have any intentions to take any further steps; they delay actions due to excuses like the lack of budget or having other priorities; and they wait for any probable triggers which can be: a mandatory law or regulation or an action from their rivals that can compel them to move further.

Data coming from the tools automatically or even those collected are most probably not appropriate for use. They might be missing values, non-relevant values, repetitive values, or other problems with the data that necessitate the step of data preparation. Therefore, the first step is to solve the problem within the data.

Sometimes, calculations are needed to reach the final data. For instance, the electricity generated from fossil combustion should be turned into the tons of carbon dioxide it contributes to. Thus, calculating the values is the second step.

The third step is to unify the units whenever possible, for the reason that understanding the data based on different units is not easy and requires more time.

As a recommendation, it is better to accompany the visual and reports with a guide on how to interpret them. This guide should be clear, understandable, and most importantly, brief.

5.2 Analyzing and Interpreting the Results

Analyzing and interpreting the results of the actions is as important as executing them. Analyzing requires knowing the data, what data were collected, how it was collected, and what they represent.

Meetings can be held with relevant departments separately to interpret the results and learn about the insights from numerous perspectives. This should be done for the reason that some topics may contain terms and require technical knowledge, which will be hard for the sustainability department to handle alone.

At the end of the process, the results can be shared in a meeting with attendees coming from all the departments involved, to evaluate the overall status, and pinpointing important issues for setting the targets for the next period.

References

Baumgartner, R. J., & Ebner, D. (2010). Corporate Sustainability Strategies: Sustainability Profiles and Maturity Levels. *Sustainable Development,* Vol. 18, pp. 76–89.

Eisner, E., Hsien, C., Mennenga, M., Khoo, Z.-Y., Dönmez, J., Herrmann, C., & Low, J. S. (2022). Self-Assessment Framework for Corporate Environmental Sustainability in the Era of Digitalization. *Sustainability,* Vol. 14, No. 2293, pp. 1–33.

Strand, R., Freeman, R. E., & Hockerts, K. (2015). Corporate Social Responsibility and Sustainability in Scandinavia: An Overview. *Journal of Business Ethics,* Vol. 127, No. 1, Special Section: Corporate Social Responsibility and Sustainability in Scandinavia, pp. 1–15.

6

Reporting

6.1 Structure of the Reports

CSR reporting bears some distinctive features, which are the variety of users and uses, objectives, measurement, and being voluntary, multi-dimensional, externality-driven, and yielding long-term benefit (Christensen, Hail, & Leuz, 2021).

Most sustainability reports start with a coverage page, followed by a page representing their content. Companies take different approaches when preparing this page; some choose to categorize the content based on the dimensions of ESG, and others prefer their very own way of classifying the subjects into different chapters. According to a paper, including pictures and unrelated information on the page containing the table of contents was not recommended, and a table of contents aiding the readers in introducing the content will suffice (Litfin, Meeh-Bunse, Luer, & Teckert, 2017). Moreover, an embedded annual-and-sustainability report was suggested over a reference report (Litfin, Meeh-Bunse, Luer, & Teckert, 2017).

Reporting a value like sustainable development indicates that the performance of a company concerning economic, environmental, and social aspects might be desired (Krajnc & Glavic, 2005). While decision makers prefer aggregated data linked to goals, individual readers favor a single value like an index (Krajnc & Glavic, 2005). This highlights the importance of knowing your audience or trying to get to know them. You might have watched movies where someone who is expected to give a presentation, first, carries out searches about the audience and their backgrounds. This is a very simple example of such a case.

Within IT companies, it is better to take a look at reports prepared so far and get inspirations from them on how to form the report. The employees who have already create reports for the company own useful insights about the audience. For instance, during a report preparation

DOI: 10.1201/9781003587583-6

process, it was recommended to utilize more visuals, and be extra cautious about the data presented via them due to the engineering background of the users.

Once, during a meeting, a question was asked about the nature of data, whether it indicates a percentage or number, and they asked for specifying the type by using a symbol on the charts.

The other example was when an employee with much more tenure in the company commented about applying soft tones of matte color.

In another situation, the position of the legends of the charts was changed, and they were located on the right of the chart to be aligned with the direction of their language. It seems to be a tiny detail; however, it demonstrates that the users of the report are of value to you, and in the case of the last example which is highly likely to happen within global and international companies, or even when the potential investors are from diverse nations, the company can leverage this chance to show that it is aware of the differences, and it embraces diversity.

6.2 Selecting the Right Style and Language for Reporting

In accordance with what was mentioned earlier in this book, there are several frameworks and standards for sustainability reporting. IT companies have the chance to select the most appropriate one and prepare their reports accordingly. However, the lack of comparability due to the plurality of guidelines emerges (Albu, Albu, Dumitru, & Dumitru, 2013). The same study pointed out that comparability rarely finds its way into the objectives of companies for sustainability reporting, and the main purposes are disclosing information, obtaining resources, and meeting key actors' expectations (Albu, Albu, Dumitru, & Dumitru, 2013).

Comparability of sustainability reporting was studied regarding its three facets of: materiality, comparability within and between companies, benchmarking, and qualitative and quantitative aspects of comparability (Korca, Costa, & Bouten, 2023). This reminds us of several important facts:

- First, knowing what is important to the company and its stakeholders and their priorities should be considered in each sustainability initiative. In a similar vein with the literature, sustainability materiality plays a pivotal role in defining the content of sustainability reports (Pizzi, Principale, & De Nuccio, 2023).

- Second, sustainability reporting is not limited to the borders of a company, but it goes beyond it due to the width of its scope. Therefore, it is better not to concentrate only on inter-organizational concepts.
- Third, operationalization should be done for each subject that is included in reporting, involving providing definition and methods for its measurement.

There are several recommendations with regard to sustainability reporting in the following:

- Since sustainability reporting is mostly and mainly voluntarily performed for the time being, companies can make their choice based on their strategies, stakeholders' expectations, and their context. Therefore, make the most of the advantage provided by the plurality of guidelines, and employ the one or ones that fit your purpose and context.
- If sustainability reporting is taking place annually, maintaining the consistency between the methodology and the standard employed can benefit companies in many ways. The first is making it easy for stakeholders to compare the data and report of different years, and this makes them assured of the transparency because applying various methods evokes this feeling that the results are no better than the results from the previous years, and the change in the method has been conducted for masking the underlying insufficiencies. Besides, since an ideal stakeholder management is one of the first priorities of sustainability reporting, keeping the methods the same simplifies their use, and they will spend less effort and time to comprehend them.
- Taking a look at the languages used in sustainability reports so far accentuates that companies have mostly preferred a formal and less formal style in their sustainability reports. The reports are not prepared completely based on the academic writing methodology, and the writer is free to use "we" and add emotions and feelings in the content. The goal is to make

the reader, primarily the stakeholders, feel that the writer or whoever prepared the report is part of the company and see himself/herself as a part of the whole, not separate from it, and portray the image that all the employees feel the same commitment and ownership with the company.

Examples are shown below:

- Use "We have invested in empowering women by investing in Project A" instead of "Company A (instead of we) has invested in empowering women by investing in Project A". The use of "we" indirectly implies that the company values its employees and their commitment a lot, and the employees believe in the power of the word "we".
- Choose "By contributing to Project A and playing an active role in it, we were able to hold software programming workshops with an attendance of 200 women, with the aim of helping them to become economically productive" over "As a partner of Project A, 200 women were educated on software programming, and were helped to become economically productive". The use of active voice in some cases is more effective than the passive voice. Furthermore, a smooth language like that of the first sentence seems more suitable for the purpose of stakeholders' engagement and involvement.
- Another suggestion is to add photos of the events to the sustainability report, both as evidence and as a transparency tool. The number of companies using real photos seems to be reducing, and they are preferring more AI-generated images. This can be the topic for marketing meetings and is dependent on the preference of the companies; nevertheless, even if real photos are not being included in the sustainability reports, they can be contained on the posts of the corporate website and social media accounts to strengthen the company's communication with its stakeholders.
- The other recommendation is to enhance the sustainability reports with hyperlinks to other documents and evidence. Since the main part of commination is performed digitally and the reports are available online, companies can gain

benefit from such digital tools. These attempts can include employing tools to make graphs and chart more readable, or links to the dynamic visual reports, or hyperlinks to other web pages on the corporate websites or social media.

- ♦ The other issue, which was mentioned earlier, is the design of the sustainability reports, which is becoming more and more attractive and professional day by day. With the use of AI, the reports have started to become more visually appealing. The role competitions play in the enhancement of the sustainability reports should not be underrated. Each year, companies enter competitions that evaluate their sustainability reports, and can use their success to raise their chance of building partnerships and networks. It is noteworthy that partnerships and networks play key role regarding sustainability initiatives in line with the 17th UN SDG. Looking at the instances in IT sector, great social responsibility projects have already been carried out by the collaboration between IT companies, the education sector, including schools and universities, the medial sector, including hospitals, charities, and social enterprises.

- Coming to the selection of the words, the words that trigger emotions in the readers are chosen. These words are directed toward creating the picture that the company put the society and environment among its priorities, and its attempts are in line with leaving positive impact on its society and environment.
- The role of social norms in structuring the societal expectations should not be forgotten, and the survival of a company relies upon meeting these expectations when considering the role sustainability disclosure plays in a society (Cho et al., 2015). Hence, it would be better to conduct research on the norms of the society the company is located at, or where the company's headquarter is based in. Because, at the end of the day, the sustainability disclosures are compared against the social forms and perceived accordingly.
- Sustainability reporting can take many forms, and there are several classifications for it. They may be integrated reports covering financial, ecological, and social dimensions or

two-dimensional (non-financial) or one-dimensional reports, while the first, which encompasses all three dimensions, can be considered as a sustainability report (either social or ecological) (Hahn & Kühnen, 2013). Indeed, not many IT companies incorporate their financial information into their sustainability reports (mainly entailing ESG dimensions). Before any preparation is made regarding the reporting, the outline should be devised for what to place or to what degree of detail. Just like any other writing task, beginning with writing the heading and sub-headings, and then trying to add clues to the content goes under each of them is a simple and efficient method.

- Taking approvals for each part of the report, which is completed from the pertinent department or individual, is much better than waiting for the report to be fully wrapped up. This saves time and effort, and inhibits the probable reworks. It reminds us of the agile methodology. Demonstrating and validating the content at specific milestones is strongly recommended. Besides, emails of the approvals should be archived for further reference because in case of any mistake or error after publishing the report, it can be referred to these emails to check the facts and correspondence.
- Sustainability reports were conveyed in relation to the company's performance, and after delving into the details, the social performance disclosure turns out to be associated with the company's performance, and not economic and environmental performance disclosures (Burhan & Rahmanti, 2012). Deciding on which section of the report to highlight more can be done by taking the stakeholders' opinions and expectations into consideration, together with the findings of studies in the literature.
- While working on sustainability reporting, it should not be forgotten that its main purpose, according to the literature, is to illustrate a company's contribution, measured and undergone independent audits, toward sustainable development rather than comparisons within and between companies (Abeysekera, 2022). Thus, the content and all the wording should be chosen with this issue in mind. To be honest,

comparisons are made by companies, reviewing their sustainability reports for different years, or comparing their reports with those of their rivals.

- It is a normal practice which is also common among the IT companies to have a glance at the IT companies' latest sustainability reports to get ideas and gain inspirations, completely within ethical framework, not with negative intentions, to get familiar with the recent trends. However, the main purpose of sustainability reporting is not comparison, of course.
- The last recommendation concerns the companies active in different countries. It was explained that these companies with regard to sustainability reporting, encounter the issue of having diverse cultural and socio-economic environments, and that they may indicate various groups of stakeholders and matters within those environments (Fifka & Drabble, 2012). Considering the environment during report preparation will be of help in selecting the right content and communication channels.

6.3 Finalizing the Reports

Like any other report, a sustainability report consists of both textual and visual parts. Since corporate reports can be employed as a legitimization tool (Cho, Michelon, & Patten, 2012), there may be attempts to disguise the facts. The findings of a study revealed that graphical presentations were manipulated in a systematic manner to build a better depiction of the companies (Cho, Michelon, & Patten, 2012). Hence, before finalizing the report and sending it for publication, it is better to review the report with a critical eye, particularly the visuals. The first thing that captures the reader's attention when turning to a page is the design and visuals included. Thus, giving extra consideration to visuals, what they represent, and how they depict it is strongly recommended. Sometimes, even not intentionally, the facts are concealed behind the colorful appearance of visuals, instead of being communicated through them. It would be good to remember that such problems will soon be revealed, and they will affect the reputation and the image of the company. Examples of similar situations can be found even through conducting a search on the internet.

Given all that has been mentioned, we know that it would not be easy for companies to disclose information when their initiatives were not satisfactory and/or sufficient, and as a result, they may try to find a way to mask the shortcomings. What should not be forgotten is that reporting and audits rely heavily on ethics, and this becomes of paramount importance in sustainability, where we are talking about corporate responsibility and our responsibilities toward our society and environment.

With regard to the sustainability report, a paper discusses the question of whether companies implement what they have incorporated in their reports, and proposes "implementation likelihood" (Kolk, 2004). Scrutinizing what has been happening in IT sector implies that the increase in the number of sustainability audits/assessments, demanding external endorsements for sustainability-related actions, such as for labor practices, and asking for certifications and awards concerning sustainability are just some of the initiatives to align actions with claims. During sustainability audits, you may be asked to provide the evidence of any external verification or endorsement in various forms such as certifications, awards, audit or control reports, and authentication document. The noteworthy issue here is that the documents should be updated, and it should be ensured that they fulfil the requirements of the auditing entity including but not restricted to the format, and the validity period.

6.4 Releasing the Reports

The appropriate time, communication channels, and their order in terms of releasing, and the content that comes along with the report for announcing the release of the report should be decided carefully with the help of the marketing and public relations department. For instance, the sustainability-related days can be a good choice, but not the days when there are other topics among the public or economic agenda. These choices can affect the readability rate and the impact of the content on the readers, shareholders, and potential investors.

References

Abeysekera, I. (2022). A Framework for Sustainability Reporting. *Sustainability Accounting, Management and Policy Journal*, Vol. 13, No. 6, pp. 1386–1409.

Albu, N., Albu, C. N., Dumitru, M., & Dumitru, V. F. (2013). Plurality or Convergence in Sustainability Reporting Standards? *Business and Sustainable Development*, Vol. 15, Iss. Special No. 7, pp. 729–742.

Burhan, A. H., & Rahmanti, W. (2012). The Impact of Sustainability Reporting on Company Perofmance. *Journal of Economics, Business, and Accountancy Ventura*, Vol. 15, No. 2, pp. 257–272.

Cho, C. H., Laine, M., Roberts, R. W., & Rodrigue, M. (2015). Organized Hypocrisy, Organizational Façades, and Sustainability Reporting. *Accounting, Organizations and Society*, Vol. 40, pp. 78–94.

Cho, C. H., Michelon, G., & Patten, D. M. (2012). Enhancement and Obfuscation through the Use of Graphs in Sustainability Reports: An International Comparison. *Sustainability Accounting, Management and Policy Journal*, Vol. 3, No. 1, pp. 74–88.

Christensen, H. B., Hail, L., & Leuz, C. (2021). Mandatory CSR and Sustainability Reporting: Economic Analysis and Literature Review. *Review of Accounting Studies*, Vol. 26, pp. 1176–1248.

Fifka, M. S., & Drabble, M. (2012). Focus and Standardization of Sustainability Reporting – A Comparative Study of the United Kingdom and Finland. *Business Strategy and the Environment*, Vol. 21, pp. 455–474.

Hahn, R., & Kühnen, M. (2013). Determinants of Sustainability Reporting: A Review of Results, Trends, Theory, and Opportunities in an Expanding Field of Research. *Journal of Cleaner Production*, Vol. 59, pp. 5–21.

Kolk, A. (2004). A Decade of Sustainability Reporting: Developments and Significance. *International Journal of Environment and Sustainable Development*, Vol. 3, No. 1, pp. 51–64.

Korca, B., Costa, E., & Bouten, L. (2023). Disentangling the Concept of Comparability in Sustainability Reporting. *Sustainability Accounting, Management and Policy Journal*, Vol. 14, No. 4, pp. 815–851.

Krajnc, D., & Glavic, P. (2005). How to Compare Companies on Relevant Dimensions. *Ecological Economics*, Vol. 55, pp. 551–563.

Litfin, T., Meeh-Bunse, G., Luer, K., & Teckert, Ö. (2017). Corporate Social Responsibility Reporting - A Stakeholder`s Perspective Approach. *Business Systems Research*, Vol. 8, No. 1, pp. 30–42.

Pizzi, S., Principale, S., & De Nuccio, E. (2023). Material Sustainability Information and Reporting Standards: Exploring the Differences between GRI and SASB. *Meditari Accountancy Research*, Vol. 31, No. 6, pp. 1654–1674.

Index

Critical Sustainability Sciences

This book explores Critical Sustainability Sciences, a new field of scientific inquiry into sustainability issues. It builds on a highly novel integration of elements from relational ontologies, critical theory, political ecology, and intercultural philosophy in support of emancipatory perspectives on sustainability and development.

The book *Critical Sustainability Sciences* begins by uncovering the weaknesses of mainstream sustainability science and debates on sustainable development. The new field of Critical Sustainability Sciences has grown out of a deep engagement with relational ontologies, which helps to overcome the dualist ontology underlying mainstream notions of sustainability and development. Dualist ontologies reinforce problematic anthropocentric divisions, for example, between humans and nature, subjects and objects, mind and matter, body and soul, etc. Examples from indigenous peoples in Bolivia, India, and Ghana – as well as integrative movements in Chile, Brazil, and Europe – show that relational conceptions of life, rooted in ecosophy and cosmosophy, can provide an intercultural philosophical foundation for Critical Sustainability Sciences. The book concludes by describing three key topics for exploration in Critical Sustainability Sciences: societal reorganization in view of emancipatory, existential, and cognitive self-determination; living labor and commons; and the development of new comprehensive relational scientific paradigms.

This book will be of great interest to students, scholars, and practitioners of emancipatory and intercultural approaches to sustainability and development.

Stephan Rist is Professor Emeritus of Human Geography at the Institute of Geography at the University of Bern, Switzerland.

Patrick Bottazzi is Assistant Professor at the Institute of Geography at the University of Bern, Switzerland.

Johanna Jacobi is Professor at ETH Zürich, Switzerland, where she leads the Agroecological Transitions Group.

Routledge Studies in Sustainability

Stories of Change and Sustainability in the Arctic Regions
The Interdependence of Local and Global
Edited by Rita Sørly, Tony Ghaye and Bård Kårtveit

Sustainable Living at the Centre for Alternative Technology
Radical Ideas and Practical Solutions
Stephen Jacobs

Rural Governance in the UK
Towards a Sustainable and Equitable Society
Edited by Adrienne Attorp, Sean Heron and Ruth McAreavey

The Environmental Impact of Cities
Death by Democracy and Capitalism
Fabricio Chicca, Brenda Vale and Robert Vale

Reimagining Labour for a Sustainable Future
Alison E. Vogelaar and Poulomi Dasgupta

Waste and Discards in the Asia Pacific Region
Social and Cultural Perspectives
Edited by Viktor Pál and Iris Borowy

Digital Innovations for a Circular Plastic Economy in Africa
Edited by Muyiwa Oyinlola and Oluwaseun Kolade

Critical Sustainability Sciences
Intercultural and Emancipatory Perspectives
Edited by Stephan Rist, Patrick Bottazzi, and Johanna Jacobi

For more information on this series, please visit: www.routledge.com/Routledge-Studies-in-Sustainability/book-series/RSSTY

Critical Sustainability Sciences

Intercultural and Emancipatory Perspectives

Edited by Stephan Rist, Patrick Bottazzi, and Johanna Jacobi

Routledge
Taylor & Francis Group
LONDON AND NEW YORK

earthscan
from Routledge

First published 2024
by Routledge
4 Park Square, Milton Park, Abingdon, Oxon OX14 4RN

and by Routledge
605 Third Avenue, New York, NY 10158

Routledge is an imprint of the Taylor & Francis Group, an informa business

British Library Cataloguing-in-Publication Data
A catalogue record for this book is available from the British Library

ISBN: 978-0-367-48941-0 (hbk)
ISBN: 978-1-032-48797-7 (pbk)
ISBN: 978-1-003-04357-7 (ebk)

DOI: 10.4324/9781003043577

Typeset in Galliard
by Apex CoVantage, LLC

Contents

Contributors

Viviana Blanco (Ph.D., Argentina), Agronomist Engineer, is Professor-Researcher in the Faculty of Agricultural and Forestry Sciences (FCAyF), UNLP, Argentina.

Patrick Bottazzi (Ph.D., Switzerland) is Assistant Professor in the Department of Integrative Geography, Institute of Geography, University of Bern.

Isis Brook (Ph.D., United Kingdom) is Visiting Research Fellow at Bath Spa University. She taught philosophy for over 25 years.

David Choquehuanca Céspedes (Bolivia) is Vice President of the Plurinational State of Bolivia since 2020. He is a world-known leader of the Aymara nation in Bolivia. He is also Lecturer of History and Anthropology at the Universidad Mayor de San Andrés (UMSA) in La Paz.

Beat Dietschy (Ph.D., Switzerland) is a Swiss philosopher and theologian. As a collaborator of Ernst Bloch, he was involved in the publication of some of his writings.

Alain Findeli (Ph.D., Canada) was professor and researcher in design at the University of Montreal, Canada (1977–2007), and the University of Nîmes, France (2007–2017).

Johanna Jacobi (Ph.D., Switzerland) is Professor who leads the Agroecological Transitions Group at ETH Zürich. She studied Geography, Biology, and Social Anthropology.

Renato S. Maluf (Ph.D., Brazil) is Professor of the Graduate Program on Development, Agriculture and Society, Federal Rural University of Rio de Janeiro, Brazil.

David Millar (Ph.D., Ghana) is Professor and President of the Millar Institute for Transdisciplinary and Development Studies (MITDS) in Ghana.

René Montalba Navarro (Ph.D., Chile) is Professor in the Faculty of Agricultural Sciences and Environment at La Universidad de La Frontera (Chile).

Jéssica Sepúlveda Pizarro (Ph.D., Chile) is a social worker, working at the John Paul II Center for Ethics and Social Responsibility, and a member of the Nucleus of Interethnic and Intercultural Studies of the Catholic University of Temuco, Chile.

Nora Räthzel (Ph.D., Sweden) is a sociologist and professor at the University of Umeå, Sweden.

Stephan Rist (Ph.D., Switzerland) is Professor Emeritus of Human Geography at the Institute of Geography at the University of Bern.

Niagia F. Santuah (Ph.D., Ghana) is an expert in communication holding a Ph.D. in Culture and Development Studies from Millar Institute for Transdisciplinary and Development Studies (MITDS) in Ghana.

Darshan Shankar is an expert in cross-cultural knowledge in Health Sciences and Vice Chancellor of the University of Trans-Disciplinary Health Sciences and Technology (TDU) in Bangalore, India.

Florencia Spirito (PhD, Chile) is a researcher and lecturer in the Department of Natural Sciences and Technology at the Universidad de Aysén, Chile.

Maxwell Ba-an Tengolzor is in the final year of the Ph.D. program. He is Lecturer in the Department of Culture and Development Studies at Millar Institute for Transdisciplinary and Development Studies (MITDS) in Bolgatanga, Ghana.

Hugo Marcelo Zunino (Ph.D., Chile) is a full professor and researcher in the Department of Social Science at the Universidad de la Frontera (Temuco-Pucón, Chile).

Acknowledgments

This book would not have been possible without the wonderfully creative and tireless collaboration of the co-authors. Special thanks are due to the Institute of Geography and the Centre for Development and Environment of the University of Bern for granting Stephan Rist's sabbatical in 2019. We are especially grateful for the generous support and hosting of Stephan Rist's sabbatical to the Universidad de la Frontera, Chile, provided through the then Dean Rodrigo Echeverría Pezoa and the current Vice Dean René Montalba. For his wonderful facilitation of the workshop in 2019, we wish to thank the director of the Pucón Campus, Rodrigo Garrido Maldonado. For his great support in editing the language of Chapters 1, 2, and 15, we especially wish to thank Anu Lannen (Centre for Development and Environment, the University of Bern, Switzerland). For his excellent coordination of the editorial process, we are particularly thankful to Matthew Shobbrook from Routledge. And last but not least, we wish to thank the anonymous reviewers of the book proposal who gave us the motivation and strength to engage in this piece of work, knowing that they considered it to be a very timely and creative way of advancing critical perspectives on sustainability and development.

Foreword

Vandana Shiva

We face an existential crisis as humanity and as a planetary civilization. Indigenous people have been uprooted, displaced, and exterminated over 500 years of colonialism. One million species are threatened with extinction, with 200 going extinct every day. The present path humanity is on is clearly non-sustainable because it is destroying life on Earth, the very infrastructure of life. Non-sustainable systems are emerging as a threat to the very survival of the human species. Humans too are a threatened species. Why has sustainability been so elusive in spite of decades of search and research?

The book *Critical Sustainability Sciences: Intercultural and Emancipatory Perspectives*, edited by Stephan Rist, Patrick Bottazzi, and Johanna Jacobi, answers these questions in novel and comprehensive ways.

The editors show that the dominant field of sustainability science is based on the same mechanistic, reductionist and use-inspired engineering paradigm that is at the root of the multiple interconnected crises we face.

Among the fundamental critiques of sustainability science, they identify four interrelated issues: a flawed ontology of dead inert nature; epistemic reductionism that ignores the diversity and plurality of worldviews; uncritical use of the concept of "development"; and the neglect of linking sustainability science with freedom, emancipation, justice, as present in critical thinking and critical theory.

The Earth is Living. We are not separate from the Earth, we are a strand in the web of life, we are members of one Earth Family. Biodiversity, the diversity of species in mutuality and interconnectedness, creates the web of life and maintains the living planet and the infrastructure of life.

The emergencies humans face in terms of hunger and thirst, disease and pandemics are rooted in the ecological crises and the crises of injustice, inequality, and inhumanity.

Mother Earth is Terra Madre, Gaia, Pachamama, Vasundhara. Emerging critical ecological and sustainability sciences recognize her as Gaia. She is living. She gives us life. We are part of her Earth Family, *Vasudhaiva Kutumbkam*, of living, creative, diverse, intelligent sentient beings.

The phrase *Vasudhaiva Kuṭumbakam* (Sanskrit: वसुधैव कुटम्बकम्) consists of several words: *vasudhā* (transl. "the earth"), *ēva* (transl. "is thus"), and *kuṭumbakam* (transl. "family")

अयं निजः परो वेति गणना लघुचेतसाम्। (Ayam Nijah Paro Veti Ganana Laghucetasam)

उदारचरितानां तु वसुधैव कुटम्बकम्॥ (Udaracaritanam Tu Vasudhaiva Kutumbakam)

(Chapter 6 of *Maha Upanishad* VI.71–73).

"This is mine, this is yours, this is a stranger, this is a relative is the thinking of a petty mind. An evolved mind sees the whole world as one Earth Family."

All beings are our relatives. There are no strangers, no enemies, no hierarchies, no superiority and inferiority. Oneness is the path to real sustainability and justice, to non-violence, peace, and harmony.

The petty-minded mechanistic and market paradigm has made us forget we are one Earth Family. It has created Eco-apartheid, the illusion that humans are separate from nature, that some powerful humans are masters and owners of the Earth, her resources, her biodiversity. Eco-apartheid goes hand in hand with anthropocentrism, the false assumption that humans are superior to our non-human relatives, who are mere objects and property to be manipulated for profits, or threats to human life that need to be exterminated. Apartheid as separation has made other people and other species strangers, a threat, an obstruction, a competitor who must be exterminated for one's existence.

This is why forests have been cleared, rivers dammed, insects pushed to extinction with pesticides, plants exterminated with Roundup, organisms genetically modified (GMOs) for patents. The petty mind is a mechanical mind, an arrogant mind that drives the hubris that the Earth is dead matter, Terra Nullius. Colonizing, industrial, mechanical man sees himself as the "creator". GMOs are an example of this hubris, allowing violence against the integrity of living systems, and using this violence to claim the right to ownership through patents and intellectual property. GMO in effect means "God Move Over", since corporations are claiming to have created life and seek to become "Life Lords", lords over life itself, collecting rents, royalties, and taxes from life's renewal and regeneration (Vandana et al. 2020).

I call this worldview Eco-apartheid.

The multiple crises and pandemics we face today – the health pandemic, the hunger pandemic, the poverty pandemic, the climate and extinction controversies, the emergency of injustice, exclusion, and inequality, dispossession, and disposability of large numbers of humanity – are all rooted in a worldview based on the illusions of separation and superiority which deny interconnectedness and oneness.

These false assumptions are as follows:

1 The transformation of *Terra Madre*, Mother Earth who is living and has rights, into *Terra Nullius*, the Empty Earth. The assumption that nature is dead inert matter, property to be owned by enclosing the commons and raw material to be extracted for profits. The denial that the Earth is Living, she is Gaia, Pachamama, Vasundhara, and that Mother Earth has no rights is the root cause of ecological destruction and violence against her.

2 *Eco-apartheid*, the assumption that humans are separate from Nature, are her conquerors, masters, and owners, and the denial of the fact that we are part of Nature, not separate from her. Apartheid is "apartness" or "separateness" in Afrikaans.
3 *Human apartheid*, based on the false assumption that colonizing man is separate from and superior to other cultures and most human beings, who are colonized, including the indigenous, the non-white and colored, women, farmers, peasants, and workers. The illusion of superiority leads to domination and discrimination based on race, gender, religion, and work. This false assumption of superiority is also used to enclose the commons that are cared for and shared by all members of a community. Enclosures allow the extraction and appropriation of resources that sustain all life, including human life.
4 *Anthropocentrism*, the assumption that humans are superior to other species which are reduced to objects to be owned, manipulated, and exploited for profits and control. And a denial that we are members of one Earth Family and all living beings are sentient beings with integrity and rights.

The anthropocentric assumption that humans are separate from nature and superior to other species who have no rights is not just a violation of the rights of our fellow beings but also a violation of our humanity and human rights. We are members of one Earth Family, and our being human is predicated on our relations with biodiversity and living seed, land and living soil, living waters and living food. Human rights defined on the basis of separation and superiority makes "dominance" and "exploitation" appear natural to being human, when they are in fact constructs based on illusion of dualistic separation and superiority which have contributed to both non-sustainability and injustice (Shiva 2016).

Denial of Nature's rights leads to destruction of Nature and a threat to the very conditions of human survival. In an ecologically interconnected world, denial of the rights of Nature translates into denial of human rights. The same constructs that lead to violence against nature and her destruction become the basis of violence against fellow human beings. Non-sustainability and injustice are part of the same process.

The petty-minded worldview of dualistic separation also engenders hierarchies and the illusion of superiority – of humans as superior to other species, men as superior to women, whites as superior to blacks and all colored people, one faith as superior to the diversity of belief systems which have nourished cultural diversity. Separation and superiority create structures of violence – violence against Nature, violence against women, violence against every "other" defined as lesser beings with the objective of colonization.

Colonization is based on separation and superiority, the construction of apartheid.

The mechanistic reductionist worldview of nature as dead inert matter and mere raw material to be extracted was constructed by Descartes and Bacon to facilitate nature's exploitation and promote colonialism and commercialization. Bacon, who is called the "father of modern science", called this shift "The Masculine Birth of Time" with a deeply patriarchal concept of the project of reductionist mechanistic science. Descartes made life and sentient beings disappear. Life is inert matter he declared: a mere machine. Locke justified the violent enclosures of the commons and the creation of private property as "improvement" of nature (Shiva 1988).

The paradigm of the "dead Earth" goes hand in hand with the scientific paradigm of mechanistic reductionism and a technological paradigm of mastery, control, and engineering instead of partnership, cooperation, and co-creativity. This ontological flaw is at the root of the failure of the dominant model of "development", "sustainable development", and sustainably science, which is more of the same.

The contributions to this volume from diverse cultures and continents, diverse disciplines and worldviews, shape Critical Sustainability Sciences, which allows us to shift from extractivism, mastery, domination, and ownership to giving, sharing, participating, and commoning the commons.

I have called this emerging paradigm and practice of participating in a living, pulsating sentient earth with all beings "Earth Democracy" (Shiva 2005), Arne Naess has referred to it as "biological equalitarianism".

Earth Democracy is a worldview, paradigm, and practice that is based on the following recognition:

1 The Earth is Living. The Living Earth is our Mother. She is *Terra Madre*, Mother Earth, Gaia, Pachamama, Vasundhara . . . Mother Earth has rights.
2 We are all members of the *One Earth Family.* We are part of the Earth, and not separate from her, not her masters. We are interconnected through the living currencies of breath, water, and nourishment. We have a duty to protect the Earth's Living Systems that provide us clean air, clean water, and clean food.
3 We are part of *One Humanity on One Planet.* All humans are equal. Our diversity enriches life and cannot be made the justification for inequality and injustice. Future generations have a right to enjoy the gifts of the Earth. Present generations have a duty of Earth Care to pass on the gifts of nature in their full diversity, integrity, and purity.
4 Earth Democracy is based on *Living Economies*, *Living Democracies*, and *Living Cultures* woven through the diversity in the web of life. Each life-form supports and sustains all others in mutuality, cooperation, and harmony. All living beings are sentient beings and have rights. All beings are creative and intelligent.

Living economies are based on co-creativity and co-production by humans as part of the Earth community, respecting the rights and integrity of all. In critical sustainability sciences, living economies are aimed at recognizing that the maintenance of the integrity of the biological basis of life at large is the precondition of any sustainable system.

This requires a shift from commodification of nature and labor to decommodification of the Living Earth and our living labor. In Earth Democracy, the economy is a subset of ecology based on the laws of Mother Earth. We share the Earth's gifts with others. Seed, biodiversity, water, and food are commons. Participation as co-creators in the Earth's ecological processes to protect the commons and defend the common good is living democracy. Cultivating the culture of oneness with the Earth is Living Culture.

Earth Democracy is a worldview, paradigm and practice that is based on the recognition that everything is interconnected, the Earth is Living, the Earth has rights, that we have duties to care for the Earth, and regenerate her soil, seeds, and biodiversity, her water and food systems. Our rights flow like a spring from our duties.

Earth Democracy recognizes that humans are part of the Earth and related to other beings. Human Rights are therefore connected to the Rights of the Earth and the Rights of other species. The path to peace, justice, and real sustainability is paved by reclaiming the commons – our common home, the Earth, and the commons of the Earth Family, of which we are a part. Through reclaiming the commons, we can imagine possibility for our common future, that we can sow the seeds of abundance through "commoning".

In the commons we care and share – for the Earth and each other. We are conscious of Nature's ecological limits, which ensure that her share of the gifts she creates goes back to her to sustain biodiversity and ecosystems. We are aware that all humans have a right to air, water, and food, and we feel responsible for the rights of future generations.

The ideas of the book *Critical Sustainability Sciences* and those of Earth Democracy are based on the creativity of ancient living cultures and emerging sciences of relationship and interconnectedness. They allow a shift from the dead Earth ontology to the ontology of a living intelligent Earth. They allow a shift from the epistemology of violence, referred to by Johann Wolfgang von Goethe as the "grim torture chamber of empiricism, mechanism, and dogmatism" to "delicate empiricism", which makes itself identical with the object thereby becoming true theory. Epistemic democracy and biodiversity of knowledges is central to true sustainability. It allows a shift from Monocultures of the Mind to Biodiversities of the Mind.

It allows a shift from linear one-way extraction from Nature and society as "development" and "growth" enabled by the mechanistic reductionist paradigm to co-creativity, sympoiesis – meaning "making with". It allows a shift from hierarchies of exclusion to biodiversity of knowledges, cultures, and economies.

The shifts emerging from Critical Sustainability Sciences are vital for our times specially when in the name of "sustainability" the Earth is being privatized and financialized, and the indigenous people who conserve 80% of the remaining biodiversity on 20% of the land that remains in their care are threatened to be colonized and displaced by false sustainability solutions, like carbon and biodiversity offsets. The legally binding regulations of the UN environmental treaties – the Biodiversity Convention and the Climate Convention – were first eroded and transformed into voluntary commitments, falling far short of actions to be taken to reverse climate change and biodiversity loss and erosion. Deregulation has now been replaced by the polluters turning the crisis they have contributed to, into a new market opportunity for profits, control, and ownership of the earth's resources, ecosystems, and biodiversity.

First the corporate industrial money-making machine mined for coal, drilled for oil, fracked for gas, burning up 600 million years of Nature's work in fossilizing her carbon, ignoring nature. Now that the ecological crises created by limitless greed can no longer be denied, they want to mine the ecological functions and services of nature as "nature-based solutions". Through the Fourth Industrial Revolution, the metaphor for mining continues to turn into reality. Data is being called the new oil. Billionaires want to mine the data from humans, including farmers, mine the data from diverse food cultures, mine the data from seeds and biodiversity, mine the data from soil and ecosystems. This is "financialization of nature", the final assault on Nature's integrity and Nature's self-organized, self-sustaining, regenerative economy. While robbing Nature of her integrity, the "financialization" of Nature is being greenwashed as protecting nature and reversing the ecological crises of biodiversity loss and extinction.

Wall street and the financial asset companies are seeing a $4,000 trillion fictitious economy of finance by extracting profits from goods and services from "Nature's assets". This commodification is an enclosure of the commons of life as whole. When the last river, forest, and land will be enclosed as an asset, the custodians will be displaced people, without access to land, forests, and water and their earth-centered livelihoods. It is misleading to call the latest stage of colonial commerce "Nature's Economy" as the Intrinsic Exchange Group (IEG) refers to it. The Natural Asset Companies (NAC) have been created by the IEG which is the Rockefeller Foundation, the New York Stock Exchange, and the Latin American Development Bank.

New proposals to address the ecological emergencies based on the old paradigm that reduces life in common to private property and money are neither sustainable nor just. They deny both the ontology of the living earth and the right of Mother Earth and human rights of free and autonomous beings and communities, indigenous people, farmers, fisherfolk, women, working people and the poor, the children, and future generations. They are blind to the diversities of ontologies and related ways of knowing that are relational and lead to living knowledges.

The rich contributions of this book about Critical Sustainability Sciences shape another worldview. It allows us to be "reborn from the death of egocentrism, anthropocentrism, eurocentrism". The contributions shine a light to help us move from darkness, hopelessness, despair, greed, and enclosures of the commons to a living future with all life on earth, cultivating hope, life, care, harmony, solidarity, and a reclaiming of the commons.

References

Shiva, V., 1988. *Staying Alive: Women, Ecology, and Survival in India.* New Delhi: Kali for Women; London: Zed Books.

Shiva, V., 2005. *Earth Democracy: Justice, Sustainability and Peace.* London: Zed Books.

Shiva, V., 2016. *Biopiracy – The Plunder of Nature and Knowledge.* Berkeley, CA: North Atlantic Books.

Vandana, S., et al., 2020. *Gates to a Global Empire: Over Seeds, Food, Health, Knowledge . . . and the Earth: A Global Citizen's Report.* https://navdanyainternational.org/wp-content/uploads/2020/10/GATES-TO-A-GLOBAL-EMPIRE-REPORT-SYNTHESIS-13.10.2020-3.pdf (Accessed 26.12.2022).

1 Why do we need critical sustainability sciences?

Stephan Rist, Patrick Bottazzi, and Johanna Jacobi

Sustainable development: between hope and despair

In the 1960s, increasing numbers of people around the world began to realize how economic growth, industrialization, and technification were associated with rapidly increasing degradation of human and environmental health. As a result, new or rejuvenated social–environmental movements emerged that called for humankind to reshape its relationship with nature. Rachel Carson's book *Silent Spring* (2015), first published in 1962, was highly influential at the time, and is often credited with spurring the rise of 'modern' environmental movements. The book begins by describing how bird populations were decreasing due to massive use of the first modern synthetic insecticide DDT. Carson goes on to trace how DDT was spreading through the food chains, also posing threats to human health and life in a widening toxic cycle. The book is considered by many to have played a key role in DDT eventually being banned for agricultural use in the early 1970s in many countries.

The debates inspired by Carson's book opened people's eyes to the rising levels of contamination of food, water, soils, and air resulting from industrial development. Corresponding movements began to demand greater restrictions on polluting practices and industries and called for recognition of the intrinsic values of nature as well as for reflection on humankind's responsibility for nature. Social–environmental movements such as the "1968" movement also articulated fundamental critiques of Western lifestyles. They argued that war, capitalism, consumerism, and "modernism" in Western countries were incompatible with peaceful coexistence between humans and all other beings forming the web of life on earth.

The environmental dimension of these critical movements became more prominent in 1972 when the Club of Rome originally published *The Limits to Growth* (Meadows 1972). While the authors of the report were not explicit on the links to capitalism, other contemporary observers put the limits to growth into a more critical perspective. Galtung (1973), for example, stated:

> Concretely, capitalism is based on the idea that capital, nature in the form of raw materials, and labor in the form of "raw" and skilled labor,

DOI: 10.4324/9781003043577-1

> are moved towards points of accumulation, "poles of growth". These accumulation points become the centers of the system; the rest becomes the periphery. Depletion, pollution, and uneven accumulation are immediate consequences.
>
> (Galtung 1973:107)

Parallel to their engagement in political activism, members of the growing social–environmental movements – mainly in industrialized countries – began striving to live according to their professed ideals, under the banner of "practice what you preach". This gave rise to the organization of rural communities, for example, composed mainly of people intentionally leaving urban areas. They sought to draw practical lessons from the radical questioning of Western lifestyles that emerged up through the early 1970s. Using terms like "back to the land" and "deep ecology", members of these movements began experimenting with new forms of life rooted in principles of self-reliance, self-sufficiency, self-determination, freedom, and fraternity (Wilbur 2013).

Meanwhile, in 1973, Arne Naess added a spiritual dimension to the definition of key claims and practices of environmental movements with his concept of deep ecology. Devall (1980) summarizes deep ecology as based on the following:

> (1) *A new cosmic/ecological metaphysics which stresses the identity (I/thou) of humans with non-human nature is a necessary condition for a viable approach to building an eco-philosophy* [italics of the original text]. In deep ecology, the wholeness and integrity of person/planet together with the principle of what Arne Naess calls "biological equalitarianism" are central and integral part of nature, not over or apart from nature. Man is a "plain citizen" of the biosphere, not its conqueror or manager. There should be a democracy of all "God's creatures" according to St. Francis; or as Spinoza said, man is a "temporary and dependent mode of the whole of God/Nature". Man flows with the system of nature rather than attempting to control all of the rest of nature. The hand of man lies lightly on the land. Man does not perfect nature, nor is man's primary duty to make nature more efficient.
>
> (Devall 1980:310)

Consequently, these movements injected a new critical, self-reflective element into the ongoing development of Western societies. There began a radical rethinking of the dominant forms of creating and distributing economic wealth through the varying forms of capitalism. This was done not only in view of distributive justice, but also by promoting a shift away from private capitalist – or centrally planned – creation of economic growth toward a "living economy" (Galtung 1986). This living economy called for wealth creation that acknowledged maintenance of the integrity of the biophysical basis of life at large as a precondition for any sustainable economic system. By the mid-1980s, an inspiring debate about the "spiritual dimension of green politics" was established (Spretnak 1986).

In summary, in the period from roughly 1960 to 1985, members of the growing social–environmental movements were engaged in highly diverse attempts at reorganizing personal, social, and economic lifestyles. A common thread running through this phase was the search for new practices and theories of life rooted in ideals of self-determination, inter- and transcultural spiritual growth, and a desire to reach beyond materialism,[1] capitalism, and other Western notions of progress or "development".

Situating the emergence of environmental movements at the start of consumer societies in the West raises the question of whether or not such movements are a "luxury phenomenon" of rich industrialized countries. At the latest since Martinez-Alier's (2002) groundbreaking work on the environmentalism of the poor, we know that the answer to this is clearly "No": Martinez-Alier shows convincingly that environmental movements led by the poor already existed at the start of the colonial period and have continued up until the present. These movements have resisted the destruction of their social and ecological foundations due to Western or imperial concepts of development.

Regarding social–environmental movements in so-called developing countries of the Global South, the "spiritual dimension" of people's view of the environment was already well established, in contrast to Western countries. Indeed, spiritual perspectives on nature formed the bedrock on which social movements of the Global South built their struggles. Among other things, it enabled them to identify capitalist modes of economic development as the root causes of poverty and environmental destruction. Persons and people demonstrating the force and the far-reaching impact of cultural and spiritual motivations and actions can be found throughout global history in many places and moments. Outstanding examples include Mahatma Gandhi from India, Nelson Mandela from South Africa, Thomas Katari or Bartolina Sisa from Bolivia-Peru, Wladimir Sergejewitsch Solowjew from Russia, or recent environmental and sustainability-related accounts of Daoist traditions in China (Miller 2017), etc. In the late 20th century, the increasingly shared concern of Northern and Southern people and nations for the "spiritual dimension of the environment" helped to build a bridge between movements, heightening the relevance of intercultural learning and cooperation among participants on all sides, including the diverse forms of approaching spirituality.

As we will see in the following sections, this open, creative, and critical attitude of social–environmental movements was gradually transformed in parallel to the consolidation and official recognition of the notion of

1 For more on our specific understanding of materialism, see page 15 of this chapter.

sustainable development by governments and influential political actors in civil society.

Notably, in this first phase, concerns about environmental consequences were not seen as an end in itself. On the one hand, environmental movements in the Global North sought to overcome ego-centered, material-growth-oriented lifestyles and economies; on the other hand, environmental movements in the Global South sought to resist the destruction of their socio-ecological and cultural foundations in the name of "modern development". In both cases, people of these movements viewed humans as holistic living beings, possessing a physical body that harbours life, soul, and spirit. In both the Global North and Global South, early environmentalists sought to harmonize the needs of their body, soul, and spirit with the material and spiritual needs of nature.

Recognition by states

In the late 1980s, the concerns of the era's social–environmental movements began to resonate with civil society, including political parties, green NGOs, and, in part, religious organizations. In 1987, the UN organized the first World Commission on Environment and Development (WCED). It was the first moment in which governmental representatives from all around the world launched the idea of sustainable development. It was presented as a global economic development trajectory aimed at balancing "the needs of humans and the environment". The so-called Brundtland Report summarized the key postulates of the WCED. This report provided the often-cited definition of sustainable development as "development that meets the needs of the present without compromising the ability of future generations to meet their own needs" (WCED 1987:6).

Official commitments made by many governments to implement novel policies for sustainable development served to inspire numerous critical social–environmental movements to participate in the joint construction of corresponding concepts and strategies with governments, business, and civil society. This process was further enabled by a new openness among UN organizations to work more closely with civil society.

The next landmark conference was the United Nations Conference on Environment and Development (UNCED) in Rio in 1992. It served as a platform for UN bodies to debate – in a rather unique broad-based collaboration with key social actors (e.g., social–environmental movements, academia, politics, and business) – about how to advance toward sustainable development, as proposed in 1987.

Carruthers (2001) took a closer critical look at the rationales behind the declaration of the Rio conference and identified three crucial elements:

- First, by relating environmental degradation to the large number of people still living in poverty, the declaration emphasized economic growth as solution to poverty and stressed poverty eradication over environmental concerns. Economic growth was represented as a precondition for mitigating poverty and environmental problems. Meanwhile, the fact that environmental problems are not mainly caused by poor people, but rather by the 20% of rich and industrialized countries, was conveniently relegated to the background, avoiding criticism of Western lifestyles embedded in capitalist, industrialized, and consumer-based economies.

- Second, emphasizing the "tradeoff" between trade and environmental protection, the latter was framed as a luxury. According to this logic, environmental protection could be afforded only when higher economic wealth was made available to the large numbers of people still living in relative or absolute poverty. Here, free trade was represented as a precondition for creation of economic growth, entrepreneurship, and competitive markets.
- Third, the Rio Declaration and corresponding action plan framed the biophysical limits to poverty alleviation, economic growth, and free trade as possible to overcome by means of enhanced administrative and technological innovation. In this way, biophysical limitations were not presented in absolute terms, but rather as a relative constraint to economic growth defined in relation to technological innovation and capacity.

In essence, the 1992 Rio conference "erased the line between 'sustainable development' and 'sustained growth' . . . sustainability had come full circle, back to an essential belief that open-ended growth will lift all boats" (Carruthers 2001:100). Against this background, it becomes clearer how concerns about sustainable development were gradually channeled into a rather neoliberal agenda. As a result, the following UN conferences dealing with sustainable development – in 1997, 2002, and 2012 – framed the challenge as that of reconciling the supposed need for economic growth with ecological and social concerns.

Until the UN conference Rio+20, in 2012, openly critical views about the contradictions between economic growth and biophysical limits were virtually absent. Instead of developing the concept of sustainable development further, policymakers, scientists, and representatives of civil society engaged in broad debates about how to define the most adequate operational objectives and indicators to measure progress toward the – now taken-for-granted – neoliberal understanding of sustainable development. In the conference report – dubbed "The Future We Want" – priority was given to issues like how to develop measurable sustainable development goals (SDGs) on behalf of policies for "green" and growth-oriented economies (United Nations Conference on Sustainable Development 2012).

In the aftermath of the 2012 conference, private and public corporations increasingly collaborated and consolidated their power more directly toward shaping policies, goals, and targets of sustainable development. This was mainly done through the establishment of Public–Private Partnerships (PPP). They served to link national governments and many international bodies of the UN system closely to transnational companies and other private actors, including NGOs, philanthropic foundations, and think tanks. They committed to define and implement concrete goals and targets of sustainable development in a common agenda. PPPs increased their influence on markets and policies, particularly by managing and steering significant financial flows to NGOs, think tanks, and like-minded governmental bodies (Elsig and Amalric 2008).

Within multilateral organizations, the PPP model went hand in hand with the emerging phenomenon of "philanthrolateralism". It refers to a new dynamic binding the agenda of sustainable development to the rise of funds flowing from private firms, individuals, and their foundations. In this way, the views and interests of private players enter the national and international policy arenas. Seitz and Martens (2017:46) observed that "private funding runs the risk of turning UN agencies, funds and programs into contractors for bilateral or public-private projects, eroding the multilateral character of the system and undermining democratic global governance".

Similarly, the trend of "philanthrocapitalism" arose at about the same time. Commonly associated with sectors like education, international development, healthcare, and agriculture, philanthrocapitalism has also gained prominence in the fields of science and sustainable development. Haydon et al. (2021:353), describes this trend as operating mainly through three cultural frames: "(1) development challenges being framed as scientific problems; (2) beneficiaries being framed as productive entrepreneurs; and (3) philanthropy being framed as social investment".

2030 Agenda: global consolidation of growth-based sustainable development

In 2015, all governments represented in the UN approved a declaration called "Transforming Our World: The 2030 Agenda for Sustainable Development". Large sections of global civil society, the business community, and religious leaders applauded the 2030 Agenda, which received the endorsement of all UN member states. Given major heterogeneity and asymmetries in the global distribution of human, technological, and natural resources – and equally diverse interests, values, and worldviews – this overwhelming agreement on the 2030 Agenda was indeed a surprising achievement.

A new feature of the 2030 Agenda was definition of 17 sustainable development goals, or SDGs, along with 169 targets and 232 indicators, ostensibly designed to monitor progress toward implementation of the SDGs. The responsibility of the private sector is repeatedly emphasized in many SDGs and targets, mainly in view of the massive investments needed to implement the 2030 Agenda.

The 2030 Agenda is based on the belief that economic growth can be made environmentally sustainable. This will supposedly be made possible by means of "decoupling" and "resource efficiency" as key technological solutions. However, according to Parrique et al. (2019:4):

> The validity of the green growth discourse relies on the assumption of an absolute, permanent, global, large and fast enough decoupling of economic growth from all critical environmental pressures. The literature reviewed clearly shows that *there is no empirical evidence for such a decoupling currently happening*. This is the case for materials, energy water, greenhouse gases, land, water pollutants, and biodiversity loss for which decoupling is either only relative, and/or observed only temporarily, and/or only locally. In most cases, decoupling is relative. When absolute decoupling occurs, it is observed only during rather short periods of time, concerning only certain resources or forms of impact, for specific locations, and with very small rates of mitigation . . . Considered all together, *the hypothesis that decoupling will allow economic growth to continue without a rise in environmental pressures appears highly compromised, if not clearly unrealistic.*
>
> (emphasis in the original)

A similar critique is made regarding the definition of goals, targets, and indicators. For example, SDG 8 seeks to "Promote sustained, inclusive and sustainable economic growth, full and productive employment and decent work for all". While essentially social and economic targets and indicators are generally well-defined, "the environmental elements in many targets were not included in indicators, or the indicators lacked ambition, or were watered down" (Elder and Olsen 2019:70). The same authors relate this to a structural difficulty in defining and articulating cross-sectorial indicators that integrate socioeconomic and environmental aspects: "Nevertheless, despite expanded stakeholder participation, the fact remained that no one had measured before what many targets sought to capture. Therefore, many indicators could not adequately capture some environmental aspects or integration among the goals" (Elder and Olsen 2019:80).

With the global endorsement of the 2030 Agenda, we face a dilemma. On the one hand, we witness surprisingly high support for the so-called sustainable development, decoupling growth from environmental externalities, and providing a basis for "green growth". On the other hand, we confront a political agenda which, from a critical scientific perspective, is not consistent or feasible in conceptual, empirical, or political terms.

Moreover, if we compare the concept of sustainable development of the 2030 Agenda with its origins in environmental movements, we must acknowledge that the gap was never wider. Previously fundamental issues that are now glaringly absent include the need to further emancipate people from oppression and domination by economic, political, social, religious, or other power groups and associated mechanisms; efforts to relate environmental issues with spiritual growth at personal and community levels; and the creation of social and economic forms of life capable of overcoming capitalist or centrally planned state-based exploitation of humans and other beings in nature.

Additional justification for our critical view of the dominant ideas, concepts, and strategies for advancing sustainable development is provided by the completely insufficient progress to date in implementing the SDGs according to the timeline of the 2030 Agenda. It begs the question: What went wrong? Why did sustainability science fail to see that the notion of sustainable development underpinning the 2030 Agenda and the SDGs is virtually impossible to achieve?

Moving backward on implementation of the 2030 agenda and the SDGs

Progress toward implementation of the SDGs has been woefully inadequate, especially considering that there are only seven years remaining until 2030. The "Sustainable Development Goals Report 2022" (United Nations 2022) shows that only three SDGs – concerning well-being for all (health), inclusive and equitable education, and safe and sustainable cities – are still likely to be met by 2030. Three goals show insufficient progress, namely, those aimed at no poverty, gender equality, and access to energy. Alarmingly, very poor progress – or even backward trends – are evidenced regarding 11 SDGs: end hunger; water availability and sanitation; inclusive economic growth; infrastructure and sustainable industrialization; inequality within and among countries; sustainable consumption; combatting climate change; sustainable use of oceans and marine resources; protection of forests and biodiversity; peace and accountable institutions; and global economic partnerships for sustainable development (United Nations 2022).

In their recent volume, Sachs et al. (2022) offer a deep dive into the possible explanations for the sobering lack of progress toward the SDGs. They conclude:

> For the second year in a row, the world is no longer making progress on the SDGs. The average SDG Index score slightly declined in 2021, partly due to slow or nonexistent recovery in poor and vulnerable countries. Multiple and overlapping health and security crises have led to a reversal in SDG progress.

Against this background, it becomes clear that the problems of implementation are not specific to just one issue or just one year, but rather represent a concerning trend occurring on various fronts – namely, that of moving *backward*.

At the same time, we witness increasingly extreme inequality in terms of the distribution of wealth and financial resources. In 2021, the poorest half of the global population owned only 2% of global wealth, while the richest 10% of the global population owned 76% of all wealth (Chancel et al. 2022).

What went wrong?

In the search for answers for what went wrong, we first look at the explanations offered by the field of sustainability science. The sustainability science community collaborated extensively with the "Open Working Group", organized by the UN Division for Sustainable Development (DSD), which was responsible for drafting the text of the SDGs beginning in 2014. We can therefore assume that they had a direct influence on key propositions of the 2030 Agenda.

More of the same

Within the sustainability science community, answers to the question of what went wrong tend to be framed pragmatically. There is little or no critical reflection on the community's apparent inability to recognize the impossibility of achieving the SDGs, in particular based on the cognitive and epistemological nature of sustainability science itself. Of course, scholars of sustainability science generally recognize that progress toward the 2030 Agenda is insufficient. But rather than critically reviewing the science–policy interactions that birthed the 2030 Agenda, prominent members of this scientific community pin their hopes on swift mobilization of unprecedented levels of funding, peace, diplomacy, and international cooperation in order to overcome delays and reboot the necessary radical transformation of the world (Sachs et al. 2022).

We do not have faith in this "pragmatic", overly optimistic perspective on sustainable development and its potential to make up for lost time. In the following sections, we therefore seek to shed light on the main features of sustainability science itself. This analysis helps to reveal specific shortcomings of sustainability science that must be addressed to move forward. Overcoming them will also delineate the topical fields that must be considered to transform sustainability science into the broader, more critical and comprehensive field of scientific inquiry; we propose – that is, *critical sustainability sciences.*

Emergence and development of sustainability science

As introduced earlier, the concept of sustainable development has its roots in the social–environmental movements of the 1960s. In the late 1980s, the concerns of these movements finally began receiving attention from certain Western governments. This led to further development of the concept of sustainable development, based on relatively close collaboration between social movements, national governments, UN institutions, and – increasingly – civil society organizations and business actors.

In this historical perspective, it is interesting to note that, despite the plethora of scientific literature on sustainable development published over the years, its core concepts are rooted in the ideas of individuals and groups who had little direct interaction with the sciences. Mitlin (1992) identifies the year 1987, marked by publication of Brundtland report "Our Common

Future", as the moment when the scientific community began to engage more significantly with topics of sustainable development. The start was made by researchers mainly from natural sciences, but this group was rapidly complemented by social scientists concerned with the "development" side of sustainable development.

From 1987 to 2000, research on sustainable development was mainly subdivided into disciplinary fields, increasingly interacting among themselves through multi- and interdisciplinary coordination. From 2000 to 2006, research on sustainable development was so abundant, and demand from policymakers was so high, that it was granted its own formal space within global academia. Under the name *sustainability science,* this type of research became an institutionalized field of scientific inquiry, consolidating itself with the establishment of the journal *Sustainability Science* in 2006 (Clark and Harley 2019).

But what are the main features, concepts, methods, practices, strengths, and limitations of sustainably science? In the following sections, we offer answers to these questions, helping to cast the contours of sustainability science in sharper relief.

The nature of sustainability science

Sustainability science emphasizes use of concepts, methods, and tools aimed at understanding the links between human societies and the ecosystems in which they are embedded; this is done with a performative goal of promoting sustainable development. Sustainability science is rooted in a rather broad basis of disciplines, often interacting in inter- and transdisciplinary research projects (Clark and Harley 2019). Sustainability science is, not only concerned with scientific theories and concepts, but also situates itself at complex interfaces between academia and actors engaged in influencing policies and politics of "sustainable development", including governments, social movements, political parties, NGOs, practitioners, think tanks, religious organizations, and a host of private and public corporations.

According to the pioneers of sustainability science (Kates et al. 2001), until the successful establishment of sustainably science in academia, the agenda of sustainable development was mainly a societal and political agenda, much less concerned with the need for empirical evidence. Sustainability scientists have tried to change this with increasing efforts toward collaboration, emerging from international scientific programs, the world academies, and independent networks of scientists (Kates et al. 2001:641).

Further, the authors identify the purpose of sustainability science as that of studying relations between nature and society, taking account of interactions across scales of time and space. Here, natural and social systems are understood as complex, self-organizing systems with high levels of uncertainty. Importantly, it is assumed that social systems belong to natural systems and cannot be considered as separate entities. Based on the high complexity, uncertainty, and "unintended consequences of scientific progress, participatory procedures involving scientists, stakeholders, advocates, active citizens, and users of knowledge are critically needed" (Kates et al. 2001:641).

According to Kates et al. (2001), sustainably science addresses the following core questions:

- How can the dynamic interactions between nature and society – including lags and inertia – be better incorporated into emerging models and conceptualizations that integrate the Earth system, human development, and sustainability?
- How are long-term trends in environment and development, including consumption and population, reshaping nature–society interactions in ways relevant to sustainability?
- What determines the vulnerability or resilience of the nature–society system in particular kinds of places and for types of ecosystems and human livelihoods?
- Can scientifically meaningful "limits" or "boundaries" be defined that would provide effective warning of conditions beyond which the nature–society systems incur a significantly increased risk of serious degradation?
- What systems of incentive structures – including markets, rules, norms, and scientific information – can most effectively improve social capacity to guide interactions between nature and society toward more sustainable trajectories?
- How can today's operational systems for monitoring and reporting on environmental and social conditions be integrated or extended to provide more useful guidance for efforts to navigate a transition toward sustainability?
- How can today's relatively independent activities of research planning, monitoring, assessment, and decision support be better integrated into systems for adaptive management and societal learning (Kates et al. 2001:642)?
- Another feature of sustainability science is that it is "primarily use-inspired, as are agricultural and health sciences, with significant fundamental and applied knowledge components, and commitment to moving such knowledge into societal action" (Kates 2011:19449).

These features of sustainably science remain valid today. In a review of the state-of-the-art and future challenges of sustainability science, Clark and Harley (2019) confirm these dominant themes. However, compared to the initial articulation of sustainability science in 2001, the focus today is slightly broader in terms of two aspects: first, sustainability science situates itself in the concept of the Anthropocene, meaning to consider the issues of uncertainty, adaptation, lack of predictability, and non-linear relations between social and ecological systems; this means, second, that sustainability research involves six key capacities, that is, the capacity to (1) measure sustainable development; (2) promote equity in the pursuit of sustainable development; (3) adapt to shocks and surprises that threaten sustainable development; (4) transform unsustainable pathways of development; (5) link knowledge with action; and

(6) devise governance arrangements that support collective action in view of sustainability (Clark and Harley 2019:3).

Sustainability and the Anthropocene

A second group of scholars dealing with sustainability has sought to go beyond the rather narrow, implicit, and fuzzy epistemological foundations of sustainability science. In recent years, the sustainability sciences have confronted two rather destabilizing concepts. The first is that of the "Anthropocene", credited to Crutzen (Steffen et al. 2007) in early 2000. The concept refers to a distinct geological epoch marked by humankind's modification of biogeochemical terrestrial processes to such an extreme extent that radical shifts in the climate, biodiversity, and Earth's habitability are imminent.

The notion of the Anthropocene became an important reference point, mainly concerning the nature and shape of the so-called planetary boundaries, beginning in 2009. A total of nine – biophysical, and hence natural science based – environmental limits were defined, which, if respected, humanity could safely operate within (Steffen et al. 2015). This concept, extensively detailed and increasingly complemented by sustainability science, received wide recognition in the global community concerned with policies of sustainable development. The main conceptual change introduced by this theory is that of systemic, irreversible human impact on the planet. Prominent proponents of sustainability science integrated the notions of the Anthropocene into a new integrative framework, thought to guide further research on sustainability (Clark and Harley 2019).

Despite its sober analysis, the concept of Anthropocene has been criticized. For example, it implicitly assumes that all humans have the same level of responsibility for today's socio-environmental problems, when in fact only part of the planet, the industrialized world – particularly the private or state-based capitalist features of these societies – are responsible for the lion's share of the problems. As a result, some observers prefer calling this new period a "capitalocene" (Moore 2016). At the same time, the concept of the Anthropocene maintains the idea that we can "purposefully manage" the crisis by developing particular capacities, albeit with great concerns about unknowns and the need for ongoing adaptation.

The limitations of current sustainability science

As the previous sections show, without a critical revision of current sustainability science, we risk misusing the massive amounts of money, goodwill, skills, and hopeful expectations of all the dedicated individuals and groups today who are eager to make changes and to change themselves. However, due to the constraints of today's prevailing systems, progress is not happening. There is an urgent need for critical reflection on the existing shortcomings of sustainability science, as outlined in the following sections.

A growing body of research is examining the weaknesses of sustainability science. These efforts generally come from research fields lying outside sustainability science, including strands of critical thinking rooted in the philosophy of science, intercultural philosophy, critical theory, post-development, decolonial studies, feminism, political economy and ecology, and others.

Among the fundamental critiques of sustainability science, we find five interrelated issues. They concern (1) lack of critical reflection on the ontology that is underpinning sustainability science; (2) insufficient engagement with critical thinking on private or state-based forms of capitalist wealth creation; (3) unreflected methods of knowledge production in contexts of plural worldviews; (4) uncritical use of the concept of "development"; and (5) neglect of linking sustainability science with the normative orientations of emancipation, autonomy, and resistance to asymmetric, oppressive power relations.

One reason why such key issues are widely absent in sustainability science concerns the lack of more systematic integration of sociological theories and concepts, for example, relating to social construction, critical realism, structure and agency, historical specificities, and more appropriate conceptualizations of collective action and issues of biocultural diversity (Longo et al. 2021). There is also a lack of integration of key debates on sustainability from other scholars and fields, including the philosophy of science (Nagatsu et al. 2020), post-developmentalism (Shiva 2018). Further, there is a lack of engagement with new notions of development, such as "Living Well", which go beyond the Western lifestyles associated with the "green economy" (Bell 2016; see also the statements about "Vivir Bien" in Chapter 3), and not enough concern for culturally sensitive notions of development and emancipation (Leff 2015). The latter includes the need to reorganize societal institutions and economic life in view of self-governance, deliberation, and the reinvention of economic activities, shifting from commodified "dead" to de-commodified "living labor" (Bond 2002).

In the following, we review these critical issues of sustainability science. We believe that identifying these limitations can provide a basis for proposing the contours and basic concepts, methods, and practices of a broader field of scientific inquiry that we call *critical sustainability sciences*. Critiquing current notions of sustainability science can enable us to identify the building blocks for our new envisioned field, drawing on relational ontologies; comprehensive research methods that consider the irreducible interactions of material, psychological, and spiritual realms of human life and the lives of other beings; intercultural notions of emancipation; and new forms of societal and economic organization.

Dualist ontology of sustainability science and disregard of critical thinking

The most fundamental critique of sustainability science concerns the lack of elucidating the limitations of its ontological, epistemological, methodological, and normative foundations. With ontology, we refer to the basic assumptions of any form of knowledge about what exists, prior to human perception. We understand epistemology as the criteria used in different fields of science to define truth claims and related standards of "scientific validity".

The dualist ontology underpinning sustainability science assumes that we live in a twofold world which is constituted, first, by material objects and their real interactions that exist independent of human perception; and, second, the universe of subjective ideas, concepts, thoughts, and feelings that are only believed to exist in the consciousness of human beings. The two systems are thought to be "coupled". However, because of the subjective or "socially constructed" character of human knowledge, we cannot know for certain to what degree we are able to capture the objective world as such.

In the context of sustainability science, adherence to a dualist ontology means viewing humans and nature as being separated, into a subjective social world and an objective natural world. Here, the relation between humans and nature is embedded in an anthropocentric worldview, in which nature is seen as instrumental for the creation of wealth and security (Kenter and O'Connor 2022)

West et al. (2020) demonstrate that sustainability science, underpinned by a dualist ontology, assumes a substantialist separation between the social and ecological dimensions of life. This dualist ontology results in the social sciences and humanities emphasizing investigation of the socially constructed, human, subjective world, on the one hand, and the natural sciences emphasizing investigation of the materiality of the supposed objective world, on the other hand. While sustainability science purports to put interaction of "coupled social and ecological systems" at the center of analysis, its scholars have yet to explain on which grounds the two opposing ontologies can be bridged or integrated. This might explain why sustainability science appears to forge ahead in an effort to overcome substantialist positions. Yet its emphasis on "coupled social and natural systems" continues to sustain, perhaps unintendedly, the separation it seeks to repair (West et al. 2020).

The dualist ontologies underpinning today's prevailing natural and social sciences also imply distinct epistemologies. Regarding the natural sciences, these epistemologies assume replicability, empirical accuracy, statistical significance, predictability, and generality of explanations of phenomena. Regarding the social sciences and humanities, they assume scientific truth claims are based on systematic qualitative research methods, according to one or another theories of interpretation (hermeneutics). Scientific validity is derived from peer-based intersubjective validations of the interpretations made by a given researcher. These peers must be able to logically follow the conceptual chain, linking the human expression that was analyzed – for example, artifacts, written documents, arts, symbolic or religious expressions – with the interpretation made by the researcher. A key objective of social science and the humanities is that of coming to an *understanding* of the world by means of the subjective reasoning of human beings; natural sciences, by contrast, generally aim at *explaining* the world through objective knowledge, codified in natural laws.

Sustainability science assumes the validity of ontology/epistemology of both the social sciences and the natural sciences. In practice, however, because of a failure to articulate the basis upon which the different ontologies are interlinked, sustainability science tends to reproduce a hierarchy, privileging the natural sciences over the social sciences: in the social organization of academia – and even more so in policymaking – the existence and use of "objective" scientific evidence, derived from positivist natural or economic science, is clearly afforded greater weight than the "subjective knowledge" derived from the social sciences and the humanities.

Moreover, in practice, the assumed equal validity of both ontologies often falls apart when inevitably pulled between two different worldviews, namely, materialism – emphasized in the natural sciences – and constructivism or idealism – emphasized in the social sciences and humanities. The natural sciences claim to produce knowledge about the "real" world, made up of matter, fitting easily with materialist worldviews. Today's more neurologically based materialism basically views the social world of ideas, feelings, and the will as resulting from material biophysical processes of the human body (Roth and Grün 2006).

Materialism is a wide field of different ontological positions that cannot be comprehensively addressed here. According to Bunge (2012), materialism is an ambiguous concept, linked at the same time to moral doctrine, a philosophy, and an entire worldview. Moral materialism is related to hedonism, playing an important role in private or state-based forms of capitalism. Philosophical materialism states that the real world is only composed of matter. From the 19th century on, materialist philosophy was embraced – in a "philosophically naïve" way (Bunge 2012) – by the "modern" natural sciences. This gave rise to a type of "scientific materialism" of natural sciences. This view assumes that all phenomena in the universe, including the human mind, have a material basis, and are therefore subject to the same physical laws, to be uncovered by scientific analysis

In this sense, when we speak of "materialism" – especially in Chapters 1 and 2 – we refer to this specific form of "philosophically naïve" ontological materialism, represented by natural science. Moreover, this type of "scientific materialism" has gained currency in society at large. In particular, this is because it can be easily linked to hedonism (important for the functioning of modern consumerism) and serves the ongoing "naturalizing" of asymmetric power relations and hierarchies, based on our supposed "egoist human nature". This also serves to turn the "struggle of the fittest" into quasi "natural laws" that cannot be changed.

More recently, this concept of "scientific materialism" and the supposed "unchangeable" nature of human beings have fed into the ideology of transhumanism: it concerns the presumed failures of human beings in addressing the converging socio-ecological and cultural crisis; however, it emphasizes "fixing human nature" by means of new neurological devices, digitalization, and artificial intelligence. However, we agree with Frodeman (2019) that the idea of "human enhancement" relates human evolution to a set of pathologies that are

> rooted in, or a reaction to, the ethos of "more". It's a cultural love affair with excess, which is prompted by the libertarian standards of our cultural productions. But the attempt to live at the speed of an electron is destined for failure.
>
> (Frodeman 2019:2)

Problematizing this specific notion of "scientific materialism" does not mean ignoring other important branches of materialist philosophy, for example, as offered by critical theory and its expansion by Karl Marx, Enrique Dussel, or Ernst Bloch, which are indeed part of critical sustainability science, as shown in Chapters 1, 2, 7, and 8. Indeed, critical sustainably science has an ontologically open and evolving character. However, valuing the contributions of critical theories based on philosophical materialism does rule out moving beyond materialism, as shown in Chapter 7 on Ernst Bloch's thinking (Dietschy 2023). Through the dialogue with relational ontologies, explicitly stating that matter is an expression of immaterial forces, critical sustainability science leaves open the possibility that materialism takes a "cosmosophic turn"; this would mean viewing the material side of things simply as the outer edge of inner forces related to spirituality (see Chapters 3–6 on indigenous ontologies and cosmosophy) or to "living ideas" as shown in Goethean science (see Chapters 12–14).

In modern dualism, the social sciences and humanities are essentially constructivist. This implies that ideas, concepts, values, visions, feelings, and related patterns of action and behavior are seen as subjectively constructed, and the degree to which they are related to an objective reality is regarded as unknowable. In this way, the constructivist ontology shares the view that the natural world is essentially separated from the human world. Social constructivism contrasts with materialism, assuming that a recognized object is socially constructed by the observer through the process of recognition. While constructivism is a diverse field of philosophical inquiry, its various strands generally assume the impossibility of definite knowledge about the relations of the social and natural world.

More recently, limitations of dualist, materialist, and constructivist ontologies in sustainability science – and elsewhere – are increasingly discussed. The tacit ignorance and ambiguity of sustainably science, in dealing with its dualist underpinnings, motivates Nagatsu et al. (2020) to identify three key areas in which philosophers of science can help to overcome the lack of clarity of sustainability science with respect to its cognitive foundations. These key areas concern the clarification of the essentially dualist ontology of sustainability sciences; the implications this has for dealing with epistemological, conceptual, and methodological criteria used for justifying specific truth claims; and the treatment of normative and ethical issues.

Hamilton (2002) offers an excellent analysis of the nature and the implications of the dualism of sustainability science. He reveals the dualist character at its core by distinguishing intuitive and rational knowledge. This results in the marginalization and trivialization of intuitive knowledge, despite its centrality to the inner worlds of human beings, birthing our original relationship to the natural world. According to Hamilton, the rise of the scientific–industrial revolution meant not only banishing intuitive knowledge, but also making instrumental scientific rationality the basis of modern development – replicated in sustainability science.

Meanwhile, the transition from intuitive to rational scientific knowledge meant shifting from a participating to a non-participating consciousness, heralding new modes "of self-awareness, an isolated ego existing 'inside' the body, and an acute psychological loss. The consciousness of non-participation meant that the natural world was now seen as essentially dead. It could therefore be exploited for the material benefit of humans" (Hamilton 2002:89).

As pointed out in the foreword of this book by Vandana Shiva (2024), the separation of humans from nature has given rise to what she calls "eco-apartheid". This phenomenon originated in

> the minds of the powerful. Its contemporary seeds were sown when the living Earth was transformed into dead matter to facilitate the industrial revolution. Reductionism replaced holism, monocultures replaced diversity and complexity, "raw material" and "dead matter" replaced a constantly renewing and vibrant Earth. As the dominant neoliberal paradigm and capitalist worldview take over the minds of governments and citizens, we give up our sovereignty and freedoms in the name of "growth", of "progress", of "development". Those who have no illusions of growth and development are the communities whose land and forests, water and biodiversity are grabbed for corporate profits, whose very sustenance is destroyed in a massive, global resource grab.
>
> (Shiva 2013:2)

Eco-apartheid is closely associated with two other macro-tendencies. On the one hand, it refers to the rise of authoritarian capitalism in the 21st century (Bloom 2016). This author points to a growing number of people adhering to the idea that authoritarian rule is required for ensuring the survival of modern forms of capitalism. This thinking is essentially based in

> an appeal to an array of affective political fantasies. Here concrete policies of privatization and corporate-led development are linked appealing narratives of progress directly championing explicitly authoritarian regimes as well as the authoritarian polices of otherwise conserved democratic governments.
>
> (Bloom 2016:vi)

On the other hand, it refers to the rise of ecologically motivated authoritarianism or the "coming authoritarian ecology" (Flipo 2018). This strand of thinking had flourished during the 1970s; under the banner of "neo-Malthusianism", scientists and politicians were claiming

> that governments should not act as central planners but should nevertheless be granted to carry out public programs and to intervene in the personal and economic activities of citizens, without having to abide by limitations emerging from citizens' private and democratic rights.
>
> (Shahar 2015:346–347)

This discourse was losing force, with the collapse of the Soviet Union. But more recently, it remerged in the shape of modern-day China, from where it slowly moves also into societies, considered as liberal democracies (Beeson 2010). Trends toward eco-authoritarian thinking are appearing in several domains of environmental policies, for example, related to the "fortress conservation of biodiversity" (Pemunta 2019), the so-called nature-based solutions in managing climate change in sustainable development (Mabon et al. 2022), or works linking the failures in tackling climate change with "the failure of democracy" (Shearman and Wayne Smith 2007).

Against this background, it becomes clear that the biggest challenge facing sustainability science today is the question of how to shift away from dualist ontologies and move on toward meaningful engagement with relational ontologies. The latter is at the center of non-dualist, holistic worldviews and can be integrated into sustainability science by means of engagement with various long-running strands of critical thinking in the social sciences and humanities. This will contribute to finding creative ways of transforming the ontological and epistemic "monoculture of the mind" of sustainability science, which is echoing "modern" sciences, through the development of *critical sustainability sciences.*

In the following chapters of this book, we demonstrate that this is not only possible, but also highly rewarding. In Chapter 2, we summarize key areas or "building blocks" for the construction of *critical sustainably sciences.* These key areas essentially concern the need for engagement with the long and rich tradition of critical thinking, hitherto neglected by mainstream sustainability science. Specifically, we demonstrate how engagement with relational ontologies, ideas of emancipation and autonomy, intra- and intercultural dialogues, political economy, and ecology can help to overcome various limitations of sustainability sciences.

Given the importance of engaging with relational ontologies, in Chapters 3–5, authors from Bolivia, India, and Africa, respectively, present the key features of the diverse relational ontologies that constitute their lifeworlds and sociocultural practices. The authors also reflect on the conditions under which these ontologies could be engaged with, and the contributions they might make to *critical sustainability sciences.*

In Chapters 6–14, we further explore key issues of *critical sustainability sciences*. Chapter 6 presents basic ideas of a cosmosophy that could orient the integration of core elements resulting from intercultural dialogues between dualist and relational ontologies. Chapter 7 expands on the prospect of a cosmosophical turn in the sustainability sciences. The author presents novel ideas for rethinking sustainability in terms of a "nature alliance", drawing on intercultural philosophy, post-development, and decolonial thinking. Chapters 8 and 9 introduce innovative ways for social movements to overcome the fictitious commodity character of labor. Chapters 10 and 11 demonstrate how struggles for "real food" in Brazil and the emancipatory concepts of agroecology build a food-, health-, and nutrition-related core of *critical sustainability sciences*, gaining expression in social movements, emancipatory sciences, and practices.

Finally, Chapters 12 and 13 present novel ways of incorporating relational ontologies and corresponding epistemologies. They serve the construction of *critical sustainability sciences* by developing key features of non-dualist – that is, *relational* – natural and social sciences. Chapter 14 presents a novel radical critique and creative expansion of the tripolar model of sustainability. It shows how relational thinking could help reorganize our societies, considering the threefold nature of a living cosmos, in which human and non-human beings are seen as possessing matter, soul, and spirit. In this perspective, societal organization would benefit from three semi-autonomous, self-determined, but interrelated realms of life: a cultural or spiritual realm of life, based on individual freedom (including creative pursuits, the arts, education, science, healthcare and medicine, media, religion, spirituality); an associative economic realm of life, based on collaboration and fraternity; and a legal realm of life, guaranteeing equality of all people, based on democratically determined political, social, and cultural rights.

References

Beeson, M., 2010. The coming of environmental authoritarianism. *Environmental Politics*, 19(2), 276–294.

Bell, K., 2016. Green economy or living well? Assessing divergent paradigms for equitable eco-social transition in South Korea and Bolivia. *Journal of Political Ecology*, 23, 71–92.

Bloom, P., 2016. *Authoritarian Capitalism in the Age of Globalization*. Glos: Edward Elgar Publishing.

Bond, P., 2002. *'Something for Nothing' in the Discourses of 'Sustainable Development'*. Paper presented to the Eighth Critical Methods Conference, Unisa Sunnyside Campus, Pretoria, September 5, 2002, Club of Rome.

Bunge, M., 2012. *Scientific Materialism*. Dordrecht: D. Reidel Publishing Company.

Carruthers, D., 2001. From opposition to orthodoxy: The remaking of sustainable development. *Journal of Third World Studies*, 18(2), 93–112.

Carson, R., 2015. *Silent Spring*. Routledge (originally published in 1962). Greenwich, CT: Fawcett Publications, Inc.

Chancel, L., Piketty, T., and Saez, E., 2022. *World Inequality Report 2022*. Cambridge, MA: Harvard University Press.

Clark, W., and Harley, A., 2019. *Sustainability Science: Towards a Synthesis*. Sustainability Science Program Working Paper 2019-01, John F. Kennedy School of Government, Harvard University, Cambridge, MA.

Devall, B., 1980. The deep ecology movement. *Natural Resources Journal*, 20, 299.

Dietschy, B., 2023. Towards a "nature alliance" – Why sustainability must be rethought in terms of relationality. In: Rist, S., Bottazzi, P., and Jacobi, J., eds. *Critical Sustainability Sciences: Intercultural and Emancipatory Perspectives*. London: Routledge.

Elder, M., and Olsen, S. H., 2019. The design of environmental priorities in the SDG s. *Global Policy*, 10, 70–82.

Elsig, M., and Amalric, F., 2008. Business and public – private partnerships for sustainability: Beyond corporate social responsibility? *Global society*, 22(3), 387–404.

Flipo, F., 2018. *The Coming Authoritarian Ecology*. Hoboken, NJ: Wiley.

Frodeman, R., 2019. *Transhumanism, Nature, and the Ends of Science*. London: Routledge.

Galtung, J., 1973. 'The limits to growth' and class politics. *Journal of Peace Research*, 10(1–2), 101–114.

Galtung, J., 1986. Towards a new economics: On the theory and practice of self-reliance. In: *The Living Economy: A New Economics in the Making*. London: Routledge.

Hamilton, C., 2002. Dualism and sustainability. *Ecological Economics*, 42(1–2), 89–99.

Haydon, S., Jung, T., and Russell, S., 2021. 'You've been framed': A critical review of academic discourse on philanthrocapitalism. *International Journal of Management Reviews*, 23(3), 353–375.

Kates, R. W., 2011. What kind of a science is sustainability science? *Proceedings of the National Academy of Sciences*, 108(49), 19449–19450.

Kates, R. W., et al., 2001. Sustainability science. *Science*, 292(5517), 641–642.

Kenter, J. O., and O'Connor, S., 2022. The life framework of values and living as nature; towards a full recognition of holistic and relational ontologies. *Sustainability Science*, 17(6), 2529–2542.

Leff, E., 2015. Encountering political ecology: Epistemology and emancipation. In: *The International Handbook of Political Ecology*. Cheltenham: Edward Elgar Publishing, 44–56.

Longo, S. B., et al., 2021. Sociology for sustainability science. *Discover Sustainability*, 2(1), 1–14.

Mabon, L., et al., 2022. Whose knowledge counts in nature-based solutions? Understanding epistemic justice for nature-based solutions through a multi-city comparison across Europe and Asia. *Environmental Science & Policy*, 136, 652–664.

Martinez-Alier, J., 2002. *The Environmentalism of the Poor: A Study of Ecological Conflicts and Valuation*. Northampton, MA: Edward Elgar Publishing.

Meadows, D., 1972. *The Limits to Growth; a Report for the Club of Rome's Project on the Predicament of Mankind*. New York: Universe Books.

Miller, J., 2017. *China's Green Religion: Daoism and the Quest for a Sustainable Future*. New York, NY: Columbia University Press.

Mitlin, D., 1992. Sustainable development: A guide to the literature. *Environment and Urbanization*, 4(1), 111–124.

Moore, J. W., 2016. *Anthropocene or Capitalocene?: Nature, History, and the Crisis of Capitalism*. Vestal, NY: Binghamton University—SUNY, PM Press.

Naess, A., 1973. The shallow and the deep, long-range ecology movement. A summary. *Inquiry*, 16(1–4), 95–100.

Nagatsu, M., et al., 2020. Philosophy of science for sustainability science. *Sustainability Science*, 1–11.

Parrique, T., et al., 2019. Decoupling debunked. *Evidence and Arguments Against Green Growth as a Sole Strategy for Sustainability. A Study Edited by the European Environment Bureau EEB.* Available from: https://www.almendron.com/tribuna/wp-content/uploads/2019/11/decoupling-debunked.pdf (Accessed 12 April 2023).

Pemunta, N. V., 2019. Fortress conservation, wildlife legislation and the Baka Pygmies of southeast Cameroon. *GeoJournal*, 84(4), 1035–1055.

Roth, G., and Grün, K.-J., 2006. *Das Gehirn und seine Freiheit: Beiträge zur neurowissenschaftlichen Grundlegung der Philosophie.* Göttingen: Vandenhoeck & Ruprecht.

Sachs, J., et al., 2022. *Sustainable Development Report 2022.* Cambridge: Cambridge University Press.

Seitz, K., and Martens, J., 2017. Philanthrolateralism: Private funding and corporate influence in the United Nations. *Global Policy*, 8, 46–50.

Shahar, D. C., 2015. Rejecting eco-authoritarianism, again. *Environmental Values*, 24(3), 345–366.

Shearman, D., and Wayne Smith, J., 2007. *The Climate Change Challenge and the Failure of Democracy.* Santa Barbara, CA: Praeger.

Shiva, V., 2013. From eco-apartheid to earth democracy. In: *Yes Naturally: How Art Saves the World*, Netherlands: Niet Normaal Foundation, 94–99.

Shiva, V., 2018. Development, ecology, and women. In: *Living with Contradictions.* London, Routledge, 658–666.

Shiva, V., 2024, Forword. In: Rist, S., Bottazzi, P., and Jacobi, J., eds. Critical Sustainability Sciences: Intercultural and Emancipatory Perspectives. London: Routledge, x–xvi.

Spretnak, C., 1986. *The Spiritual Dimension of Green Politics.* Santa Fe: Bear & Company.

Steffen, W., Crutzen, P. J., and McNeill, J. R., 2007. The anthropocene: Are humans now overwhelming the great forces of nature. *AMBIO: A Journal of the Human Environment*, 36(8), 614–621.

Steffen, W., et al., 2015. Planetary boundaries: Guiding human development on a changing planet. *Science*, 347(6223), 1259855.

United Nations, 2022. *The Sustainable Development Goals Report.* Available from: https://unstats.un.org/sdgs/report/2022/The-Sustainable-Development-Goals-Report-2022.pdf (Accessed 12 April 2023).

United Nations Conference on Sustainable Development, 2012. *Declaration of Rio – The Future We Want. Outcome document.* Rio de Janeiro, Brazil.

West, S., et al., 2020. A relational turn for sustainability science? Relational thinking, leverage points and transformations. *Ecosystems and People*, 16(1), 304–325.

Wilbur, A., 2013. Growing a radical ruralism: Back-to-the-land as practice and ideal. *Geography Compass*, 7(2), 149–160.

World Conference on Environment and Development (WCED), 1987. *Report of the World Commission on Environment and Development: Our Common Future.* United Nations, New York.

2 Key areas for critical sustainability sciences

Stephan Rist, Patrick Bottazzi, and Johanna Jacobi

Relational ontologies: a core of critical sustainability sciences

West et al. (2020) suggest that currently established sustainability science must take a "relational turn" to overcome its dualist ontological foundations. This means shifting from a substantialist analysis – of ostensibly separate entities of humans and nature – to an analysis that sees both humans and nature as living and, thus, continually changing entities. To understand these living entities, a relational ontology requires putting the continually unfolding and changing *relations* between these interacting entities at the center of scientific inquiry.

West et al. (2020) conclude that a critical discussion about relational approaches in sustainability science

> might contribute to (i) a paradigm shift in sustainability science, and (ii) transformations towards sustainability. Relational approaches foster more dynamic, holistic accounts of human – nature connectedness; more situated and diverse knowledges for decision-making; and new domains and methods of intervention that nurture relationships in place and practice.
>
> (West et al. 2020:304)

Our concept of relational ontology does not derive directly from the more common understanding rooted in mainstream philosophy. Rather, we share the definition deriving from critical anthropology that views ontology as an individual and social process aimed at advancing cognitive self-determination based on people's lived experiences (de Castro 2003). We believe that this existential definition better reflects the purpose and perspective of relational ontologies, namely, exploring the lived experience – including thoughts, feelings, and actions – of people who are involved in the creation of diverse counter-hegemonic worlds and in resistance to imposed concepts of "nature", "truth", and "reality", etc. (Bertelsen and Bendixsen 2016).

Consequently, the scope of relational ontologies is much broader than that of dualist ontology. Many relational ontologies consider the cosmos as part of an emergent process of communication among all the entities that compose it – including human beings and entities "other than human beings" such as plants,

DOI: 10.4324/9781003043577-2

animals, water, minerals, winds, clouds, rain, stars, the sun, planets, and related notions of time and space, etc. This also implies that the future – rather than following predefined paths of evolution, development, or revolution – results from emerging patterns of what Siena (2005) refers to as a cosmosophical process of communication between the members of a "planetary community of destiny".

This definition of relational ontology also assumes that nothing is defined a priori (Escobar 2016). Instead, it emphasizes the emergent, negotiated nature of entities and their meanings. The process is an outcome of ongoing dialogue between personal introspection and social communication. The forms of life from which relational ontological conceptualizations emerge assume that "nothing is outside" (Blaser 2009; Choquehuanca Céspedes and Rist 2024). As a result, the task of the human community is not to impose itself above all other living entities of the cosmos. Rather, related ontologies are rooted in the lives of all existing entities. In this context, the human community must grow to understand itself as born as an "organ" of the cosmos, with the cosmos creating, for itself, the faculty of human thought, intuition, self-awareness, and emotions.

Choquehuanca Céspedes and Rist (2024) expresses this cosmosophical dimension of relational ontologies in Chapter 3 as follows:

> We are the cosmos that thinks, we are cosmos made flesh and spirit. In the community no one has to feel free to destroy the other, be it animal or plant, because we all complement each other, because we are all sisters and brothers raised by Mother Earth, and because we all drink from the milk of Mother Earth, which is water. Human beings are similar to other living beings because human beings are part of the natural and cosmic totality and because in cosmic life we are energy and life.
>
> (Choquehuanca Céspedes and Rist 2024:49)

This implies that the entity comprising the cosmos collaboratively co-create its structure, dynamics, and tendencies through semiotic communication (Moser 2018). Relational ontologies lay the foundation for processes of self-education that enable people to see, hear, touch, and feel, in order to create thoughts, emotions, and words as precursors to actions. In this way, human activities are but one aspect of the living cosmos.

In line with relational ontologies, the concept of the "Universe" is replaced by the concept of a "Pluriverse" or a "multi-verse" in the words of the Zapatistas of Mexico (Kothari et al. 2019). Here, human thought is seen as reaching beyond mere reason and formal logic, encompassing what is expressed as "pensa-siento" in Spanish – or "think-feel", as roughly translated into English. This implies that human thought "requires moving out of the epistemic space of Western social theory and into the epistemic configurations associated with the multiple relational ontologies of worlds in struggle" (Escobar 2016:16).

At the same time, we agree with one critical point raised in the context of post-development, namely, that of the impossibility of understanding relational ontologies outside their historical context. In this way, relational ontologies are also marked by the long history of colonization of lifeworlds based on capitalist expansion and reorganization. This means that critical sustainability sciences must enable collaborative identification and review of any residues, internalizations, or other hidden forms of domination of lifeforms, so as to support resistance to ongoing processes of colonization.

In this way, overcoming the dualist ontology of sustainability science appears possible to the extent it starts engaging with relational thinking. Adopting the idea of relational ontologies would have three advantages. First, it would help to restore a more comprehensive view of humans to research, similar to that originally proposed by deep ecology and the environmental movements of the 1960s and beyond, with human beings considered as living entities that are co-constituted by matter (body) and the soul/mind and spiritual faculties. As observed by Hamilton (2002:89): "While recognizing the value of reason, the well-spring of environmentalism is intuitive knowledge. Environmentalism seeks to transcend duality, reassert the cosmic unity and breathe life back into Nature".

Second, adopting relational ontologies would enable sustainability science to engage in meaningful dialogue and processes of mutual learning with other worldviews that traditionally go beyond the dualist separation of nature and society. In this realm, we find many non-Western, often indigenous groups (Grindsted 2018) as well as a growing number of environmental groups concerned with "subtle spirituality" (Becci et al. 2021). These groups employ varying traditional and new spiritual practices – such as meditation, yoga, contemplation, rituals, or cosmocentric celebrations – to transform purely rationalist thinking, feeling, and actions into "living thoughts" concerned with intuition and caring actions.

Third, the engagement of sustainability science with relational ontologies implies reconnecting with a variety of vibrant, long-running traditions of critical thinking, around since the dawn of the "modern age of sciences". This enables integration of key questions raised by critical sciences – hitherto largely neglected by sustainably science – including issues identified by political economy, political ecology, integral psychology, feminism, decolonial and post-development studies, liberation theology, ecosophy, and Goethean sciences and anthroposophy. These, in turn, can reveal new paths of evolution that expand emancipation and autonomy at the individual and societal levels, supporting the development of individuals, communities, and institutions that promote forms of living based on existential and cognitive self-determination.

Intra- and intercultural dialogue

Overcoming the monoculture of mind, life, and development

In sustainability science, researchers have begun to recognize that the neglect of engagement with its own and other philosophies of science also complicates interaction with the huge range of biocultural diversity, often expressed in non-Western worldviews (Caldas et al. 2015). As sustainability science is based on ambiguous and implicit dualist (materialist or constructivist) views of the world, it can easily enter into tension or conflict with other non-dualist worldviews. This is especially important as many worldviews

> particularly those in indigenous cultures – place a higher value on the spiritual and nonmaterial aspects than what is often represented by the discourse surrounding Western conservation policy. Alternative framings of the biosocial complex that recognizes nature's intrinsic value can be powerful motivators for social change and for local-scale conservation efforts.
>
> (Kohler et al. 2019:1014)

This perspective acknowledges, for example, that local-scale conservation efforts of indigenous people are typically rooted in worldviews in which the spiritual and non-tangible aspects of nature are decisive. However, the following question usually goes unexplored: Are indigenous people willingly agreeing to reduce the issue of conservation of nature to mere material parameters, as expressed in numbers of threatened species on a red list elaborated by materialist ecologists? Or, do indigenous people instead view conservation of nature as an end in itself, as part of a wider practice of "Living Well in reciprocity with nature"? In our view, these are important questions that should be addressed based on intra- and interepistemological conversations (Rist et al. 2011).

Despite the growing recognition of the value of indigenous knowledge for sustainability science, efforts toward facilitating authentic and respectful interaction between differing worldviews remain in their infancy (Ruwhiu et al. 2022). Sustainability science tends to consider other non-dualist, relational ontology-based worldviews through an instrumental lens: indigenous knowledge or institutions are recognized to the degree they are compatible with the dualist ontology of sustainably science; relational ontologies are thus becoming instrumental in achieving goals of sustainability science, defined without an authentic inclusion of knowledge created in non-dualist worldviews (Odora Hoppers 2021).

More generally, on the one hand, we find worldviews ranging from dualist to monist ontologies (materialism, realism, idealism, spiritism, constructivism, anthroposophy, ecosophy, etc.). On the other hand, we find numerous religious and spiritual traditions, ranging from monotheist to polytheistic belief systems. Ontologies are not always expressed through rationalist and explicit reasoning, as in today's Western philosophical practice. Ontologies can also be based on rather existential perceptions or feelings, emerging from rituals, meditation, prayers, and other moments in people's lifeworlds. From the resulting existential basis, the persons of these communities translate their perceptions, visons, and associated feelings into thoughts and words. This permits them to communicate among themselves, and with people belonging to other belief systems (Brook 2021).

The huge trove of different worldviews is linked with equally high cultural diversity. When it exercises an instrumental relationship with indigenous worldviews, sustainability science risks aiding the violation of cognitive justice. This occurs when science is used to justify state violence and technological interventions that invade and disrupt the private life of citizens in the name of progress and development (Nandy 1988).

Taken together, this increases the marginalization of the multiple ways of knowing, feeling, and understanding the world, which are at the heart of the rich cultural diversity that humanity has created. Cultural diversity is put under pressure, mainly by the expansion of what Shiva (2001:199) calls a "monoculture of mind", leading to

> disappearance of local knowledge through its interaction with the dominant Western knowledge takes place at many levels, through many steps. First, local knowledge is made to disappear by simply not seeing it, by negating its very existence. The Western systems of knowledge have generally been viewed as universal. However, the dominant system is also a local system, with its social basis in a particular culture, class and gender. It is the globalized version of a very local and parochial tradition.
>
> (Shiva 2001:199)

This monoculture of mind especially compromises the diverse ways of seeing and feeling the world still present in many indigenous groups, as well as many social, cultural, religious, and spiritual movements that are concerned with building ways of living and "re-linking" the material world, human beings, and other beings – although the spiritual dimensions of existence often continue to be discredited and ignored (Gambon and Bottazzi 2021). These ways of living are based on divergent, subaltern, alternative or post-materialist, relational ontologies. They demonstrate that it is possible to build individual and community lives, as well as corresponding societal structures that go beyond seeing humans as selfish beings who supposedly cannot save themselves from the "struggle between all, and everything" without oppressing, competing, or annulling their alter-egos.

Additionally, it is important to note that the monoculture of mind also implies, on the one hand, a supposedly universal understanding of life and living, as determined by the materialistic explanations of "natural laws" in prevailing natural sciences. On the other hand, monocultural "modern universality" also asks us to accept that the "laws of the social world" are given by the currently dominating constructivist social sciences and humanities. However, this implies accepting that human beings cannot know to what degree they are part of reality, and opens the door to ego-centered behavior, competition, and the reduction of social relations to maximization of tangible and intangible utilities negotiated in corresponding marketplaces.

Toward an intra- and interontological dialogue

In Chapter 4, Shankar (2024) summarizes some important starting points and key conditions for intra- and interontological dialogue. The currently dominanting scientific ontology has its roots in the renaissance European culture. Domination and marginalization of ontologies is generally an expression of politics of knowledge which in turn is a function of the geo-politics that governs national and international, economic, social, and technological processes. It is important in public interest to analyze and review relational ontologies specially when one observes large scale adverse effects of a dominant ontological system. Promotion of multicultural knowledge especially in university and knowledge institutions, instead of monoculture is critical for advancement of civilizational evolution. The review of ontologies underlying indigenous health sciences and practice is urgent. This is because firstly they are effective and secondly because they are holistic. They should be assessed based on their actual performance and in terms of their own intellectual framework and not by an external politically dominant ontological yardstick.

Further, from the perspective of the age-old Indian Ayurvedic knowledge system, Shankar (2024) pointedly observes:

> Western Science is a recent functional ontological framework for comprehension of nature. It enjoys the well-deserved credit of liberating Europe from the dark ages (from dogma to reason). But it's super-imposition on the non-Western cultures with the claim that it is the only credible way of interpreting nature has perhaps given it the role of a cultural oppressor.
>
> (Shankar 2024:71)

Against this background, epistemic and geopolitical issues may partly explain why Western-oriented mainstream sciences have not embraced relational ontology-based indigenous knowledge traditions. While the sovereignty of previously colonized nations has made advances,

> the governing elite and sections of intelligentsia due to their deep engagement with a complex international ecosystem of networked economic, scientific, educational, and political institutions, appear to have got trapped in a powerful centralized global system of knowledge and practices (with their underlying ontology) from which, for fear of isolation, most of them do not have the courage or funding opportunities to explore with self-respect, self-confidence and non-chauvinist intent, knowledge derived from their own cultural roots.
>
> (Shankar 2024:70–71)

The praxis of communication between different ontological communities makes clear that such dialogue must be understood as a process of personal and collective reflection. It constantly oscillates between internal moments, in which specific ontological communities review their own thoughts through introspection, and external moments in which they share the results of internal inquiry with other ontological communities with which they engage in dialogue. It is therefore crucial to be aware that it is in fact an ***intra- and interontological dialogue*** (Rist and Dahdouh-Guebas 2006).

In addition, the practice of interaction between ontological communities (relational or not) shows that recurring questions in intra- and interontological dialogue revolve around the relationships – manifesting in each ontological community – between spirit, mind, life and matter, life and death; between the social and the natural; and between object and subject (Delgado et al. 2012; Rist et al. 2011).

Understanding life and living as more than socio-materialist determination is a central issue of the manifold social–environmental and justice movements enacting them in their everyday lives, all around the world. They are countering the classical Western-based notions of "development" as promoted by conventional sciences, policies, forms of organization, and technologies. These movements sometimes "revive or creatively re-interpret long-standing indigenous movements; yet others revisit older philosophies and religious traditions. All of them ask: What would a better life look like, and how do we get there?" (Kothari et al. 2019:xxviii).

This means that we must go further than just looking at how to achieve predefined visions, goals, or targets of sustainable development. Before coming to these – indeed, vital issues – it is more important to begin reflecting on how to live sustainably by first answering the fundamental questions posed by Vandana Shiva (2019:vii): "What is it to live, to be alive? What is it to live well, to be well? What is knowledge, what is intelligence? What is ecology, what is economy? What is freedom, what is democracy? What is our future?" This means that "development" is much more than defining forms of economic growth, distribution of wealth infrastructure, healthcare facilities, etc.

Indeed, the very notion of "development" should be replaced by the diverse answers to Shiva's fundamental questions, emphasizing practices and skills aimed at emancipation and everyday experience, considering the expressions of spiritual entities in the continuity of social to natural realms of life – referred to as "cosmopraxis" by Munter and Note (2009). These new forms of enacting human beings in the Pluriverse replaces classical concepts of "development". Post-development thinking is therefore focusing on the highly diverse notions of coevolution of humans and other than human beings, for example, expressed in ideas relating to Suma Qamaña, Sumak Kawsay, Sansaeng, Ubuntu, or Waniambi to Tobati Engros in East Papúa (Dietschy 2015).

Thus, interontological dialogue becomes a process of social learning within and among ontological communities that interact and continually review their own conceptions in order to relate them to possible external contributions. Interontological dialogues enhance existing networks of thought and foster new ones in a continual process. Questions of core cognitive essences differentiating one ontological community from another are not grounds for conflict, but rather key sources for exploration of opportunities for complementarity, synergy, and co-creation of common ground on behalf of jointly developed, mutually shared notions of "sustainability" and "development" as well as the institutions needed to realize them.

Against this background, in Chapter 5, Millar et al. (2024) calls for a "transcendental philosophical approach to research" to guide the search for interontological exchange and cooperation within the community of critical sustainability science. He further observes:

> Transdisciplinarity would then provide the basic methods, tools and techniques for conducting research and development in an emancipatory manner – getting all stakeholders (including rural ontologically-rich, but material resource-poor farmers) included and involved so as to own the processes in a collegial manner. The ontological unpinning of the Dagara phenomenology of "Vurr" makes it imperative to put spirituality at the center of human and environmental systems.
>
> (Millar et al. 2024:90)

Emancipation and autonomy: engagement with critical theory

Critical theory

As laid out earlier, an important task of critical sustainably sciences consists in making visible the inherent links between the dualist ontology of the currently dominating sciences and the consequences this has for the stabilization and enhancement of capitalist appropriation of "human and natural resources". This implies to link capitalist relations with issues of oppression, coercion, manipulation, exclusion, and injustice among large parts of the global population. This means to relate critical sustainably sciences with the larger body of critical theory.

Critical theory began emerging at the beginning of the last century. It struggled to free itself from previous rather orthodox approaches of Marxism. It sought to understand the oppressive nature of the regimes of accumulation, rule, and "truth" in modern Western societies. These elements appeared to contradict professed Western enlightenment values of individual freedom, self-determination, representative democracy, and free trade. In its narrowest sense, critical theory refers to several consecutive generations of German philosophers and social theorists in the Western European Marxist tradition. They started to appear as a scientific strand of independent thinking from the 1930s onward.

Though a very broad field of scientific inquiry, critical theory is usefully summarized as follows by Bohman (2005:1):

> According to these theorists, a "critical" theory may be distinguished from a "traditional" theory according to a specific practical purpose: a theory is critical to the extent that it seeks human "emancipation from

> slavery", acts as a "liberating . . . influence", and works "to create a world which satisfies the needs and powers of" human beings ([Horkheimer 1972b]:246, cited in Bohman 2005:1). Because such theories aim to explain and transform all the circumstances that enslave human beings, many "critical theories" in the broader sense have been developed. They have emerged in connection with the many social movements that identify varied dimensions of the domination of human beings in modern societies. In both the broad and the narrow senses, however, a critical theory provides the descriptive and normative bases for social inquiry aimed at decreasing domination and increasing freedom in all their forms.

The first wave of critical theory initially focused on the human side of modernity as experienced in the mid-20th century. Before the ecological contradictions of modernization became a scientific and political issue, critical theory mainly emphasized human emancipation. This was seen as an avenue for overcoming all forms of oppression and alienation, embraced by modern forms of economic, social, political, and cultural organizations.

In more recent accounts of critical theory, the relationship between oppressive, alienating institutions and environmental crises has gained increasing attention. A good example is Fraser and Jaeggi (2018). Based on a synthesis of thoughts from the various fields of Marxism, feminism, ecology, and postcolonialism, they distinguish between foreground and background conditions in the reproduction of capitalism. *Foreground conditions* are those that are explicitly included in the pricing mechanisms of goods and services, such as the cost of labor, natural resources, or rent on capital invested. Background conditions refer to those inputs into the capitalist economy that are not priced in the production processes, such as the deterioration of nature, human health, education, as well as the need to care for the young or the sick people who are not accounted for in the "productive sector". These unaccounted social and ecological costs are shouldered by workers and people in general. They constitute an increasingly heavy burden, putting the human and natural base of our societies under enormous pressure.

Background conditions are non-commoditized and concern unwaged social reproductive labor, land, seeds, non-human nature, public and legal powers, as well as the confiscated labor and resources of racialized groups. Capitalist regimes of accumulation depend on the continuous flow of "free gifts" from the realms of social–ecological reproduction. As a result, capitalism is doomed to crisis and eventual failure; its maintenance is only possible insofar as it continues to consume the non-economic background conditions on which its apparently successful development is based. Unceasing depletion of nature is part of an accelerating, self-destructing process of the very social–ecological basis upon which capitalist wealth accumulation currently rests (Fraser and Jaeggi 2018).

In their critical analysis of SDG implementation, Menton et al. (2020) adopt the concept of environmental justice – a more recent concept emerging from critical theory. They move environmental justice beyond its four classical principles, namely, distribution, procedure, recognition, and capabilities. They propose an intersectional, decolonial approach to environmental justice. They propose that the use of environmental justice for implementation of the SDGs is useful to the extent it makes explicit "power dynamics, complex interactions among injustices, and listens to the different 'senses of justice' and desires of theorists, activists, and other stakeholders from the Global South" (Menton et al. 2020:1621).

Gottschlich and Bellina (2017) link critical theory with the discourse of sustainability, emphasizing that sustainable development must be rethought beyond its rather neoliberal conceptualization. They propose beginning by questioning the asymmetric power relations of the dominant economic system; only on this basis can sustainability become part of the wider, longer-term aspirations of human beings, centering on emancipatory, non-oppressive struggles rooted in democratic, solidarity, associative, and self-determined reform of the socioeconomic realms of life.

Linking critical sustainability sciences with critical theory enables the identification of new concepts – including concepts that can help to explain why the 2030 Agenda and SDGs have been doomed to fail. By now it is clear: the 2030 Agenda is overly instrumental. According to its logic, any form of social or economic organization is acceptable as long as it enables one or another sustainability objective or target to be met. This bluntly pragmatic rationale turns society – and eventually life on earth – into an object of planning, monitoring, correcting, educating, and improving. Private and public actors are given – or assume – the power to reorganize our societies in the service of their techno-materialist–instrumental rationales (Telleria and Garcia-Arias 2022; Luke 2005).

Emancipation and liberation

The engagement of critical sustainability scientists with key questions of critical theory allows the scientists to contribute with critical reflections and knowledge to the learning process with people living based on relational ontologies. These emancipatory issues were initially linked to Western-borne struggles to free societies from slavery, serfdom, and religious coercion; this was achieved mainly through the spread of and access to rational thinking among a small proportion of clerical and monarchist elites and eventually among the majority of "normal" people.

It is true that the process of emancipation gained momentum with the rise and spread of Enlightenment values – individual freedom and self-determination – and the principles of the French Revolution, including societal organization based on fraternity, freedom, and equity. On the one hand, however, it is important to remain aware and critical of the Euro- and anthropocentric liberal origins of these values and challenge their universal validity. On the other hand, we should acknowledge how these values have affected –

in varying forms and degrees – the evolution of societies even beyond those considered constitutive of the dominant "Western world". In this way, the fundamental value of liberation and emancipation from oppression and coercion has fueled both social–environmental movements in the West (Leff 2015; Brulle 1995) and struggles for decolonization in the Global South.

Against this background, Doyle and Doherty (2006) note that many social–environmental movements carved out agendas which situate the protection of ecosystems in a perspective of emancipation that concerns the liberation of humans and non-humans from degradation and subjugation by non-sustainable forms of socio-ecological development. As they observe (Doyle and Doherty 2006:881):

> "Environment", therefore, can be either a symbol for liberation or repression; emancipation or conditionality. It can be used to support democracy or, alternatively, to support authoritarianism; it can be used to attack neoliberalism and corporate-controlled globalization, and it can be used to support it; it can be used to lionize concepts of "the local", and it can be utilized to denigrate local systems of meaning in a neocolonial fashion.

This raises the question of the experiences and advances that could be adopted as starting points in this endeavor. We therefore summarize some key reflections that emerge from the attempts to understand emancipation and liberation beyond its supposedly universal character. Our definition of ontology as the search for higher levels of existential and cognitive self-determination contains a strong component of emancipation: communities whose lives revolve around relational ontologies aspire to liberate themselves from the multiple forms of domination emerging from educational, academic, social, economic, and cultural institutions that express dualistic, materialistic, and ego- and anthropocentric values.

Consequently, we understand emancipation as a double strategy: on the one hand, the concept of emancipation enables us to become aware of power and its concomitant mechanisms of oppression. This includes revealing the conditions under which subaltern people are alienated by power structures in such a way that it is difficult for them to recognize the forms of domination to which they are exposed. On the other hand, emancipation implies the development of a new language to express not a linear and inevitable evolution, but rather aspirations based on the actions of human beings capable of resistance and using their capacities of reasoning to become aware of their alienation and revolt against the forces of domination (Boltanski 2011).

Thus, critical theory provides an indispensable element for existential and cognitive self-determination: it represents the conviction that the critique of societal conditions must come from the immanent critique of human groups living the contradictions of dualist, capitalist, and/or socialist "modernities".

Against this backdrop, critical theory must pursue a double task: (1) illuminating the changes that occur throughout historical contexts, in relation to the constellations of the individual, nature, and society; (2) seeking to go beyond Marx, in the sense of understanding the social and cultural human being, not merely the human being as determined by material and economic conditions.

Another critical aspect from the perspective of relational ontologies concerns the anthropocentric and Eurocentric nature of mainstream analyses. In his "philosophy of liberation", Dussel (2011) adopts a decolonial, pragmatic, and global perspective in an attempt to overcome Eurocentric notions of liberation.

However, from the point of view of relational ontologies, the concept of liberation particularly falls short as a result of its anthropocentric bias. Dussel recognizes the importance of religious life – even though it is a product of colonization – as a vital force for advancing counter-hegemonic agendas. However, he does not critique its strongly anthropocentric perspective, which runs counter to the existential and conceptual self-determination of the victims of modernity, that is, those who are oriented according to relational ontologies.

In other words, Dussel fails to grasp the cosmosophic dimension that inspires the religious–spiritual life of the victims of Western modernity. We suggest integrating this broader perspective with that of Boff (2011): he promotes an "instaurative rupture" in order to concretize a new social pact that, from the perspective of excluded and oppressed communities, should not only be made between peoples, but also between human beings and other beings of nature. In this way, overcoming the separation between matter and spirit becomes part of a transition from a dualistic ontology to a holistic vision (*visão holística*). This holistic vision places all existing entities in relation to one another in order to show that human life is an integral – but not a superior – part of a cosmic process. Further, this holistic vision emphasizes the fallacy of separating and hierarchizing the fundamental ontological categories of matter, spirit, life, and death, as doing so contradicts the lived experiences and aspirations that emerge from relational ontologies.

We propose to address this critical element by developing an interontological notion of emancipation. This could enable identification of the possible limits of domination inherent in social hierarchies and related forms of social and political organization. Notably, the development of an intercultural concept of emancipation underlying the basic principles of subaltern relational ontologies raises the problem of representation. We agree with Asher and Wainwright (2019), who, drawing on Spivak (1988), assert that representation of subaltern knowledge cannot be done by those who do not belong to it. This must be acknowledged to avoid recreating intellectual domination or paternalism, based on essentialisms, romanticism, or skepticism.

In response to this challenge, critical sustainability scientists – who do not necessarily belong to the communities possessing relational ontologies – have proposed transdisciplinary modes of knowledge co-creation in collaboration with those groups who live according to relational ontologies. This creates space for open exchange and cooperation, in which learning from each other is more important than any other outcome. As shown in the practice of transdisciplinary work, critical sustainability scientists can become part of movements that are based on relational ontologies. And they can do so without losing their role as "reflective scientists" who help broaden the scope and objects of joint reflection, including issues of power relations or oppression as well as options for overcoming them through emancipatory action (Pohl et al. 2010).

Of course, this is only possible to the extent that those possessing relational ontologies are willing to participate in such a process. In this way, critical sustainability scientists avoid the pitfalls trying to speak *for* those who hold relational ontologies, and instead speak *with them* about their views on the imposition and effects of dualistic ontology, as well as related forms of domination, colonization, and emancipation.

A concrete example of the notion of freedom or liberation, emerging from indigenous Andean relational ontologies, is provided by Choquehuanca Céspedes and Rist (2024) in Chapter 3. They observe that Living Well:

> goes beyond the mere search for freedom and is oriented towards the search for complementarity among all beings. When we speak of freedom we are only talking about the individual and collective freedom of human beings, people and society. However, when we talk about the search for complementarity we are talking about the life and freedom of all living beings raised by Mother Earth and not only about the freedom of human beings. In Living Well, everyone is the size he/she is, no more and no less, and thus must respect and complement each other.
>
> (Choquehuanca Céspedes and Rist 2024: 55)

Autonomy

Closely related to the issue of emancipation is autonomy. Esteva (2019) differentiates between three interrelated concepts. He first points to ontonomy, representing traditional, endogenous norms and values, still enacted everywhere; second, he refers to autonomy, as the process through which a group or community creates and develops new norms and values; third, he describes heteronomy, meaning the imposition of rules on others by those with the power to do so. Autonomic movements, for example, related to the creation or revitalization of relational ontologies and forms of life, strive to widen the spheres of ontonomy and autonomy as much as possible. Such movements, of course, confront the manifold attempts of heteronomic forces aimed at homogenizing and erasing the values, norms, and attitudes embraced by ontonomy and autonomy. Autonomic movements, also present in the sphere of social–environmental movements, create room for an emancipatory horizon aimed at life and living beyond capitalisms, industrial modes of production, wage labor, and patriarchy.

Against this background, it becomes clear that sustainability science could benefit greatly from linking research and action with the aspiration of emancipation. Moreover, combining emancipation with the understanding of autonomy of Esteva (2019) would help to avoid holding up emancipation as a transcultural universality; instead, in the context of autonomic movements, authentic ideas of emancipation could be sown based on meanings and experiences of people engaged in the multiple movements of the Pluriverse. Finding ways out of domination, without necessarily following Western-based notions of liberation and emancipation, is enabled through the engagement of critical sustainably sciences in transdisciplinary co-creation of knowledge, based on respect for the normative, institutional, and structural ontonomy and autonomy of both critical sustainability scientists and the people living based on relational ontologies.

Key topics for the construction of critical sustainability sciences

Surveying the specific contents presented in the following chapters, we have identified three key topics that can provide the foundations for construction of critical sustainability sciences. They are (1) the search for new forms of societal organization, (2) efforts to overcome the "fictitious" commodity character of human labor and the "natural resources" exploited in the production, trade, and consumption of goods and services; and (3) the task of revitalizing existing – often marginalized – holistic paradigms and creating new holistic paradigms that are needed to enable critical sustainability sciences, based on relational thinking, emancipation, and interculturality.

Chapters 3–14 contribute to these key topics in diverse ways. Chapter 15 summarizes the three key topics and how critical sustainability sciences could be further developed based on relational ontologies, intra- and intercultural dialogue, and emancipation.

References

Asher, K., and Wainwright, J., 2019. After post-development: On capitalism, difference, and representation. *Antipode*, 51(1), 25–44.

Becci, I., Monnot, C., and Wernli, B., 2021. Sensing 'Subtle Spirituality' among environmentalists: A Swiss study. *Journal for the Study of Religion, Nature & Culture*, 15(3).

Bertelsen, B. E., and Bendixsen, S., 2016. Recalibrating alterity, difference, ontology: Anthropological engagements with human and non-human worlds. In: *Critical Anthropological Engagements in Human Alterity and Difference*. Cham, Switzerland: Palmgrave McMillan, Springer, 1–40.

Blaser, M., 2009. Political ontology. *Cultural Studies*, 23(5–6), 873–896.

Boff, L., 2011. *Ecología: grito de la tierra, grito de los pobres*. Trotta Madrid, Traducido: Herranz, Juan Carlos Rodríguez.

Bohman, J., 2005. Critical theory [online]. In: Zalta, E. N., ed. *The Stanford Encyclopedia of Philosophy* (Spring 2021 ed.). https://plato.stanford.edu/archives/spr2021/entries/critical-theory/ (Accessed 12.12.2022).

Boltanski, L., 2011. *On Critique: A Sociology of Emancipation*. Malden, MA: Polity.

Brook, I., 2021. A new science from a historical figure: Goethe as holistic scientist. In: *Subtle Agroecologies*. New York, NY: CRC Press, Tailer & Francis, 71–79.

Brulle, R. J., 1995. Environmentalism and human emancipation. In: *Social Movements*. London: Palgrave Macmillan, 309–328.

Caldas, M. M., et al., 2015. Endogenizing culture in sustainability science research and policy. *Proceedings of the National Academy of Sciences*, 112(27), 8157–8159.

Castro de, E. B. V., 2003. *After-Dinner Speech Given at Anthropology and Science, the 5th Decennial Conference of the Association of Social Anthropologists of the UK and Commonwealth*. Department of Social Anthropology, University of Manchester.

Choquehuanca Céspedes, D., and Rist, S., 2024. The great disruption – A culture that understands that everything is interrelated, that nothing is divided, and nothing is outside. In: Rist, S., Bottazzi, P., and Jacobi, J., eds. *Critical Sustainability Sciences: Intercultural and Emancipatory Perspectives*. London: Routledge.

Delgado, F., et al., 2012. Knowledge dialogues for sustainable endogenous development: Reforming higher education and research in Bolivia. In: Haverkort, B., et al., eds. *Towards Co-Creation of Sciences. Building on the Plurality of Worldviews, Values and Methods in Different Knowledge Communities*. Bangalore: CAPUTRED, IAMI, UDS, UMSS. Nimby Books, India, 186–233.

Dietschy, B., 2015. Hacia una ecología de saberes¿ Cómo superar el colonialismo epistemológico? *Pasos*, 166, 20–30.

Doyle, T., and Doherty, B., 2006. Green public spheres and the green governance state: The politics of emancipation and ecological conditionality. *Environmental Politics*, 15(5), 881–892.

Dussel, E., 2011. *Filosofía de la liberación*. Mexico City: Fondo de cultura económica.

Escobar, A., 2016. Thinking-feeling with the Earth: Territorial struggles and the ontological dimension of the epistemologies of the South. *AIBR. Revista de Antropología Iberoamericana*, 11(1).

Esteva, G., 2019. Autonomy. In: Kothari, A., et al., eds. *Pluriverse: A Post-Development Dictionary*. Delhi, India: Tulika Books and Authorsupfront, 99–102.

Fraser, N., and Jaeggi, R., 2018. *Capitalism: A Conversation in Critical Theory*. Cambridge: Polity, John Wiley & Sons.

Gambon, H., and Bottazzi, P., 2021. The political ontology of protected area co-management: Worlding and nature perceptions among stakeholders. *Journal of Political Ecology*, 28(1), 646–662.

Gottschlich, D., and Bellina, L., 2017. Environmental justice and care: Critical emancipatory contributions to sustainability discourse. *Agriculture and Human Values*, 34(4), 941–953.

Grindsted, T. S., 2018. Geoscience and sustainability – In between keywords and buzzwords. *Geoforum*, 91, 57–60.

Hamilton, C., 2002. Dualism and sustainability. *Ecological Economics*, 42(1–2), 89–99.

Kohler, F., et al., 2019. Embracing diverse worldviews to share planet Earth. *Conservation Biology*, 33(5), 1014–1022.

Kothari, A., et al., eds., 2019. *Pluriverse: A Post-Development Dictionary.* Delhi, India: Tulika Books and Authorsupfront.

Leff, E., 2015. Encountering political ecology: Epistemology and emancipation. In: *The International Handbook of Political Ecology.* Cheltenham: Edward Elgar Publishing, 44–56.

Luke, T. W. 2005. Neither sustainable nor development: Reconsidering sustainability in development. *Sustainable Development,* 13(4), 228–238.

Menton, M., et al., 2020. Environmental justice and the SDGs: From synergies to gaps and contradictions. *Sustainability Science,* 15(6), 1621–1636.

Millar, D., Santuah, N., and Ba-an, M., 2024. Cosmovisions and critical sustainability sciences – an African ontology of "Vurr" (an energy) amongst the Dagara of Southwest Burkina Faso and Northwest Ghana. In: Rist, S., Bottazzi, P., and Jacobi, J., eds. *Critical Sustainability Sciences: Intercultural and Emancipatory Perspectives.* London: Routledge.

Moser, K., 2018. Rethinking the essence of human and other-than-human communication in the anthropocene epoch: A biosemiotic interpretation of Edgar Morin's "complex thought". *Humanities,* 7(2), 57.

Munter, K. D., and Note, N., 2009. Cosmopraxis and contextualising among the contemporary Aymara. In: *Worldviews and Cultures.* Frankfurt: Springer, 87–102.

Nandy, A., 1988. *Science, Hegemony and Violence: A Requiem for Modernity.* Delhi: Oxford University Press.

Odora Hoppers, C., 2021. Research on Indigenous knowledge systems: The search for cognitive justice. *International Journal of Lifelong Education,* 40(4), 310–327.

Pohl, C., et al., 2010. Researchers' roles in knowledge co-production: experience from sustainability research in Kenya, Switzerland, Bolivia and Nepal. *Science and Public Policy,* 37(4), 267–281.

Rist, S., and Dahdouh-Guebas, F., 2006. Ethnosciences – A step towards the integration of scientific and non-scientific forms of knowledge in the management of natural resources for the future. *Environment, Development and Sustainability,* 8(4), 467–493.

Rist, S., et al., 2011. Endogenous knowledge: Implications for sustainable development. In: Wiesmann, U., Hurni, H., and with an international group of co-editors, eds. *Research for Sustainable Development: Foundations, Experiences, and Perspectives. Perspectives of the Swiss National Centre of Competence in Research (NCCR) North-South.* Berne: University of Bern, Bern, Switzerland: Geographica Bernensia, 119–146.

Ruwhiu, D., et al., 2022. Enhancing the sustainability science agenda through Indigenous methodology. *Sustainability Science,* 17(2), 403–414.

Shankar, D., 2024. Relational Ontologies in health sciences and practices in India. In: Rist, S., Bottazzi, P., and Jacobi, J., eds. *Critical Sustainability Sciences: Intercultural and Emancipatory Perspectives.* London: Routledge.

Shiva, V., 2001. Monocultures of the mind. *Creative Management and Development,* 199.

Shiva, V., 2019. *ONENESS vs the 1%: Shattering Illusions, Seeding Freedom.* New Internationalist.

Siena, S. D., 2005. Cosmosophy. *World Futures,* 61(6), 409–440.

Spivak, C., 1988. *Can the Subaltern Speak? Marxism and the Interpretation of Culture.* Urbana: University of Illinois Press.

Telleria, J., and Garcia-Arias, J., 2022. The fantasmatic narrative of 'sustainable development'. A political analysis of the 2030 global development agenda. *Environment and Planning C: Politics and Space*, 40(1), 241–259.
West, S., et al., 2020. A relational turn for sustainability science? Relational thinking, leverage points and transformations. *Ecosystems and People*, 16(1), 304–325.

3 A culture that understands that everything is interrelated, that nothing is divided, and nothing is outside

*David Choquehuanca Céspedes and Stephan Rist**

Introduction

In this section, we expose the key characteristics of the indigenous Aymara worldview, based on testimonies given by Choquehuanca Céspedes, a member of the Aymara nation, raised in an indigenous community on the shores of Lake Titicaca. He does so through a speech delivered on November 8, 2020, the moment he assumed the Vice Presidency of the Plurinational State of Bolivia (Choquehuanca Céspedes 2020). We consider that this speech summarizes fundamental aspects of the Aymara indigenous cosmovision in a historical moment in which men and women reconquer – after a brutal dictatorship of almost 12 months – the democratic instances of the plurinational state. As a result of a clean and transparent election, held on October 18, 2020, and legitimized with a support of more than 55% of the electorate to the "Political Instrument for the Sovereignty of the Peoples – Movement Towards Socialism (IPSP-MAS)", their elected representatives started again, to govern Bolivia.

We see that the concepts, perceptions, and feelings expressed in this discourse – are full of elements of a relational ontology, which are unique in the world of high level politics. This allows us to explore central aspects that emerge when we assume not only a relational ontology (Aymara in this case) for personal and community life, but also what it means when relational ontologies break into the collective planes of the societal, political, economic, cultural, psychological, and spiritual spheres of life.

We also hope that the readers will feel how a relational ontology contradicts or enters in tension with and opposes those who have assumed the Western "thought style", be it in terms of societal, or in the form of scientific knowledge, related to social, natural sciences, or humanities. We invite to *sentipensar*[1] the fundamental differences that are popping up, when

* Compilation, footnote explanations and translation.

1 The notion of "sentipensar" expresses the idea of thinking as happening unseparated from feeling; it could be translated as "think-feel". According to Escobar (2016) the terms sentipensar

DOI: 10.4324/9781003043577-3

we approach or become part of the existential and intellectual universe that emerges from the many communities and nations that have assumed – many times since centuries ago – relational ontologies.

We invite with this chapter to follow, feel, and reciprocate the movements of ***sentipensar*** allowing us knowing the forces, tensions, and contradictions with which the indigenous nations of Bolivia defend, adapt, reinvent, complement, and expand their ontological bases in a relational and cosmosophical perspective. Despite the difficult conditions, which are expressed in the multiple forms of domination, oppression, coercion, and instrumentalization, exercised by the different forms of private, state, and mixed capitalism, all of them are founded on materialistic ontologies derived from dualism.

After the reproduction of the speech, we add some excerpts from another key text by Choquehuanca Céspedes (2017), "Manifiest of Vivir Bien/ Living Well:[2] Our World Is Possible". This work was presented when Choquehuanca Céspedes was acting as Chancellor of the Plurinational State of Bolivia from 2006 to 2017, which allows us to complement some aspects, not sufficiently thematized in the inaugural speech as Vice President.

In order to draw attention to the aspects of the discourse that are directly or indirectly related to the central aspects of this book, that is, the need to advance a reconceptualization of sustainability from a critical, emancipatory, and intercultural perspectives, we include some footnotes, written by Stephan Rist that summarizes the meaning of notions expressed in indigenous languages, used in documents of Choquehuanca Céspedes.

Inaugural speech of David Choquehuanca Céspedes as vice president of the Plurinational State of Bolivia, Plurinational Legislative Assembly, November 8, 2020

> With the permission of our gods, of our older brothers and sisters and of our ***pachamama***,[3] of our ancestors, of our ***acha***

and sentipensamiento are used by the Colombian sociologist Orlando Fals Borda (Fals-Borda and Moncayo, 2009), for describing the living principle of the riverine and swamp communities of Colombia's Caribbean coast. They imply the art of living based on thinking with both heart and mind. See http://www.youtube.com/watch?v=LbJWqetRuMo. Sentipensamiento was later on popularized by the Uruguayan writer Eduardo Galeano as the ability found among popular classes to act without separating mind and body, reason and emotion (see, e.g., http://www.youtube.com/watch?v=wUGVz8wATls).

2 We will translate this concept as "Living Well" in English throughout this chapter More details on what "Living Well" embraces are found at the end of this chapter..

3 Pachamama: Mother Earth, which also means Mother *of* the Earth; the original meaning, however, refers to "paya ch'a mama"; this expression refers to two (paya) opposed forces (ch'amak), i.e., a cosmic and telluric force; that - while "interacting repeatedly" (expressed as "mama") - give birth to everything on earth; mama refers to concretization of the two eternal forces in a concrete geo-historic circumstance; in everyday language "paya" is pronounced as

chilas,[4] with the permission of our *patujú*,[5] of our rainbow, of our sacred coca leaf. With the permission of our peoples, with the permission of all those who are present and not present in this hemicycle. Today I want to share our thoughts in a few minutes. The obligation to communicate, the obligation to dialogue, is a principle of Living Well.[6]

We, the people of millenary cultures, of the culture of life, have maintained our origins since the dawn of ancient times. We children have inherited a millenary culture that understands that everything is interrelated, that nothing is divided, and nothing is outside. That is why they tell us that we should all go together, that no one should be left behind, that everyone should have everything, and no one should lack anything.

And the welfare of all, is the welfare of oneself, that helping is a reason to grow and be happy, that giving up for the benefit of others makes us feel strengthened, that uniting and recognizing ourselves in the whole is the path of yesterday, today, tomorrow, and always, from where we have never strayed.

The *ayni*, the *minka*, the *tama*, the *tumpa*, our *colka*[7] and other codes of millenary cultures are the essence of our life, of our *ayllu*.[8] *Ayllu* is not only an organization of society of human beings, *ayllu* is a system of organization of life of all beings, of everything that exists, of everything that flows in balance in our planet or Mother Earth.

For centuries, the civilizing canons of *Abya Yala*[9] were deconstructed, re-semanticized and many of them exterminated, the original thoughts were systematically subjected to colonial thoughts. We are alive, we are from *Tiwanacu*,[10] we are strong, we are like a stone, we are *cholke*, we are

"pa", "ch'amak" turns to "ch'a" and "mama" remains in its original pronunciation; putting these concepts into one expression becomes "pachamama"; it expresses the "two key creative forces (cosmic and telluric) that are the source of what takes existence in concrete spatial-temporal circumstances. (Source: Personal communication by Ing. Juan San Martin, researcher and philosopher of Andean culture, La Paz, Bolivia, 11.5.2023). . Additional details see also: https://www.facebook.com/vivecandelariacom/videos/la-dualidad-andina/366891494903363/?locale=hi_IN.

4 Achachilas: Invisible guardians and protectors, but present in the sacred hills, of the spirits of the ancestors of the Aymara nation.

5 The national flowers that represent the different regions of Bolivia are the tricolor Kantuta, which symbolizes the human–nature unions of the valleys and the altiplano, and the Patujú flower, which represents the Amazon region.

6 This is the translation of "Vivir Bien" or "Buen Vivir", which became well-known concepts in South America, and even used in other languages using this Spanish expressions.

7 Various forms of reciprocal work in different regions of Bolivia and in neighboring countries.

8 Means community of beings (human and other living entities).

9 In the decolonial perspective, Abya Yala is a concept of the Kuna language of Panama for the Latin American continent; this concept is assumed by many politically organized indigenous groups in Latin America.

10 Tiwanaku is a ceremonial center in Bolivia that had its heyday between 100 BC and 1000 AD). The ceremonial center has been revitalized through the ritual possession of the presidency of

sinchi,[11] we are *rumy*,[12] we are *jenecherú*,[13] "the fire that was never extinguished", we are from Samaipata, we are Jaguar, we are *katari*,[14] we are *Aínos*, we are *Maoris*, we are Comanches, we are Mayas, we are Guaranies, we are Mapuche, Mojeños, we are Aymaras, we are Quechuas, we are Jokis, and we are all the peoples of the culture of life that awaken our *larama*, which means to be rebellious with wisdom.

Today Bolivia and the world live a transition that repeats itself every 2,000 years, in the context of the cyclicality of time; we move from no time to time, starting a new dawn, a new *pachakuti*[15] in our history. A new sun and a new expression in the language of life, where empathy for the other or the collective good replaces selfish individualism. Where Bolivians look at each other as equals and know that united we are worth more. We are in times of becoming *jiwasa* again, it is not me, it is us.

Jiwasa is the death of egocentrism, *jiwasa* is the death of anthropocentrism and the death of Eurocentrism. We are living a time of return to being *jisambae*; *jisambae* is a code that has protected our Guarani brothers, and *jambae* is the same, a person who has no owner, no one in this world has to feel owner of no one and nothing.

Since 2006, we started in Bolivia a hard work to connect our individual and collective roots, to be ourselves again, to return to our center, to the *taypi*,[16] to the pacha,[17] to the balance from where the wisdom of one of the most important civilizations of our planet emerges.

Evo Morales and Álvaro García Linera in 2006, which was repeated in the spiritual possession of the presidency of Luis Arce and David Choquehuanca.

11 In Quechua, it means "the strong, brave and resistant".

12 In Quechua "rock, stone".

13 In the Guaraní language, "my father of fire"; the ember to light the fire that never goes out.

14 In Aymara "snake". It is also a historical character of great importance: Tupac Katari led an army of some 40,000 soldiers fighting against the European colonizers. The use of the symbolic name "Katari" (snake) instead of using his real second name, indicates that he went through spiritual initiation, yet snake is the Andean symbol of initiation.After encircling and besieging the city of La Paz in 1781, he was captured and dismembered. Before dying he shouted, "they can kill me, but millions will return". For this reason "we return million" was the main campaign slogan of Luis Arce and David Choquehuanca of the IPSP-MAS in 2020.

15 "It is the time of pachakuti (return to the path of equilibrium) and the awakening of the stone beings. With this awakening, the spirit, the thought, the knowledge, the feeling, and the 'cosmobiovision' of the millenary peoples are depetrified. It is the time of the awakening of the sleeping giant. It is the awakening of the millenary peoples and cultures with all their codes and knowledge protected throughout the years of resistance, so that they can guide us on how we can once again reach the culture and the path of life and Living Well" (Choquehuanca 2017:28).

16 "Consensus is the agreement that is established among the members of a community in search of the axis of a consensus or the center (taypi). Consensus does not impose what one thinks, feels, wants and says, but rather it gathers and organizes what everyone thinks, feels, wants and says. Consensus is not unanimity, consensus is not what one wants, consensus is what we all want" (Choquehuanca 2017:56).

17 Means in Aymara: earth (living), cosmos, universe, union of time–space.

We are in the process of recovering our knowledge, the codes of the culture of life, the civilizing canons of a society that lived in intimate connection with the cosmos, with the world, with nature and with the individual and collective life of building our *sumak kamaña*, our *aillincausay*,[18] our *sumajakalle*,[19] which is to guarantee the individual good and the collective or community good.

We are in times of recovering our identity, our cultural roots, our sake, we have cultural roots, we have philosophy, history, we have everything, we are people, and we have rights.

One of the unbreakable canons of our civilization is the inherited wisdom around the *pacha*, to ensure balances in all times and spaces, means to know how to manage all the complementary energies, the cosmic forces that come from the sky and those are emerging with the earth from below the earth. Two main forces, the cosmic and telluric, interact creating what we call life as a visible totality *pachamama* and spiritual *pachakama*.[20]

By understanding life in terms of energy we have the possibility to modify our history, matter, and life as the convergence of the *chachawarmi*[21] when we refer to the complementarity of opposites. The new time we are beginning will be sustained by the energy of the *ayllu*, the community, consensus, horizontality, complementary balances and the common good.

Historically, revolution is understood as a political act to change the social structure, in order to transform the life of the individual; none of the revolutions has managed to modify the existance of power, to eliminate control over people. It has not succeeded in changing the nature of power, but power has succeeded in distorting the minds of politicians. Power can corrupt and it is very difficult to change the nature of power and its institutions, but it is a challenge that we will assume from the wisdom of our peoples. The revolution of ideas is the revolution of balances because we are convinced that in order to transform society, government, bureaucracy, laws and the political system, we must change as individuals.

We will promote opposing coincidences to seek solutions between the right and the left, between the rebelliousness of the young and the wisdom of grandparents, between the limits of science and unyielding nature, between creative minorities and traditional majorities, between

18 Sumak kamaña (Aymara), aillincausay (Quechua) mean "to live well" in Spanish.
19 Action to ensure a good complementarity between individual and community good life.
20 Creator of the (living) earth; spiritual universe representing the invisible face of pachamama.
21 In Aymara, it means everything is male–female. It is used to express the complementarity of opposites that unite to give life to a new cycle of separation–union of opposite poles.

the sick and the healthy, between the rulers and the ruled, between the cultured leadership and the gift of serving others.

Our truth is very simple, the condor takes flight only when its right wing is in perfect balance with its left wing. The task of forming us as balanced individuals was brutally interrupted centuries ago, we have not concluded it, and the time of the era of the *ayllu*, community, is already with us. It demands that we be free and balanced individuals to build harmonious relationships with others and with our environment. It is urgent that we be beings capable of sustaining equilibrium for ourselves and for the community.

We are in times of the brothers of the *apanaka pachakuti*,[22] brothers of change, where our struggle was not only for us but also for them and not against them; we seek brotherhood, we do not seek confrontation, we seek peace, we are not of the culture of war or domination, our struggle is against all kinds of subjugation and against the colonial, patriarchal, single thought, wherever it comes from.

The idea of the encounter between spirit and matter, heaven and earth, of *pachamama* and *pachakama* allows us to think that a new woman and man will be able to heal humanity, the planet, and the beautiful life on it and restore beauty to our Mother Earth. We will defend the sacred treasures of our culture from all interference, we will defend our peoples, our natural resources, our freedoms, and our rights.

We will return to our *kapak ñan*,[23] the noble path of integration, the path of truth, the path of brotherhood, the path of unity, the path of respect for our authorities, our sisters, the path of respect for fire, the path of respect for the rain, the path of respect for our mountains, the path of respect for our rivers, the path of respect for our Mother Earth, the path of respect for the sovereignty of our peoples.

Brothers, sisters, in order to finish, Bolivians must overcome division, hatred, racism, discrimination among compatriots, no more persecution of freedom of expression, no more judicialization of politics.

No more abuse of power. Power has to be there for helping, power has to circulate, power, just like the economy, has to be redistributed, has to circulate, has to flow, just like blood flows within our organism, no more impunity, justice brothers.

But justice must be truly independent, let's put an end to intolerance, to the violation of human rights and of our Mother Earth.

The new time means listening to the message of our peoples that comes from the bottom of their hearts; it means healing wounds, looking at each other with respect, recovering the homeland, dreaming together,

22 In Aymara: To understand the brothers who oppose each other.

23 In Quechua, the true pathway of us, the right pathway, the pathway of learning. It is also used to designate the famous network of roads that crossed the Inca Empire and the astronomical milky way.

building brotherhood, harmony, integration, hope to guarantee peace and happiness for the new generations.

Only in this way can we achieve good living and govern ourselves. *Jallalla.*[24]

Cosmos, living beings, and humans

Choquehuanca Céspedes (2017:20–21) deepens the relationship between cosmos and human living beings as follows:

> The path of life is the path of integrality between human beings and nature, living in the totality of the cosmos and Mother Earth as part of a great telluric–cosmic family. Nature is the primordial substance, the principle of all being. Human beings are an intrinsic part of nature and nature is a constitutive part of us.
>
> The path of life is the path of the "cosmobiocentric" vision (which has the cosmos and life as its center) of millenary societies, which contradicts the anthropocentric vision of the West. The peoples of the world, and especially the indigenous peoples, have resisted the Western anthropocentric vision in many ways, sometimes violently, sometimes peacefully and often clandestinely.
>
> In the "cosmobiocentric" vision of millenary and ancestral cultures, the human being coexists in harmony and balance with the *taqpacha* (all that exists), where human beings and nature constitute a single living being.
>
> The path of life is the path of integrality between human beings and nature, living in the totality of the cosmos and Mother Earth as part of a great cosmic–telluric family.
>
> The human being assumes the sense of belonging to nature as a whole and forms a single unit with Mother Earth and the cosmos. The native peoples of the Andes have protected the vision of the *Pachamama* (Mother Earth in equilibrium), conceiving that all beings circulate in each other, and in her the birth, living, and dying are simply changes of form in a community of stellar energy. It is a permanent being and always just happening.
>
> Human beings in Mother Earth are part of the *tama* (big family), we are an organized community, we are mountain, we are plants, we are jaguar, we are eagle, we are *katari* (snake), we are breeze, we are water, we are nature, we are a community of life.

Choquehuanca Céspedes (2017:39) continues:

> In the view of millenary peoples and cultures, everything corresponds to each other, there is correspondence at all levels and in all categories.

24 Exclamation in Aymara meaning "for life".

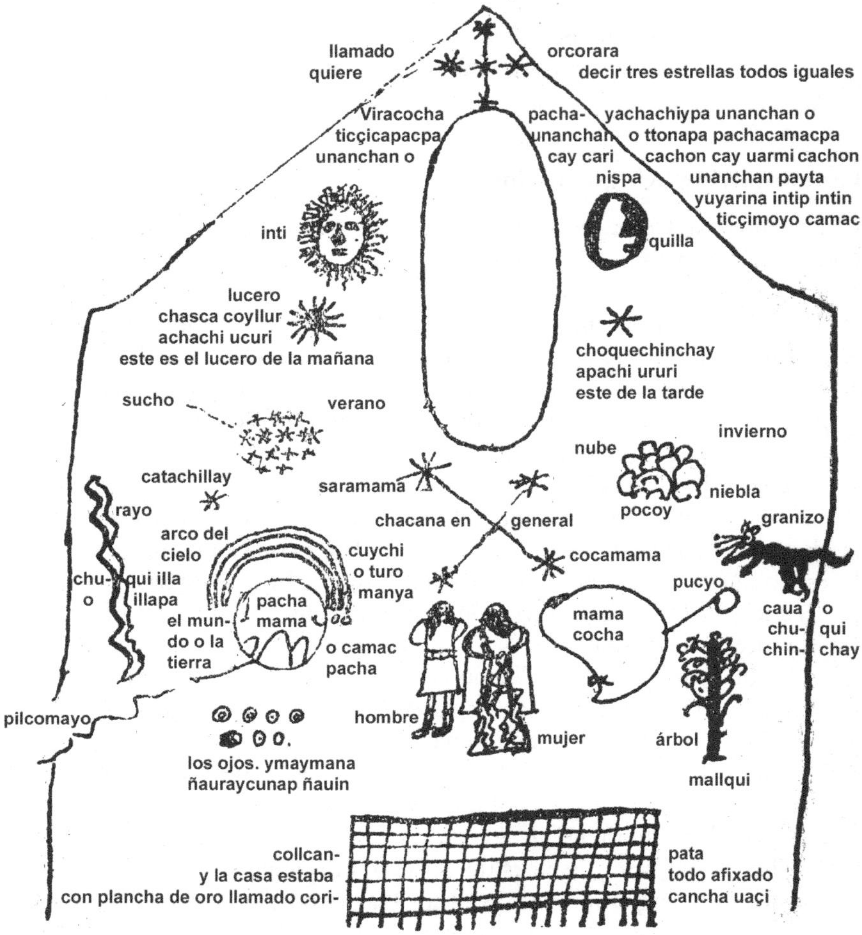

Figure 3.1 The Andean cosmovision and in the center the chakana (southern cross). It is found in the Inca temple of Koricancha in Cuzco, Peru (Juan de Santa Cruz Pachacuti Yamqui Salcamaygua, 1613) *Source:* Public domain, via Wikimedia Commons, https://commons.wikimedia.org/wiki/File:Santa_Cruz_Pachacuti_Yamqui_es.gif#/media/File:Santa_Cruz_Pachacuti_Yamqui_es.gif)

There is correspondence between the *alaxpacha* or *alaypacha* (world above) and the *manqhapacha* or *ukhupacha* (world below), between the flows of our material world and the cosmic flows, and between life and death, because death is part of life.

Harmony with wholeness also tells us that everything is connected and linked to the whole. Reality is a network of relationships, and these relationships are derived from a holistic coexistence with the cosmos. Reality is objective and subjective, material, and spiritual, matter and energy.

> Everything we live, feel, do and think is part of multiple relationships with other experiences, feelings, facts, and thoughts. That is why everything has to flow in harmony and be part of the harmony of the cosmos.

In the perspective of a relational ontology, we note that Choquehuanca Céspedes presents cosmos as a living being that, in union with Mother Earth, forms an entity which – instead of having been created – contains within itself its own creative forces and in which all beings circulate *in* each other; in this vision, birth, life, and death are simply changes in the ways of living in a community of stellar and telluric energy. There are no impenetrable boundaries between all the entities that make up the union of cosmos and Mother Earth, implying that all that exists is a result of the relationships of the cosmos and Mother Earth, the constituent forces. The two entities form a totality that in its visible and living dimension is *pachamama* and in its spiritual dimension, it is the source of life, *pachaqamac.*

The human community at the pachamama–pachqamac union

Choquehuanca Céspedes (2017) situates the organized human community in the current millenary peoples and cultures of the South.[25] He indicates:

> The culture of life in order not to be devoured by the West, and by the monster of capitalism, we have opted for various forms of resistance. We have protected ourselves as stones, as surnames, as codes and as words in our native languages. For a long time, we have been apparently stony, asleep, waiting for the time of awakening to save life and *Pachamama* (Mother Earth in equilibrium). We have never moved away from our Tiwanaku (culture of the Andean region before the Quechuas with its center in this locality). We have never strayed far from the pyramids of Egypt, the Mayan and Aztec pyramids, the citadel of Machu Picchu, the sacred temples in China, India and Iraq, and other sacred sites of the culture of life.
>
> The West and the monster of capitalism have tried to silence us, dominate us, destroy us, and eliminate us, but we are still alive and resisting. We are like the wind that cannot be bent, we are like the fire that cannot be subdued, we are like the sound of water that cannot be silenced. We are a stone that cannot be broken and a mountain that cannot be destroyed. We are rebellious, stubborn, and revolutionary. We are the

25 The Commission of Millenarian Cultures brings together nine countries in the world with millenarian cultures – China, India, Iran, Iraq, Egypt, Greece, Italy, Mexico, and Bolivia, countries that have millenarian cultures and decided to organize themselves to take a new look at our past, asserting that "we have much to learn from those millenarian cultures that have many coincidences and seek harmony" (Choquehuanca 2017:25).

> warriors of the rainbow. Much death and destruction has been committed against us, the ancestral and millenary peoples, but thanks to the strength of our cultures that knew how to resist, we have not been exterminated. They have imposed on us foreign languages, customs and clothing, calendars, and religions, they have persecuted our spirits and our leaders, they have tried to dominate us physically and ideologically and they have occupied our lands and territories, but they have not defeated us. They have made many of us feel ashamed of being indigenous and natives and fearful of being daughters and sons of millenary peoples, but they have not subdued us.
>
> Today, the indigenous peoples and the daughters and sons of millenary cultures are millions, and we continue to speak our languages, we continue to maintain our festivals, dances and rituals, our music, our religions and cosmobiovisions, and we continue to live in community in rural and urban areas around the world.
>
> Choquehuanca (2017:25–27)

In this sense, human communities live the change of epoch and the beginning of a new time, the *pachakuti*. Choquehuanca Céspedes (2017) continues:

> The change of epoch and the new time is based on the vision of the pachakuti, on the prophecy of the Lakota people and on the prophecy of the end of the world of the Mayas. In the Andean region the great change of time and space is designated as *pachakuti* (the return to order and equilibrium). It is necessary to order and return to equilibrium in our communities, in our countries and in the entire planet. This great change must take place, considering that human beings are part of an ancestral society and a current society, which coexist as two realities that must be maintained in complementarity and balance. The path of the new time is based on the philosophy of *pachakuti*, *suma qamaña* or *allin kawsay* (concepts of Living Well protected by the Quechua and Aymara) and *jallalla* (exclamation that means "for life" in Aymara).
>
> It is the time of the fulfillment of the prophecy of our Lakota brothers, who have preserved for hundreds of years the message of their ancestors, which says that when the world is on the brink of the abyss, when there is chaos in humanity, when there is crisis, from the South of the continent will emerge vigorous rainbow warriors who will return harmony to the planet.
>
> It is the time of the prophecy of the Mayans, who predicted that the former world ends at midnight on December 21, 2012, when the sun would receive a strong synchronizing ray from the center of the galaxy that would change its polarity, and produce a gigantic radiant flare. The world of "chaos-mos" ended and *pachakuti* began, the construction of the world of Living Well.
>
> (Choquehuanca 2017:28–29)

Which are the actions for recovering and revealing the millenarian codes of ancestral cultures? Choquehuanca Céspedes (2017) responds this question, presented in the following, quote:

- Reading the wrinkles of our grandmothers and grandfathers.
- Awaken our "cosmocimiento" (expression meaning the knowledge of the cosmos) and "pensasiento" (expression meaning the conjugation of thought with feeling).
- Recover our energetic centers and our natural "multiverse".
- Recover and dialogue with our *wak'as* (sacred places).
- Return to our *puma punku* (door to be puma) and recover all the lost energy.
- Transit through our *chakana* (square cross) that is the bridge to the "multiverse" and stellar energy.
- Feeling part of the whole and respecting cosmic order and the laws of nature.
- Back to nature (tiger, jaguar, ant, plant, breeze, puma, snake and walking mountain).
- Recover our mantras (prayers for life).
- Awaken our *qhawana* (look beyond what our eyes see).
- Awaken our *larama* (to be rebellious with wisdom).
- Return to being *iyambae* (being a person who has no owner).
- Return to our *qhapaq ñan* (the pathway of life).
- Back to loving life.

This awakening means to go beyond the knowledge of the laws made by humanity to dominate nature, because these are exclusive, since they have been made only for the interests of human beings. To awaken the *"cosmoser"* (to become cosmos again) means to know the laws of the totality of nature from the depths of Mother Earth. Cosmoser is to be and feel like the plant, jaguar, snake. With the wisdom and richness of the "cosmoser" we begin the construction of this epochal change towards Living Well.

(Choquehuanca 2017:31–32)

Choquehuanca Céspedes (2017) elaborates on what "*comoser*" means:

We are the cosmos that thinks, we are cosmos made flesh and spirit. In the community no one has to feel free for destroying the other, whether animal or plant, because we all complement each other, because we are all sisters and brothers, raised by Mother Earth, because we all drink from the milk of Mother Earth, which is water. Human beings are similar to other living beings, because human beings are part of the natural and cosmic totality, and because in cosmic life, we are energy and life.

Identity is built based on differences and similarities. We have an individual identity, an identity as family and community, an identity

> as countries, an identity as continents and a cosmic–telluric identity in Mother Earth. We are a universal family, each with our identities, with our differences and similarities. We are the totality of the "multiverse".
> (Choquehuanca 2017:38)

We conclude that in a perspective of the relational ontology of Choquehuanca Céspedes, the human community is an integral part of the union of *pachamama* and *pachaqamac*, being part of this organism. It has the particularity that the cosmos *pachamama–pachaqama* can come to think of itself, to the extent that the human community awakens the faculties to create, deepen, and expand the cosmoser and cosmoscience. In this sense, the present world is the result of emerging patterns that Siena (2005) called a "cosmosophical process" of communication among the members of a "planetary community of destiny".

The relational ontology presented here reveals an important epistemological dimension. As opposed to dualistic ontologies, which separate the material from the social and spiritual, the objective from the subjective, the transcendent from the immanent, etc., a relational epistemology cannot generate knowledge by widening the gap between the material, the social, and the spiritual, or between the objective and the subjective. Thus, Western research inserts between the observed and the observer, ever greater obstacles, to better "measure" the quantitative, material nature of the objects of inquiry. The measurements are then fed into increasingly abstract statistical models or mathematical models that have become so complex that it is difficult to know to what extent their results represent real features of life, as lived by us the human beings.

The alienation between the ideals of a free and self-determined daily life and the materialistic and dualistic scientific knowledge, assumed in all current societies, also requires new ways of constructing knowledge.

Choquehuanca Céspedes (2017) points in this direction by recalling that Western knowledge (materialistic, dualistic, and instrumental) has to be expanded taking into account that as a human community, we also have the capacity to feel that "we are the cosmos that thinks, we are cosmos made flesh and spirit". In epistemological terms, this implies reversing the construction of knowledge: instead of widening the distance between observing subject and the observed object, we have to shrink this distance to such a degree that the distance can collapse and the borders between the subjective and the objective disappear; this might facilitate that from our inner world of "think-feeling" (senti-pensar) a cosmism can emerge that we can share collectively. This cosmism is free of alienation and the result of an emancipatory process, controllable by each one, who becomes part of the same cosmosophical process.

Choquehuanca Céspedes (2016) expands on a relational epistemology as follows:

> Our wise ancestors inherited us the codes, ways of acting to find balance in interpersonal relationships and achieve harmony in coexistence

with Mother Earth, they always agreed to establish dialogue (*Aruski pasipxañanakasakipunirakispawa*)[26] with their similar "western scientists", although for many centuries we were excluded, our knowledge was stolen and then patented to enrich other nations. Today the world is reflecting on this mistake and proposes to facilitate spaces and instances to promote inter-scientific-transdisciplinary dialogue; it seems to us a very good initiative that we accompany, as a government with our laws, the creation of conditions permitting that, both sciences are recognized with the same level and hierarchy, for their contribution to the integral development of humanity. This accompaniment goes further, with the facilitation of spaces for dialogue at the level of the United Nations, and other regional and local instances. It is intelligent that we rebuild the y*atichaui*, which means we need to promote more inter-scientific gatherings, to share with the old wise men; we must return to "read the wrinkles of our grandparents", because in them this knowledge is kept.

Our grandparents taught us to do science with the heart, which we call *con-ciencia* (*chuymanpi*), the heart is the place where we keep our own ethical codes, fundamental principles of our *cosmocimiento* of our *cosmoser*, because the Being, for us, is not reduced to the individual; it begins in the couple (*chacha- warmi).* and consolidates in the family (*jatha*) and is built in the community (ayllu) and coexists in the *pacha* (Totality). Thus, the integral and ample BEING, is the community. We reason with the heart, that is why we say "corazonada" (a combination of reason and heart in Enghlish); we understand that due to its linguistic limitations, Spanish cannot adequately express the categories of our feelings and thoughts, therefore, we must recover terms that help to re-signify and approach the richness expressed in our native languages. Recently, a Peruvian researcher and linguist has confirmed that the Quechua language, in its formulation, is created in the sense of including the other, and in this regard, it is no different in Aymara or Guarani; our languages were created imitating nature, to dialogue with her; therefore, the first tasks to be done are to investigate, deepen, and disseminate the ancestral languages.

(Choquehuanca Céspedes 2016:13–14)

Living Well as a new horizon for civilization

It has been mentioned several times that life in a relational ontological perspective has to be oriented in the principles of Living Well. The current time of *pachakuti* is the beginning of the new dawn, which is

> [a] new sun and a new expression in the language of life where empathy for the other, or the collective good, replaces selfish individualism.

26 It is Aymara and means "Necessarily we must always communicate with each other".

> Where Bolivians look at each other as equals and know that united, we are worth more. We are in times of becoming *jiwasa* again; I am not I, we are us.
>
> (Choquehuanca Céspedes 2020:2)

But it is important to point out that *jiwasa* means "reborn from the death of egocentrism, anthropocentrism, and Eurocentrism".

Although the notions of Living Well were pronounced from the Andean nations of South America, it is interesting to note that they are no longer restricted to the Andean nations, but rather we find proposals for thinking that point to the same new civilizational horizon, such as the *iyambae* (*person who has no owner*) and the *Teko Kavi* (to be well with nature, with the spirits, with the elders, with the children, and with everything around you in the Guarani culture, *Ñuke Mapu* (Mapuche in Chile and Argentina), *Lekil Kuxlejal* (to be well and at peace with oneself, our family and community, and with Mother Earth) of the *Tseltal* Maya, etc. (Quiroga Canaviri and Zaiduni Salazar 2022).

From the Latin American context, the ideas of Living Well began to travel around the world, finding echoes and dialogue with thoughts in Buddhist philosophies (happyness), Greek philosophies (eudaimonia of Aristotle), ecosophy (of Felix Guattari and Raimon Panikkar), anthroposophy (Rudolf Steiner), or in the emerging thoughts from the West, such as quantum physics. The latter proves, according to Hans-Peter Dürr (2009), that

> [m]atter is not composed of matter! If we break matter down more and more, in the hope of finding the smallest, most amorphous, and pure matter, in the end, there is nothing left that reminds us of matter. In the end, there is no more substance, only form, aspect, symmetry, relation. This conclusion has been, and still, is very puzzling. If matter is not composed of matter, this means, that the primacy of matter and form is reversed: The primordial is the relation, and the substance the secondary. According to the new physics, matter is a phenomenon that only appears in a rough contemplation. Matter/substance is coagulated form. Perhaps we could say: At the end of all the fragmentation of matter there remains something that is more akin to spirit-holistic, open, alive: the potentiality, the possibility of a realization. Matter is the dross of this spirit – dismountable, separable, determined reality.
>
> (Dürr 2009:86)

Choquehuanca Céspedes (2017) places the struggle of the indigenous nations of Latin America in this broader global context. He points out that from the perspective of Living Well, the specific contributions and reflections to the new civilizational horizon revolve around the following principles:

> Our struggle goes beyond the simple division between capitalism and socialism. While for capitalism the most important thing is capital,

surplus value, and money, for socialism the most important thing is the human being and guaranteeing the ever-growing needs of societies. Living Well fights against capitalism and goes beyond socialism, since for Living Well the most important thing is to build a culture of life.

Living Well has coincidences with socialism because it also seeks to satisfy the needs of the human being, but it is a struggle that goes beyond socialism and the human being. For Living Well, the most important thing is life. Living Well fights against capitalism because it has enslaved Mother Earth and all human beings living on it. Subjugating others and nature is not in line with Living Well. Living Well fights for the rights of Mother Earth and all beings that coexist with Mother Earth.

Living Well fights against the capitalist development model implemented by Western society, which is the model of better (material and purely economic) living; this model has brought imbalances, crises and disasters to humanity and Mother Earth. The construction of the horizon of Living Well is based on the system of life of the *ayllu* and the community life. The *ayllu* is a true system of organization of life that goes beyond the organization of society. That is why we propose a communitarian socialism for life, based on the *ayllu* system.

Practices to build the civilizational horizon of Living Well

- Communicate and dialogue with ourselves, with others, with nature, and with the sacred.
- To guarantee individual well-being and collective well-being.
- To achieve that everything is in its place and in balance.
- Strengthen the government of all and of the whole, based on consensus.
- Back to *tama* (our big family).
- Recover the *tumpa* (reciprocal control among all of us).
- Cooperate with each other.
- Perform our actions with communal energy and love.
- Ensure healthy food in good times and bad.
- Recover our *saphi* (cultural roots).
- Respect the similar and the different.
- Recovering the balance of Mother Earth.
- Back to being *jiwasa* (meaning it's not me, it's us).

In order to achieve Living Well, there must be an individual *pachakuti* and a collective or community *pachakuti*. First, it is fundamental to achieve a change within each and every one of us, in order to change our community. If we do not change individually, we will not be able to change others. We have to return to our *qhapaq ñan*, to our *amuyu and yuyay*, to our *sarawi* or *tape* (the path of our life horizon).

In Living Well, based on the knowledge and codes of our millenary peoples and cultures, we have to transcend from the government of the majorities to the government of all and of the whole, from the search for freedom to the complementarity between the different and the similar, from the struggle for justice to the reestablishment of equilibrium among all living beings, from the struggle for the recovery of ancestral knowledge to the establishment of an inter-scientific civilization with inter-civilizational dialogue, and from the struggle for equity to the search for a decolonized and depatriarchalized society.

Our struggle goes beyond the quest for democracy and advances to the search for consensus

Western democracy has worked an ideal of government which is democracy based on representative democracy, where people vote to elect our representatives and where political power is separated from society. In democracy, the majorities subjugate the minorities, and to subjugate the neighbor is not Living Well, just as it is not Living Well to exploit the neighbor, to work, to lie, and to steal.

Subordinating the minorities is not Living Well. In Living Well, we are not against democracy, and we support it, but we consider that it is not enough. In Living Well, we seek the government of all, so it is not enough to make decisions through democracy, it is important to move toward decision-making through consensus.

While democracy is the government of the majorities, consensus is the government of all. That is why in Living Well, our struggle goes beyond the democracy of the majorities that subjugate the minorities. From the government of democracy, we have to advance to the government of consensus, where the rulers rule by obeying the people and where political power is not separated from society.

Consensus is a function of what other people (men and women) think, but it is also a function of *taqpacha*. Consensus allows us to listen and express different thoughts and feelings until we reach an agreement between different contradictory positions. In decision-making, within the framework of consensus, we not only think about human beings, but also think about everything that exists. In the search for consensus, it is important to listen to everyone, including sacred sites, nature, all living beings, and all of us who are different but similar in the view of Mother Earth. Consensus is the agreement that is established among the members of a community looking for the axis or center (*taypi*). Consensus does not impose what one thinks, feels, wants, and says, but rather it gathers and organizes what everyone thinks, feels, wants, and speaks. Consensus is not unanimity, consensus is not what one wants, consensus is what we all want.

In Living Well, governing is a communal action, and it is not left to only a few to solve our problems. In Living Well, to know how to govern is to return the capacity to solve problems to the organizations, neighborhoods, and communities. Living Well is not waiting for others to solve our problems; on the contrary, it is to solve the problems for ourselves. Living Well is to govern ourselves by awakening our capacity to organize and to know how to solve our problems.

Our struggle goes beyond the search for social justice and advances to the search for equilibrium

Colonialism has divided and annulled us, individualism has dehumanized us, and capitalism has denaturalized us. We are dismembered and alienated. The continents have been dismembered, the countries have been divided, communities have been weakened, and people are enslaved by consumerism and markets. In Living Well, our struggle goes beyond social justice because when we speak of justice, we speak only of human beings, and we do not consider the whole. That is why in Living Well, we fight for justice, but we go beyond that, we seek balance among all beings of Mother Earth. That is why we speak of *pachamama* as Mother Earth in balance. Therefore, the struggle of Living Well goes beyond justice toward the search for a world in balance.

Our struggle goes beyond the search for freedom and advances to the search for complementarity between similarities and differences

In the Western civilizational model, the struggle for freedom is the supreme objective of social struggles. The French Revolution established the motto of the struggle for "liberty, equality and fraternity". The Universal Declaration of Human Rights established individual and collective freedoms, defining liberty as the sacred and imprescriptible right that all human beings possess, where freedom is the faculty to act according to their will, respecting the law and the rights of others.

In Living Well, we are not against the struggle for freedom, and we share it. In Living Well, we demand respect so that no one can take the freedom to make decisions and think for us; because no one has the right to feel that he or she is the owner of anyone and anything. No one can make decisions or think for the other, we must all make our own decisions and be considered.

But the struggle for Living Well goes beyond the search for freedom and is oriented toward the search for complementarity among all beings. When we speak of freedom, we are only talking about the individual and collective freedom of human beings, people, and society. However, when we talk about the search for complementarity, we are talking about the life and freedom of all living beings raised by Mother Earth, and not only about the freedom of human beings. In Living Well, everyone is the size he or she is, no more and no less, and thus must respect and complement each other.

Our struggle goes beyond the pursuit of human rights and advances to the pursuit of the rights of Mother Earth and cosmic rights

Living Well does not only seek the recognition and exercise of human rights and the rights of indigenous, native, and peasant nations and peoples. Living Well also seeks the interdependence and articulation of the rights of peoples with the rights of Mother Earth, and, from there, also including the cosmic rights, as a full expression of the articulation of human beings with nature.

Living Well establishes the recognition of the rights of Mother Earth, within the framework of cosmic rights, without neglecting the respect and defense of human rights, as well as the rights of indigenous nations and peoples, in their different dimensions of individual, collective, economic, cultural, and political rights.

Rights have to be thought for the totality of Mother Earth, and not only for human beings. Everything that exists must be considered, be included, and be present, with its rights in full exercise, and in articulation with the duties and obligations, or demands of life, which are the commitment of all to advance on the path of the culture of life.

Our struggle goes beyond the search for the revalorization of ancestral knowledge and advances to the search for inter-scientific and intercultural dialogue

Throughout history, Western thought has tried to position itself as superior to that of non-Western peoples. Western thought has been designated as scientific knowledge, while the knowledge of millenary peoples and cultures is designated as indigenous and local knowledge systems as traditional and ancestral knowledge.

The knowledge of millenary peoples is not considered as science. In this hierarchization of knowledge, Western sciences are placed in the foreground, and the rest of the knowledge that does not have the rank of science is placed as local knowledge in the background.

The power of the West is nourished by this colonial vision of science. For the West, Western science explains and projects the model of civilization of the entire universe, excluding local knowledge. The Western scientific principle of the excluded third party does not allow the integration of the knowledge of millenary peoples and cultures into its civilizing model. For the West, local knowledge is localized, specific, and does not have the capacity to generalize and, therefore, continues to project a universal model of civilization.

From Living Well, the first thing we work for is equality between the modern sciences of the West and the sciences of the millenary peoples and cultures. We also seek inter-scientific and intercultural dialogue. In this dialogue, we recover the best of Western scientific knowledge and articulate it in a complementary way with the sciences of millenary peoples. This powerful science with an inter- scientific outlook is the scientific basis of Living Well.

Our struggle goes beyond the search for equity and moves forward in the search for decolonized and depatriarchalized societies

It is important to build just, equitable, and supportive societies, where the differences between social classes, between rich and poor, and between castes have been eliminated; but this is not enough. If we do not eliminate the structures that sustain these divisions, we will, again, establish populations that are differentiated into classes, castes, and social strata. For avoiding this, we have to eliminate all the structures, built historically to differentiate us, and that make some people first class, and others, second class.

These structures are based on the discrimination of some groups over others, on religious structures, skin color, money, cultural origin, and other factors that enslave us. In Living Well, the construction of equitable societies is important. But it is not enough, we also have to advance in the processes of decolonization and depatriarchalizing of societies.

Decolonization means breaking the mental structures that think that there are absolute truths and superior thoughts, discriminating against the knowledge and thoughts that come from other realities. In the processes of colonization, mental models were imposed that discriminated and marginalized the native and indigenous population, establishing hierarchies between societies and people. Decolonization means putting all things on the same level, so that there are neither superior nor inferior people, societies, and beings. Decolonization means to begin to see reality with our own eyes, and no longer with the eyes of the West, to think with our own minds, and no longer with the mind of the West, to recover our ideologies, our knowledge, and codes of life, and to learn to move according to the laws of nature and our ancestors.

Depatriarchalizing means breaking and de-structuring the structures of domination of men over women in the whole of social, political, and economic relations, building structures of equality and complementarity between men and women. It is to become whole again, in the framework of the polarity of the *chacha-warmi* (man and woman who complement each other) and of the tetralectic of the *chacha-warmi*, because each masculine being has a part that is feminine, and each feminine being has a part that is masculine" (Choquehuanca Céspedes 2017:53–61).

References

Choquehuanca Céspedes, D., 2016. Presentación. In: Delgado, F., and Rist, S., eds. *Ciencias, diálogo de saberes y transdisciplinariedad. Aportes teórico-metodológicos para la sustentabilidad alimentaria y del desarrollo*. La Paz, Bolivia: CLACSO, AGRUCO-UMSS, CDE University of Berne, 13–16.

Choquehuanca Céspedes, D., 2017. *Manifiesto del Vivir Bien. Our World is Possible*. La Paz, Bolivia: Cancilleria del Estado Plurinational State of Bolivia.

Choquehuanca Céspedes, D., 2020. *Speech of Assumption of the Vice Presidency of the Plurinational State of Bolivia*. La Paz: Vicepresidency of the Plurinational State of Bolivia.

Dürr, H. P., 2009. *Warum es ums Ganze geht: Neues Denken für eine Welt im Umbruch* [Why Everything is at Stake: New Thinking for a Changing World]. München: oekom Verlag.

Escobar, A., 2016. Thinking-feeling with the Earth: Territorial struggles and the ontological dimension of the epistemologies of the South. *Revista de Antropología Iberoamericana*, 11(1), 11–32.

Fals Borda, O., and Moncayo, V. M., 2009. *Una sociología sentipensante para América Latina*. Bogotá: CLACSO, e-libro, Corp.

Juan de Santa Cruz Pachacuti Yamqui Salcamaygua, 1613. *Relación de las Antigüedades de Este Reino del Perú (1613): Cosmology of the IncaJuan Santa Cruz Pachacuti*. Source: Public domain, via Wikimedia: Commons, https://commons.wikimedia.org/wiki/File:Santa_Cruz_Pachacuti_Yamqui_es.gif#/media/File:Santa_Cruz_Pachacuti_Yamqui_es.gif (Accessed 15.12.2022).

Quiroga Canaviri, León J., and Zaiduni Salazar, Eduardo M., 2022. La métrica del vivir bien: Construcción de un índice de reconciliación y equilibrio. *Revista Enfoques*, 6(21), 79–105.

Siena, S. D., 2005. Cosmosophy. *World Futures*, 61(6), 409–440.

4 Relational ontologies in health sciences and practices in India

Darshan Shankar

Author' orientation

I must begin this chapter by informing readers that I have no academic training in Western philosophy, biology, or medicine. My specific interest in the subject of India's cultural heritage (Shankar and Unnikrishnan 2004), health in particular, was awakened by firsthand social interactions with villagers and genuine traditional knowledge practitioners in several parts of India. In the healthcare space, I have interacted over the last 40 years with dozens of both the folk healers and physician-scholars who are living carriers of our health heritage. I was fortunate to closely observe by living for 12 years in tribal communities, their joyful outlook on life. I am familiar with the incredible knowledge that local communities possess about ecosystem-specific medicinal plants and their reliable uses for human, animal, and crop health. I observed this in many parts of the country, north, south, east, and west, including in the Trans-Himalayas. I had the opportunity to see the reflections of centuries-old indigenous knowledge encounters between India and Thailand, which must have been intimate because they had percolated into Thai villages where through personal interactions, I learnt with surprise, the folks knew technical Ayurveda terms of *simha* (*kapha*), *vata*, and *pitha* (Ayurveda terms for key biological functions) used in their description of local plant properties. I also had the privilege of friendship and mentorship with some of the leading Ayurveda physician-scholars in Maharashtra, Gujarat, Rajasthan, Uttar Pradesh, Kerala, and Tamil Nadu.

I was so struck by the extraordinary functionality and deep knowledge of the living health traditions that I felt inspired to *do* something to revitalize the widespread treasure of indigenous knowledge and practice that I witnessed. I noted with anguish that despite their functionality, the indigenous health knowledge and practices were being neglected by the mainstream due to sheer ignorance of the Western educated intelligentsia about the value of their epistemological and ontological foundations.

Thus, I founded in the mid-1980s along with the active involvement of competent colleagues a grassroots national organization of folk healers called the *Lok Swasthya Parampara Samvardhan Samiti* (Committee for

DOI: 10.4324/9781003043577-4

revitalization of local health traditions). This movement continues to grow today under the direct leadership of folk health practitioners. Subsequently, in the early 1990s, I founded a Foundation for Revitalization of Local Health Traditions (FRLHT), which 20 years later was awarded a university status. This university, named "The University of Trans-Disciplinary Health Sciences and Technology" (TDU) (www.tdu.edu.in), today is perhaps the only university in the country working on revitalization of multiple dimensions of the indigenous knowledge of healthcare and for the conservation of its medicinal plants resource base.

Introduction to the epistemic history of India's medical heritage

At the outset of this introduction, I would like to share with the reader my understanding of tradition and modernity. This is because particularly in colonized societies (Nandy 1983), sociologists have created a false dichotomy between them. Due to this dichotomy, the epistemic history of indigenous knowledge gets distorted. Tradition and modernity are seen by most modern sociologists as being located in two polar opposites when in fact they lie on a continuum. Just as the past, present, and future lie on a continuum, the present being derived from the past and the future from the present, so also the roots of modernity lie in tradition. The appropriate sociological description of modernity in my reckoning ought to be "evolving tradition" (Darshan Shankar 2014). A society that does not have roots in its own traditions is a lost society. A culture as it evolves modernizes by building on its traditions. This adjustment to temporal change may include discarding, modifying, building upon, and innovation. It may even borrow and integrate knowledge from other cultures. Despite such changes, the reference always is to its own past. Across centuries, there are hundreds of examples of transmission of knowledge across cultures. This transmission has resulted in "cross-cultural fusion" of knowledge, techniques, and resources, oftentimes transforming the original knowledge to create new cultural ownerships and societal impacts (Latour 1984; Lipphardt and Ludwig 2011). The history of evolution of knowledge informs us that transmissions and fusions across cultures resulted in significant advances in fields as varied as mathematics, philosophy, logic, medicine, architecture, agriculture, material sciences, military, culinary practices, fine and performing arts, and metaphysics (Harari 2015). While the cultural origins of knowledge reflect indigenous epistemologies and ontologies, there are instances where cross-cultural transfusion may erode the roots or modify the expressions of the original culture. A striking example of erosion in a practical dimension of the Indian health knowledge systems is in the field of surgery. It is well-established that Ayurveda (surgical text called *Sushrut Samhita*)[1]

1 A rare manuscript of *Sushruta Samhita* is available in the Burrough Wellcome collection, in 183 Euston Road, London, UK.

transmitted surgery to the world of medicine. Indians practiced surgeries, including those related to ophthalmology, orthopedics, brain, and Cesarean section four millennia ago. Even as recently as the 18th century, the British learnt plastic surgery from India (Dharampal 1971; S Saraf 2006), but during the last three centuries, India has made no major surgical innovations. The leaders of global surgery today are located in Europe, the United States, and other nations, and Indian health workers need to learn from them. Such cultural shifts in knowledge leaderships occur in several domains. They could even take place in the 21st century.

With this perspective on tradition and modernity, let me begin the narration on the epistemic history of India's Medical Heritage. It may surprise non-Indians to know that health knowledge and practices in India have an unbroken, evolving history of over three millennia (Meulenbeld 2017). They continue to be living traditions. Even today, they are socially visible in two interconnected streams as was the case perhaps 3,500 years ago, and this observation is recorded in a verse in a foundational medical text (Charaka n.d.), originally composed around 1500 BCE. The first stream comprises ecosystem and ethnic community-specific folk knowledge. This community-based knowledge is of simple but effective methods of diagnosis, processing of natural products (over 5,000 species (FRLHT and NMPB 2010) across India), and treatment with them. The knowledge is rooted in healthcare practices of village communities from the Trans-Himalayas to the seacoast (Shankar and Unnikrishnan 2004). It is based on keen observations of nature and intelligent inferences. The knowledge and practices of this stream are not codified. They are passed down the generations in a highly decentralized manner from hundreds of experienced teachers to their students. The transmission takes place through a remarkably efficient and a sacred system of oral transmission that does not appear to have suffered distortion during oral transmission over centuries. Its reliability is perhaps due to the sacred value placed on such knowledge so that it is learnt diligently. Today, this oral tradition is eroding due to indiscriminate destruction of ecosystems and their associated bio-cultures. The second stream comprises codified knowledge called Ayurveda – the knowledge (Veda) of changes in life processes (Ayu). This knowledge is much more comprehensive in scope as compared to the oral traditions. It is codified in thousands of medical manuscripts on a wide range of subjects ranging from epistemology and ontology to pharmacology, pharmacy, medicinal flora, fauna, metals and minerals to diagnosis and treatment. The most comprehensive literary history of Ayurveda has surprisingly been written by a Dutch scholar (Meulenbeld 2017) who dedicated several decades of his life to the mission. His volumes reveal that there are texts on diagnosis, treatment, epistemology, pharmacy, materia medica (which includes flora, fauna, and medicinal metals and minerals), and pharmacology. This literature is in the form of foundational texts, commentaries, critiques, and lexicons, written up to the 17th century CE. Throughout their evolution, the oral and codified streams have enjoyed a symbiotic relationship.

A unique and significant fact about the epistemic history of Ayurveda particularly relevant for the theme of this chapter is the fact that despite its antiquity, Ayurveda (1500 BCE) is a living tradition even today. Ayurveda holds the unique record of being perhaps the world's longest, unbroken, and evolving knowledge which still stands on its original foundations embodying principles and strategies discovered during its inception that can be used for contemporary application (Darshan Shankar 2021).

The foundational Ayurveda texts are practical and realistic. They explicitly pronounced that all possible diseases can never be named (Charaka n.d.). Only typical diseases were described in terms of etiology, symptoms, treatment principles, and strategies. Ayurveda emphasized that it is impossible to enumerate all possible disease manifestations or predict those that may occur in future. The Ayurveda knowledge system trains physicians to apply a dynamic (non-temporal) framework to observe and classify human phenotypes, and their health conditions in terms of the degree of perturbation of the physiological functions (doshas), body tissues (dhatu), metabolism (agni), the sensory organs, mind, and other inherent biological factors (Valiathan 2013). This framework helps physicians to discover new etiologies and treatments for new diseases that may occur from time to time. Ayurveda in a similar realistic vein explicitly recognizes clinical limitations. It acknowledged that not all types of biological imbalances can be restored by a physician. It advises physicians on types of imbalances that can be reversed (sadhya), others that may be difficult to restore fully (kruch sadhya), and certain imbalances that are not curable (asadhya). It is this brilliant and realistic framework that is the core feature of the epistemology of classical Ayurveda. Its relevance is evident today from the fact that Ayurveda in the 21st century remains a functional healthcare knowledge system resting on its original foundations (Hankey 2005). This unique feature gives rise to a seemingly ridiculous question about epistemology of knowledge: Can principles (Siddhantas) underlying healthcare knowledge generated 3,500 years ago actually possess functionality today? This in fact is the case. Hence a sharper question arises: What is the relationship of knowledge and time? How have aspects of Ayurveda knowledge system become timeless (*shashwat*). The author assures the reader that no indigenous Indian magic or miracle is involved here. The answer clearly is that Ayurveda knowledge is derived from a holistic perspective of nature (Darshan Shankar 2021). This perspective is perhaps made possible in a sensitive intuitive state of mind. It is a holistic perspective from which, if any culture were to view nature, the knowledge generated in any time or space may not become obsolete with time because it experiences nature in its intrinsic completeness. This kind of perspective is perhaps not possible in an analytical state of mind because analysis, although useful for understanding mechanisms that underlie whole phenomena, dissects and breaks the whole into parts and thus yields a reduced and partial view. Over time with improved methods and techniques, the partial view may keep getting broadened. The functionality of Ayurveda provides testimony that such a holistic epistemology and ontology exists. So, what is this perspective?

In Ayurvedic epistemology, the relationship between the observer (*purush*) and the observed (*prakriti*) does not separate the two as in modern science. One observes nature by becoming one with it. This is the tenet of the experiential indigenous school called "sankya darshan" (Sarvepalli Radhakrishnan 1923). The term "Darshan" implies that it is not a speculative school but an experiential one. Ayurveda further uses Yoga Darshan (again experiential) to actually help seekers to experience the inherent interconnectedness or oneness in all manifestations of nature. Its propositions are founded on a contextual multivalued scheme of indigenous logic (*Nyaya*) which is not an abstract, idealized two-valued logic (yes/no) like Aristotolean logic. The Ayurvedic texts are written in Sanskrit language which has a profound grammar. The etymology of the word Ayur-Veda is itself very significant. The root meaning of the word Veda is knowledge, but the meaning of Ayu is superficially understood. The root meaning of Ayu is not merely life processes but is derived from *gachati*, *gaman*, which means "change". The term Ayurveda thus refers to dynamic knowledge of *biological changes* in life processes. The dynamic scheme for understanding *change* in the world/universe/nature is also apparent in Indian cultural and intellectual traditions across all domains of knowledge. This is evident from the deeply insightful etymology of the Sanskrit word "Jagat", which in English language is translated as a static term, namely, world or universe. Etymologically however, the root meaning in Sanskrit of the word *Jagat* implies a "changing" (*gat*) phenomena (*jag*). A Sanskrit grammarian would spell out its etymology as *gacha ti iti jagate*, the universe is that, which is *constantly changing*. While the vastness and infinite diversity of forms in the universe was known, Indian epistemology of knowledge realized that underlying the vastness and diversity of the manifest physical and biological world was the constancy of change at multiple levels.

Relational ontologies an Indian perspective

While the epistemic history is outlined earlier as a contextual background before focusing on relational ontologies associated with health sciences, it may be useful to state the position and perspective from which the author will view relational ontologies. The author draws upon epistemology and ontology derived from the foundations of the Indian yoga knowledge system. While readers may be familiar with elementary yoga practices which are in public domain, they constitute only the tip of an iceberg with very deep foundations. The source of the yoga knowledge system's understanding of the reality of nature and the universe (ontology) is in principle similar to the source of any other knowledge tradition, across space and time. It is derived from the sensitive use of the six natural human instruments of knowing, namely, the five senses and the mind. All cultures use these very same instruments in order to discover the world because they are the only tools for knowing and discovering the world available to humankind. The cultural differences in perception and experience of nature arise because they use these instruments differently and in different social and natural environments.

The author has very little scholarship in Western philosophy, hence with some diffidence, I outline subsequently the difference between yoga foundations and contemporary Western philosophy, particularly as reflected in modern neurobiology.

The sharp point of departure in the yoga epistemological and ontological framework from the modern Western tradition is that in yoga knowledge system, the brain (*mastishk*) and mind (*Manas Antarkaranam*) are seen to be two different and distinct entities, although they are known to work in conjunction (Swami Vivekananda 2015). The mind is experienced to be a metaphysical entity, distinct from the brain which is a biological entity. The term metaphysical implies that it is not an entity that can be observed through the sensory apparatus (e.g., modern imaging instruments). The mind cannot be directly observed. It can only be experienced. Its multidimensional existence can only be verified by experiencing it. The mind is experienced by students of yoga to exist in multiple states, including thoughtless states and deeply interconnected states with all in nature. The yogic intuitive states of mind can be experienced by any disciplined seeker. The mind (called *sookshma sharir*) can even be separated for short periods from the body. The rational state of mind with thoughts and emotions (trackable through neuroimaging) is certainly recognized and respected as it is a commonly available state to communicate coherently and effectively. The yoga knowledge tradition, however, opens to seekers via disciplined training to multiple other intuitive states of mind. All these states enhance the connectivity of the individual to the essence in other forms of natural manifestation. They create a relationship of oneness between the self and the universe and thus promote a natural feeling of existential harmony with all forms of animate, inanimate, and even non-sensory expressions of nature (e.g., spirits). The separation of brain, a biological entity, and the mind, a metaphysical entity, is the fundamental point of ontological departure between the indigenous Indian yoga tradition and the Western knowledge traditions.

The mind as is well-known in modern Western neuroscience is synonymous with neural brain functions. It is only a notional term. It refers to intellectual and emotional activities that reflect as neural changes in the brain. It has no separate existence from the brain.

The author reiterates that the essential foundations for epistemological and ontological systems in any culture arise from the way in which the six instruments of knowing (are trained to) experience nature and the universe. The differences in mind–body experiences lie at the center of the cultural divide across societies. In African tradition, for example, the culture claims to experience the existence of ancestral spirits and communicates with them. Acceptance of the reality of spirits is an ontological shift and obviously assumes the actuality of existence of spirits (a metaphysical entity) experienced by the trained African mind. The modern European scientific mind only admits sensory observations analyzed by reason and logic (both seen as neural brain functions) as real. In the scientific ontology, both the yogi's and the African ontologies are unreal. A liberal view by the scientific community could concede that there may be other non-scientific (not necessarily false) ontologies, but would assert that within the limitations of the scientific framework, neither the metaphysical minds nor the ancestral spirits exist. An intolerant view assumes that no state of mind beyond the rational state interpreting sensory data is available to human minds and would label non-sensory experiences as those held by traditional cultures as rubbish or superstition. Indigenous cultures, on the other hand, would regard scientific ontology to be only one of the many expressions of nature's reality. At this juncture, it is important to point out that the question regarding which ontology is true is irrelevant. They both appear to be true within their own epistemological frameworks.

To sum up, the indigenous Indian general perspective on relational ontologies is derived from the *sankhya* and yoga knowledge systems. In the author's view, they have their roots in the differences related to cultural experience and understanding of both the brain and mind or more generally the body and mind. The modern Western view accepts as real only reasonable interpretation of what the senses or the extremely sophisticated scientific instruments that extend their range perceive. It also accepts theoretical extrapolation of sensory observations, not actually experienced, like the quantum world, as plausible, within the analytical framework of advanced reason expressed in the language of mathematics and logic. Several non-Western cultures depart from this scientific ontological framework because of their different experiences of mind. Apart from this fundamental instrumental (the way of using five senses and the mind) point of departure between the indigenous Indian and Western perspectives on nature and society, there also appear to be several proximate natural and social environmental factors which arise from natural ecosystem, history, economic, sociological, and politics that affect epistemological and ontological perceptions.

It is the non-academic perception of the author that the scientific and non-scientific ontologies can in principle coexist and be mutually supportive of each other. This is observed in several societies where in private life, religious and spiritual traditions based on non-scientific ontologies are practiced, but in professional life a scientific work ethic is adopted.

Relational ontologies in the context of health sciences in India

Shifting the focus on health sciences, I think from the reader's point of view, communicating on a relatively abstract subject like "relational ontologies of health sciences and practices" in India, may be better illuminated if one takes recourse to presenting a brief account of the recent post-18th-century history of medicine. The history may lend real social context to the evolution of ontological changes that have actually taken place in the Indian health space. In fact, a major ontological and epistemological change did occur in health sciences and practices in India only post–18th century. This change happened after the intimate colonial encounter between India and Europe. This recent

Indo-European encounter took place in the backdrop of significant cultural, economic, science and technology, and geopolitical changes. The historical fact of the ontological shift in India's health culture, on account of the European encounter, is elaborated subsequently. The 18th century is thus an ontological milestone in India, not only in the health space but perhaps in multiple knowledge and social dimensions of India's contemporary development. The reader will need to go back and forth in this chapter, into the long epistemic history of medicine in India, to understand the nature of dramatic ontological transformation brought about in health science and practice by recent encounter with the European world.

The change in health knowledge and practice in the 18th century marks the advent of what may be termed as a "scientific ontology" into the health culture of India. Its impact in the health space is narrated subsequently. Western medicine was introduced in India by the Europeans in the 18th century. The highlight of early European medicine was the germ theory which has graduated in the 20th and 21st centuries into advanced molecular and cell biology, genetics, and, more recently, the role of stem cells and the microbiome. The entry of scientific ontology in the late 18th century was the period when European medicine was relatively limited in its scientific scope and essentially comprised of empirical observations. Despite its elementary knowledge and capacities, it was legitimized in the colonial world, including India, not on merit of the depth and value of its understanding of health (or rather of specific diseases) but solely by political imposition. This kind of free entry into a country is unlike these days when permission for a new medical practice and even more so for an entire system of medicine would require the proponents to place on the table a substantial body of researched knowledge, stringent proof of efficacy, and a scheme of evidence. In 1835, in Eastern India, the Calcutta Medical College (CMC) was established (Kiran Kumbhar 2019). The CMC marked the official entry of European medical education into the country. From this time, the Colonial Government, while it could not extinguish indigenous knowledge and practices in the country, offered no state patronage to the indigenous health knowledge system despite its widespread acceptance by civil society. In 1896, the British formed an institutional mechanism for health delivery, namely, the Indian Medical Services (IMS) (Crawford 1901). The officers of the Indian Medical Services were initially mostly military surgeons of European origin who were selected in England. With the opening of Calcutta Medical College, IMS was also gradually opened to the natives of India who were selected to serve in Subordinate Military Medical Services or as Assistant Civil Surgeons in subdivisional civil hospitals.

An important landmark in the health sector in 1946 was the "Health Survey and Development Committee report", also called the Bhore Committee. This committee was appointed by the Government of India. Its mandate was to survey the existing health structure in the country and make recommendations for future development of healthcare in India. The Committee submitted its report in the same year (NHP CC DC 2015). This report although written 75 years ago introduced scientific ontology into India's Health System. I quote subsequently relevant excerpt from this report because they clearly spell out the underlying epistemological and ontological underpinnings of new health culture that was seeded then and continues to grow in India. The recommendations of the Bhore Committee were accepted by Government, and they continue to prevail. The ontological roots of India's modern healthcare system can be traced to the recommendations of the Bhore Committee. I urge readers to read the following brief excerpt because it has a striking similarity, even after 75 years, to the present-day policy and practice discourses on healthcare, medical education, and research.

The epistemic character of the recommendations rested on the assumption that the scientific epistemology and ontology introduced by Europeans was the most credible basis for reliable health delivery in India. It rejected knowledge or practice which was not derived by the scientific method even if the knowledge and practice was observed to be perfectly functional.

> In considering the question of the place which the indigenous systems of medical treatment should occupy in any planned organization of medical relief and public health in the country, we are faced with certain difficulties. We realize the hold that these systems exercise not merely over the illiterate masses but over considerable sections of the intelligentsia. We have also to recognize that treatment by practitioners of these systems is said to be cheap, and it is claimed that the empirical knowledge, that has been accumulated over centuries has resulted in a fund of experience of the properties and medicinal use of minerals, herbs and plants which is of some value. Further, the undoubted part that these systems have played in the long distant past in influencing the development of medicine and surgery in other countries of the world has naturally engendered a feeling of patriotic pride in the place they will always occupy in any world history of the rise and development of medicine. This feeling has not been without its effect on the value which is attached by some to the practice of these systems.
>
> We are unfortunately not in a position to assess the real value of these systems of medical treatment as practiced today as we have been unable, with the time and opportunities at our disposal, to conduct such an investigation into this problem as would justify clear-cut recommendations. This in no way reflects upon these systems. It has, however, to he recognized that great improvements have taken place in the field of public health as the' result of the many discoveries of science which are and can be implemented only through the scientific system of medicine and through personnel trained in such a system. No system of medical treatment, which is static in conception and practice and does not keep pace with the discoveries and researches of scientific workers the world over, can hope to give the best available ministration to those who seek its aid.

> We feel that it would be unfair and unjust merely because some other (indigenous)method of treatment is said to be cheaper, to deny to anyone in this country the benefit of the scientific system and of the daily growing volume of research and achievement in the wide world of science.
>
> (Bhore 1946:455–457)

The most recent National Health Policy released in 2017 (Ministry of Health and Family Welfare 2017) echoes the same ontological and epistemological orientation that was seeded in 1946. So huge has been the effect of this ontological shift that even in 2022, the Government of India, led by a right-wing political party, allocates around 95% of the health budget to Western biomedicine and a small 5% or so to the indigenous health systems.

Analysis of relational ontologies in healthcare space

The setting for the analysis of relational ontologies in the Indian health landscape has been laid in the previous section. The health policies and budgetary allocations made by the Government of India in the last 75 years[2] reflect the fact that the new scientific ontology seeded by the 1946 recommendations of the Bhore Committee took effect.

Surprisingly, this ontological dimension of change in health sciences and practice is hardly ever discussed in active public discourses on healthcare. This is perhaps because policy analysts have never been challenged to delve deep into the ontological and epistemological roots of praxis, although it underlies the changes that appear on the surface.

The usual explanation for social changes in all dimensions of knowledge, including health sciences, is attributed to proximate causes, namely, changing political, scientific, and economic factors, and naively in the author's view to Eurocentric notions of modernity and progress.

Let us look in some detail at the scientific ontology introduced into Indian health science and practices (Darshan Shankar, et al. 2022). The modern European scientific knowledge system is based on an atomic and cellular perspective of physical (chemistry) and biological reality alongside technology. It is the most recent functional innovation for study of biological change at the microlevel. In the context of civilizational evolution, modern science indeed constitutes a milestone. Without discounting its extraordinary strengths, it is evident that the epistemology and ontology of science, based on its observer–observed framework (that separates the two), gives scientists a reduced and partial view of the complexity of both natural substances and processes.

2 The budget for indigenous systems in 2022–2023 is on the order of US$428 million. The average allocation has been around US$50 million per year. This constitutes 1–3% of the National Health Budget.

In biology, the field of experimental pharmacology is employed as an ontological tool in medicine. It is well-established that biological models used in labs to design potential medical interventions do not capture the complexity and interconnectedness of the human biome. Their results from changes observed at cellular and molecular levels cannot be extrapolated to the whole living systems (Houghton et al. 2007; Hopkins 2008). The laboratories with the best scientific infrastructure do not have "study models" sophisticated enough to mimic the complexity of life. Futuristic projects like "body on a chip" (Wake Forest School of Medicine 2022) are trying to create complex interconnected bioassays that can mimic the human biome. But such advancements are still at least two decades away.

Similarly, the limitations of biomedicine are observed in clinical pharmacology which tests lab results on humans. Statisticians are aware that clinical evidence is a nuanced term. This is because there are grades in quality of evidence. It may surprise readers to know that extant practices are based on low quality of evidence (Tricoci et al. 2009). Today, at the frontiers of evidence theory, there is a growing rejection of the assumption that randomized clinical trials (RCTs) represent the golden standard of evidence (Longhurst et al. 2014). This reconsideration about the value of RCTs is first due to their small sample sizes (due to their prohibitive cost), which do not sufficiently represent the target population segment. Even more importantly, clinical trials throw up doubtful conclusions due to the recent realization that the inclusion criteria and disease classification system used in the best designed trials do not capture the genetic (intra-specific) variability within a given health condition. This limitation may have its roots in epistemology of health sciences. Thus, the extant schemes of international disease classification used in the RCTs are gross and therefore erroneously accept heterogeneous populations to be homogeneous. The error in design of clinical trials gets further compounded because the trials test the effectiveness of uniform interventions on heterogeneous groups. The results from such trials are therefore naturally incomplete and in fact misleading (Murray et al. 2022).

What about Ayurveda ontology? Its epistemic history is already narrated. While it is etymologically clear that Ayurveda and biology both carry equivalent meaning and possess exactly the same goal, namely, the study of "changes" in life processes, their difference lies in their perspectives. Ayurveda has a systemic perspective for observing biological change. In Indian knowledge systems, biological change is viewed in terms of changes that occur in the five natural states or fields inherent in nature (*panchmahabhutas*) (Shankar 2020). These are five fields that exist in the human being and in every animate and inanimate forms, in particular configurations. The fields are symbolically called space (*akasha*), wind (*vayu*), fire (*agni*), water (*aap*), and earth (*prithvi*). The explanation of the properties and behavior of these fields in the biological system will not be attempted here.

It is natural to expect that knowledge systems like biology and Ayurveda constructed on completely different perspectives (foundations) will not look alike in terms of principles, structures, concepts, categories, and methods. Only at the level of goals, they are in fact essentially similar. The rejection of Ayurveda in policy and profession as circles is apparently on the ground that it appears to be unscientific. It is puzzling however to observers as to why and how it is clinically proficient because in India, Ayurveda annually treats at least a hundred million customers. Ayurveda is observed to manage competently neurological, musculo-skeletal, gastrointestinal tract (GIT), respiratory skin, non-communicable and chronic diseases, and even serious infections. However, it cannot deal with emergencies, acute conditions, and advanced surgery. A trial-and-error explanation for its functionality and competence appears unviable to explain consistently effective application of principles and practices across multiple domains ranging from Ayurvedic pharmacology to pharmacy and diagnostics to clinical management. However, Ayurveda is certainly not at all a science in the Western sense of the term because unlike Western science, it is not founded on a molecular view of nature.

The question to be considered and pondered over is as follows: Is the molecular perspective the only way of discovering the truth of nature? Scientists, including those working in the colonized and Western worlds, due to their education in Western sciences perhaps hold an erroneous belief that modern European sciences are the only universal way of understanding biological and physical change (Twitchett 1955). However, knowledge is not a function of time but in author's view, a function of a unified view of nature. This unified view has been discussed above in the discussion on ontology. Therefore, the latest scientific advances perceived from a reductionist perspective, even if they be highly illuminating need not be the complete or the only way of knowing. Science can certainly learn from profound systemic perspectives of Ayurveda because they can advance its own partial observations. It may be further argued and demonstrated that other cultural knowledge systems also possess universal attributes and capacity to study and apply their knowledge of biological change effectively, even though the knowledge is derived using nonmolecular frameworks. The test of a non-mainstream knowledge system in the author's view should not be in its conformity to a singular method, principles, or set of concepts, however profound, but by evaluating if its theory and practice (in its own framework) can consistently be applied to transform and solve real-life problems

Concluding remarks

The author believes the reason why mainstream health sciences in India and globally have not supported indigenous knowledge traditions is essentially epistemic and geopolitical. Despite the national sovereignty of previously colonized nations, the governing elite and sections of intelligentsia, due to their deep engagement with a complex international ecosystem of networked economic, scientific, educational, and political institutions, appear to have got trapped in a powerful centralized global system of knowledge and practices (with their underlying ontology) from which, for fear of isolation, most of them do not have the courage or funding opportunities to explore with self-respect, self-confidence, and non-chauvinist intent the knowledge derived from their own cultural roots.

Western science is a recent functional ontological framework for comprehension of nature. It enjoys the well-deserved credit of liberating Europe from the dark ages (from dogma to reason). But its superimposition on the non-Western cultures with the claim that it is the only credible way of interpreting nature has perhaps given it the role of a cultural oppressor. In the context of this chapter, the author points out that Ayurvedic ontological perspective is currently functional and has been so for the last several millennia. It has unique epistemological and ontological features that have kept it alive. Therefore, the claim of monopoly of Western medicine as the most effective health science, despite its impressive technologies, certainly appears to be unjustified.

A more complex sociological question may be asked. Why have native scientists in previously colonized countries, even after sovereignty, not asserted the value of their own intellectual and cultural traditions: an assertion which is not based on chauvinism but proven merits of indigenous knowledge systems. The analysis of this complex sociological phenomenon is outside the competence of the author. One infers that the interlinked national and international networks of educational, scientific, economic, technology, and political institutions established during the last 200 years and seeded by the Europeans have thus far been too powerfully a force to challenge. It may need a new form of non-mainstream, international knowledge cooperation to initiate changes. The idea is not to reject any ontology but to establish cooperative links across knowledge cultures.

The scientific ontology which only accepts the coherent (rational) interpretation of sensory reality of the natural world as real perhaps also extends to the social "sciences", which have also appended the term science to social studies (Sarukkai 2005). The author assumes that modern ontology in the social sciences is also guided by sensory perceptions. The modern Western ontological framework for creation and interpretation of social institutions and processes appears to accept, justify, and promote unlimited economic growth of individuals and of societies as an acceptable ontological goal as long as such growth takes place within a presumed democratic framework of equal opportunity and a system of rule of law. The author is ignorant of the epistemological arguments justifying the concept of growth-oriented socioeconomic and political systems because of his incompetence in social studies. However, judging from the levels of disparity and tensions that have been seen during the last two centuries within and across nations, through colonialism, two world wars, the period of Cold War, the creation of centralized international institutions for governance and finance, the continuing imbalanced trends in the global world order, the scale of human poverty and the unprecedented pace of destruction of nature, and its consequences like climate change, suggests that the modern epistemological and ontological foundations that guide social development are also in need of urgent review.

The time to reimagine society may be right, because of the growing realization of the limits to growth and the incompleteness of modern knowledge systems in many domains not only in healthcare. Developing relational ontological analytical frameworks to promote international knowledge collaboration could be the approach adopted by sensitive sections of the knowledge community that are seeking constructive solutions for serious societal problems that are beginning to surface.

References

Bhore, J., 1946. Indigenous systems of medicine. In: Bhore, J., ed. *Report of the Health Survey and Development Committee ('Bhore Committee Report')*. Delhi: Govt. of India Press

Charaka, n.d. *Charaka Samhita*. Varanasi, India: Chowkhamba Publication.

Charaka, n.d. Vimana sthana (Cs.vI.6.4) (tatra vyādhayō' aparisaṅkhyēyā bhavanti). In: Acharya, Y. T., ed. *Charaka Samhita*. Varanasi, India: Chaukhamba Prakashan.

Crawford, D. G., 1901. Notes on the history of the Bengal medical service. *The Indian Medical Gazette*, 36(1), 1.

Darshan Shankar, A. 2021. Roadmap for Ayurveda education in modern India. In: Sarangapani, P. M., and Pappu, R., eds. *Handbook of Education Systems in South Asia*. Singapore: Springer.

Darshan Shankar, et al., 2022. Breaking silos: Can the emerging field of Ayurvedic biology contribute to the advancement of Indian health science. *Current Science*, 122(3).

Darshan Shankar, Vijay B, Singh, G., Vishnuprasad, C. N., Godbole, A., and Kumar, S., 2022. Breaking silos: Can the emerging field of Ayurvedic biology contribute to the advancement of Indian health science. *Current Science*, 122(3).

Darshan shankar. The Unfinished Agenda of Modernization. In: Claude Alvares, editor. Multicultural Knowledge and the University [Internet]. Other India Press; 2014 [cited 2022 Feb 17]. p. 120–7. Available from: https://otherindiabookstore.com/books/1-multicultural-knowledge-and-the-university (Accessed 12 April 2023).

Deutsche Nationalbibliothek,

Dharampal, 1971. *Indian Science and Technology in the Eighteenth Century: Some contemporary European Accounts*. Other India Press Mapusa 403 507 Goa, India Available from: https://shikshantar.org/sites/default/files/PDF/1DharampalJiCollectedWritings-Indian-science-and-technology.pdf (Accessed 12 April 2023).

Escobar, A., 2016. Thinking-feeling with the Earth: Territorial struggles and the ontological dimension of the epistemologies of the South. *AIBR Revista de Antropologia Iberoamericana*, 11(1), 11–32.

FRLHT and NMPB, 2010. *Indian Medicinal Plants Database* [online]. http://medicinalplants.in/ (Accessed 17.12.2021).

Hankey, A., 2005. CAM modalities can stimulate advances in theoretical biology. *Evidence-Based Complementary and Alternative Medicine: eCAM*, 2(1), 5–12.

Harari, Y., 2015. *Sapiens: A Brief History of Humankind*. Random House UK, Penguin (Vintage).

Hopkins, A. L., 2008. Network pharmacology: The next paradigm in drug discovery. *Nature Chemical Biology*, 4(11), 682–690.

Houghton, P. J., Howes, M. J., Lee, C. C., and Steventon, G., 2007. Uses and abuses of in vitro tests in ethnopharmacology: Visualizing an elephant. *Journal of Ethnopharmacology*, 110(3), 391–400.

http://www.stsinfrastructures.org/content/visualization-and-cognition-thinking-eyes-and-hands

Kumbhar, K., 2019. *How the 1835 Calcutta Medical College Catalyzed the Ascendancy of Biomedicine in India*. Caring for India.

Latour, B., 1984. Visualisation and cognition: Thinking with eyes and hands. In: *Knowledge and Society: Studies in the Sociology of Culture Past and Present*. JAI press INC

Lipphardt, V., and Ludwig, D., 2011. Knowledge transfer and science transfer. *Notes*.

Longhurst, C. A., Harrington, R. A., and Shah, N. H., 2014. A "green button" for using aggregate patient data at the point of care. *Health Affairs (Project Hope)*, 33(7), 1229–1235.

Meulenbeld, G. J., 2017. *History of Indian Medical Literature* (Groningen: Forsten, 1999–2020). ISBN 9069801248 9789069801247. Five volumes in three parts, 1A, 1B, 2A, 2B, 3.

Ministry of Health and Family Welfare, 2017. *National Health Policy, 2017*. https://www.india.gov.in/national-health-policy-2017

Murray, C. J., et al., 2022. Global burden of bacterial antimicrobial resistance in 2019: A systematic analysis. *The Lancet*, 399(10325), 629–655.

Nandy, A., 1983. *The Intimate Enemy: Loss and Recovery of Self Under Colonialism*. Oxford: Oxford University Press.

NHP CC DC, 2015. *Bhore Committee, 1946 – National Health Portal of India 2015*.

Patwardhan, B., 2015. In search of roots: Tracing the history and philosophy of Indian medicine. *Indian Journal of History of Science*, 50(4), 629–641.

Sarukkai, S., 2005. *Indian Philosophy and Philosophy of Science*. Motilal Banarsidass Publishers

Sarvepalli Radhakrishnan, 1923. *Indian Philosophy*, 2nd ed. Oxford: Oxford University Press.

Shankar, D., 2020. Roadmap for Ayurveda education in modern India. In: Sarangapani, P. M., and Pappu, R., eds. *Handbook of Education Systems in South Asia*. Singapore: Springer, 1–20.

Shankar, D., and Unnikrishnan, P. M., 2004. Challenging the Indian medical heritage. (book) In: *Challenging the Indian Medical Heritage*, 1–241. Chapter Alternative Macro Visions. 10.1017/UPO9788175968752.005.

S Saraf, R. P., 2006. Sushruta: The first Plastic Surgeon in 600 B.C. *The Internet Journal of Plastic Surgery*, 4(2).

Swami Vivekananda, 2015. Raja yoga – appendix. In: *Patanjali Yoga-Sutra*. Mumbai: Bharati Vidya Bhavan.

Tricoci, P., Allen, J. M., Kramer, J. M., Califf, R. M., and Smith, S. C., 2009. Scientific evidence underlying the ACC/AHA clinical practice guidelines. *JAMA*, 301(8), 831–841.

Twitchett, D. O., 1955. Joseph Needham: Science and civilization in China. Volume 1. Introductory orientations. xxxviii, 318 pp, 36 illustrations. Cambridge: University Press, 1954, 52s. 6d. *Bulletin of the School of Oriental and African Studies*, 17(2), 383–385.
Valiathan, M. S., 2013. *An Introduction to Ayurveda*. Delhi: Orient Blackswan.
Wake Forest School of Medicine, 2022. *Wake Forest Institute for Regenerative Medicine (WFIRM)* [online]. https://school.wakehealth.edu/Research/Institutes- and-Centers/Wake-Forest-Institute-for-Regenerative-Medicine (Accessed 18.2.2022).

5 Cosmovisions and critical sustainability sciences

An African ontology of *"Vurr"* (an energy) among the Dagara of Southwest Burkina Faso and Northwest Ghana

David Millar, Niagia F. Santuah, and Maxwell Ba-an Tengolzor

Introduction

People all around the world are bombarded by a flurry of subjects and ideas from scientists and science. Not all ideas are necessarily true, which means the undercurrent supporting the opinion is not always necessarily in conformity with reality. The undercurrent is the worldview or cosmovision. As people try to understand phenomena around them, worldview provides the needed guidance as the broad conceptual synthesis that forms one's perception on the whole of reality (Dewitt 2004). Since the human being's senses are not perfect and human ability to interpret what is perceived by the senses is imperfect, it stands to reason that human beings need to look to a higher being as the only true source of knowledge about reality (Chalk 2006).

The reality as perceived by the dominant conceptual Western worldview, which has been fundamentally responsible for the greatest improvements in science in the last 200 years of human history, may have been altered, but its core concept that existence is explainable solely in material terms, with no accounting of the metaphysical, still dominates global thinking. Admittedly, the mechanistic and materialistic view of nature brought us the Enlightenment, the Industrial Revolution, and the Green Revolution, which have witnessed quantitative leaps in human progress. But as Laszlo (2006) points out, the "unrestrained, purely quantitative growth in energy and material production and consumption, which has been the main focus of theories of the western worldview, is not possible on a finite planet with delicately balanced biosphere" (Laszlo 2006:96). This has implications for sustainability science. As former United Nations Secretary-General, Ban Ki-moon, succinctly puts it at a press conference in Marrakech, Morocco on November 15, 2016, "we don't have plan B because there is no planet B". We need resilient local societies and economies developing at the rhythm of nature and sustainably meeting the needs of the past, current, and future generations in innovative ways that are grounded in pristine ontologies.

DOI: 10.4324/9781003043577-5

This chapter juxtaposes contemporary sustainability science discourse with African ontologies from indigenous case studies. This anecdotal piece: "Vurr" (an Energy) and Ramifications of "Vurr" (a Life), discusses the various life-forms the concept could assume; the gnoseology of which can be ascribed to the Living, the Dead, and the Yet Unborn.

Background

Western ideologically skewed research has always been the prerogative of an exclusionist club of stakeholders who call themselves "the Researchers". Hence science, which claims to be the dominant product of research, has also fallen victim to the same mode of exclusion. For instance, research is tilted in favor of the material at the expense the immaterial such as spirituality. But "scientists should not be afraid to investigate spirituality and spiritual experiences since they represent a central aspect of human existence" (Beauregard et al. 2014:273). Millar (2018) has amply demonstrated the research power and ability of smallholder farmers in northern Ghana. Millar has benchmarked formal scientific principles (that which is done in laboratories) with Farmers' Experimentation (that which Farmers do in their daily livelihood activities on their farms), with the latter leaning more toward sustainable development than the former, by drawing on their relationship with nature and the environment and incorporating lived spiritual realities.

The World Commission on Environment and Development defines sustainable development as "development that meets the needs of the present without compromising the ability of future generations to meet their own needs" (Brundtland and Mansour 2010:43).

> In essence, sustainable development is a process of change in which the exploitation of resources, the direction of investments, the orientation of technological development and institutional change are all in harmony and enhance both current and future potential to meet human needs and aspirations.
>
> (Brundtland and Mansour 2010:46)

According to Lütteken and Hagedorn (1999:1), there is a close relationship and a continuous interaction between people and ecology, which means

> sustainability is a relationship between dynamic human economic systems and dynamic, but slower, ecological systems, in which: (a) human life can develop indefinitely; (b) human individuals can flourish; (c) human culture can develop and (d) effects of human activities remain within bounds so as not to destroy the diversity, complexity and functioning of the ecological life-support system.

Sustainability science then is a transdisciplinary research field that draws on a myriad of related disciplines for understanding how human–environment interactions and systems affect the prospects for attaining sustainable development (Kabisch et al. 2015). Sustainability science promotes scholarship to analyze multidisciplinary information and data to provide informed decision-making through resource management and integrate various sources of knowledge to support the process of achieving economically profitable, socially scrupulous, and ecologically sound development. Sustainability science is grounded in an open worldview – one that embraces and celebrates diverse sources of knowing while leaving room for the unknown, unknowable. This is how an open worldview provides the needed guidance as the broad conceptual synthesis that forms one's perception on the whole of reality (Dewitt 2004). In this regard, sustainability science was supposed to be the crucible for science and nature to interact in meaningful ways that mimic reality, which is itself shaped by our perception of reality.

However, by the same argumentation, what Western scientists do in sustaining science has also made sustainability science fall victim of exclusionism on all pillars of science (ontology – origin and nature of knowledge, gnoseology – proof and validity of knowledge, epistemology – knowing/enquiring and learning of knowledge, axiology – ethics and values of knowledge, knowledge community–a community of users of that knowledge). Though African ways of knowing fit well into the Eurocentric notions of the pillars of science, these are still not considered as science, at best just signs. So far sustainability science is seen through monodisciplinary, multidisciplinary, and

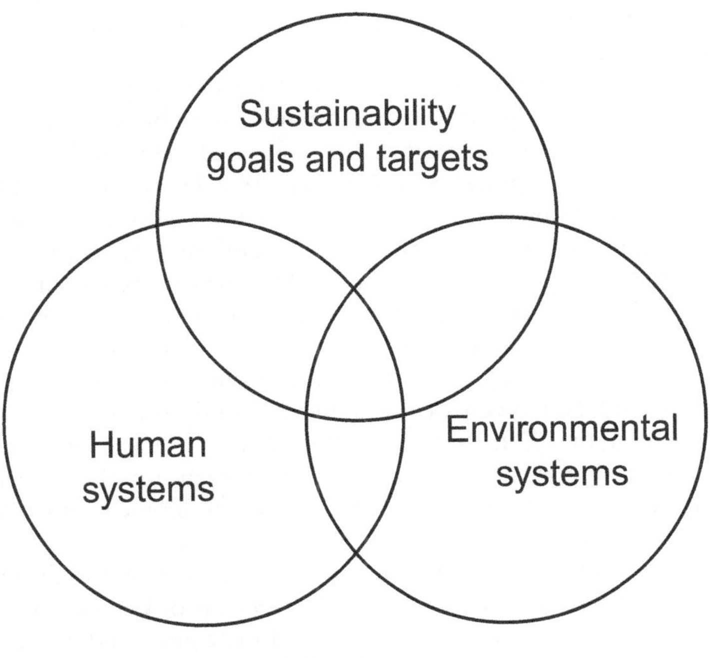

Figure 5.1 The domains of sustainability science.

Source: Levin and Clark 2010, modified by authors.

interdisciplinary lenses, that is, viewpoints that satisfy the dictates of the exclusive club of scientists alluded to earlier.

A new platform for scientific discourses into sustainability issues, called Critical Sustainability Sciences (CSS), builds on a highly novel integration of elements from critical theory, intercultural philosophy, and a relational concept of emancipatory science. According to Rist et al. (2019), Critical Sustainability Sciences are based on overcoming three basic flaws that render the current mainstream debates on sustainability ineffective. "[These] concern [1] it's essentially apolitical use-inspiration, [2] the lack of ontological reflection, and [3] the reluctance to integrate the root causes of the devastations in the ecological, social, economic and cultural realms of life" (Rist et al. 2019:3).

With Critical Sustainability Sciences, space has now been created for actors from outside academia to get involved in research processes in order to integrate the best available knowledge, reconcile values and preferences, and in so doing create shared ownership for scientific problems and or preferable solutions (Kates 2011; Filho 2018). It should eventually be transformed into transcendental sustainability science. Spirituality is a network linking humans to an AllFather, the universe, and to each other. In indigenous societies, spirituality often infers that people do not own land, but must care for the land as part of their sacred stewardship within the purpose and direction of the cosmic order (Champagne 2015). Therefore, in addition to the economic, environmental, and social aspects of a people being undeniably key, for a people's true sustainability, it is important to reflect a transdisciplinary approach to actions and inactions, which are underpinned by a people's spirituality (Lang and Wiek 2012), that is, their feeling of connection to the unseen, immaterial forces of nature.

The term *transdisciplinarity* was coined "to give the expression to the need that was perceived – especially in education – to celebrate the transgression of disciplinary boundaries, an act that far surpassed the multidisciplinary and the interdisciplinary approaches" (Nicolescu 2002:2). *Transdisciplinarity* was meant to help us embrace the extraordinary cultural, historical, religious, and political diversity of the many different peoples of the earth, permitting an intuitive understanding of the ideas that might otherwise be incomprehensible or inaccessible. But against all our fears, transdisciplinary research has been co-opted as a new tool for domination, having been "[adopted] as a fashionable slogan devoid of any content, pouring from emptiness into the void" (Nicolescu 2002:2). However, comparatively recently, proponents of transdisciplinarity (Rigolot 2020; Fann et al. 2018) have made a strong and spirited case for broadening science and the key actors in science, and for the inclusion of the so-called non-scientist as collegial to the development of science. This position leverages Critical Sustainable Science to a higher order discourse or higher pedestal of sustainability science – Transcendental Sustainability Science. This chapter calls for a review of the domains of Critical Sustainability Sciences to include an overarching context of the culture and hence the spirituality of a people, explored in a transdisciplinary and transcendental manner.

Bridging sustainability science and emancipatory science

Sustainability Science and Emancipatory Science are two schools of thought which are quite similar in their objectives – with Sustainability Science focusing on "sustainability" and Emancipatory Science emphasizing "citizen participation and liberation". We posit that transdisciplinary research concentrates on making sustainability science more sustainable.

Earlier scholars like Cornwall and Jewkes (1995) were already querying participatory research and asking critical questions for rethinking participation from a more sociocultural perspective. From an ideological viewpoint, Wright (2007) brings a Marxist perspective to bear on science: "Emancipatory Social Science seeks to generate scientific knowledge relevant to the collective project of challenging various forms of human oppression and creating the conditions in which people can live flourishing lives". In comparatively recent times, Hunt and Thornsbury (2014) re-echoes the crucial role of cultural sensitivity in transdisciplinary research and sustainability of a people.

Although transdisciplinarity seems to provide a window in the search for sustainability and emancipatory science, critical questions still remain: (1) What are the ramifications of transdisciplinarity in the different world cultures? (2) How does spirituality position itself in all these discourses? (3) As human beings, what is our responsibility toward nature? (4) What opportunities exist for breaking down walls and building bridges across worldviews?

The African worldview[1] is expressed in their proverbs and parables, religious myths and mythologies, rites and rituals, social order, and life events. It is a triune of a cosmovision in which the natural, the human, and the spiritual interconnect and interface with a Supreme Being – a *Transpirihumanatural* entity. This worldview, or philosophical position, is that human beings are not just part of nature; human beings are nature itself. This worldview engenders an ethical responsibility toward nature and all forms of life, and responsibility toward an AllFather up high in the firmaments. This explains why "sustainability transformations" must be connected to the immaterial and values of care and spirituality" (Apusigah and Millar 2010:10).

Western scientists are genuinely concerned with sustainable development also in and for Africa, a commitment exemplified in the Sustainable Development Goals (SDGs). But the SDGs will not be achieved, and sustainable development will never happen if the aforementioned values and spirituality are not part of sustainability. But concepts such as values and spirituality will remain

1 Mentioning "African Worldview" in singular is often questioned. However, given the number of diverse tribes and languages and the geographical diversity throughout sub-Saharan Africa, the issue of any unity in that diversity can justifiably be raised. But this issue has been addressed by many authors (Chalk 2006 cited Mbiti 1969; Parrinder 1969; Richmond and Gestrin 1998; Gehman 1999). Imasogie (1983:53) says, "in spite of the differences, there is a core of Africanness that runs through their cultures and religions. In view of this, one may speak legitimately of an African worldview, the local peculiarities notwithstanding".

abstract ideas unless they can be brought home for people to touch and feel them. So in this chapter, we explore the manifestation and practical application of these concepts through the African ontology of *Vurr* (an energy) among the Dagara of southwest Burkina Faso and northwest Ghana. We will briefly introduce the Dagara, describe our methodology, ground the ontological debate by addressing key ontological questions in CSS, and elucidate the African worldview and cosmovisions, before focusing on Dagara worldview and constellations of knowing. We then examine the Dagara ontology and implications for CSS. We conclude by calling for a coevolution of sciences, inviting a respectful dialogue to be carried out under equal conditions, leaving open the ontological grounds from which might emerge new ways of knowing.

Methodology

The Dagara in space and time

The Dagara of southwest Burkina Faso and Dagara of northwest Ghana see themselves as one and the same people – a nation-state. We can easily notice these linkages in their "Yirlu" (paternal linkages) and the "Belu" (maternal linkages). This positioning deepens the debate between "The State – Geopolitical" and "The Nation-State – Sociocultural", raising the question: Which is their true identity? It also raises the question whether these people are patrilineal or they have vestiges of matriclans (also matrilineal). Though the Dagara are known by different names such as "Lobrr, Lobi, Dagarti, Degabe, Dagao, Eiyaa", they share a common belief and strong identity with the living, the dead, and the yet unborn. They see the act of living as a continuum: that transitions between the living, the dead, and the yet unborn. And all of them are *ancestor-centric* by nature. Hence *ancestorcentricism* is more like a cult that binds them all together; the Dagara consult the ancestors before they plant seed in the soil, consult the ancestors as they weed in the plants, and thank the ancestors for a successful harvest and beg for good health up to the next cropping season. On all these occasions, there are communal meetings, joint sacrifices, and sharing of resources. How does this translate into their cosmovision? How do ontologies of this nature influence in changing the narrative about sustainability science? The Dagara concept "Vurr" (Energy) allows us to study the science of being.

Sources and analyses of field information

Self-introspection (Creswell and Creswell 2005)*:* All the researchers (and co-authors of this chapter) come from the parts of northern Ghana that share borders with Burkina Faso. The lead, David Millar, is a Dagara himself of Ghana and Burkina Faso extraction. Niagia F. Santuah is a Kasena from Chiana in the Upper East Region. Maxwell Ba-an Tengolzor is from Bolgatanga, also in the Upper East Region. Though the authors are of different ethic groupings, the differences in viewpoints of the authors allow for triangulation of findings since the knowledges of the area are fairly similar, thus significantly reducing the element of extreme subjectivity.

Interrogating selected elders: With a checklist or discussion guide, Key Informant Interviews were conducted. Sampling of "who-next" was based on snowballing sampling procedure (Creswell and Creswell 2005).

Phased assertion (Kirk and Miller 1986) *with selected Dagara scholars*: This aspect of intellectual investigation into ontologies is rare in Ghana. In order to authenticate the emerging positions, a series of dialogues were organized with a purposive sample (Patton 2002) of Dagara scholars in Ghana and Burkina Faso.

Monitored radio discussions: Ghana – *Lawra Radio*, 101.1 MHz; *Radio Freed*, 92.3 MHz; and *Radio Upper West*, 90.1 MHz, and Burkina Faso – *Radio Goulou*, 99.5 MHz.

Vocabulary mapping (Millar and Derbile 2018)*:* This was a major tool used for the analyses, which involved tracking the common local vocabulary use by gender and by generation to establish meaning and import.

Grounding the ontological debate

Ontological questions and critical sustainability science

Ontological questions begin with who am I, or what am I? But they do not end there. These questions lead to an endless list of other questions that extend, but not limited to, what is my relationship with fellow humans, other forms of life, and non-living forms? Where did I come from? Where do I go from here? Does human's existence terminate at death? Are heaven and hell real places where anyone would wish to go or avoid? If I am a being, is there a Supreme Being? If so, what are Its/Her/His attributes? It is the answers to these questions, or the failure to find answers to these questions, or even our inability to appreciate the need to ask these questions, which serve as the compost from which a worldview sprouts (Apusigah and Millar 2010:10).

Whether we are born with a worldview, or socialized into one, "different beliefs about the universe lead to different behavior and [actions]" (Lewis 1952:43). From an African perspective, a worldview determines beliefs, and beliefs determine behavior (praxis). Thus, a worldview is universally regarded as

> a coherent set of bodies of knowledge, concerning all aspects of the world. This coherent set allows people to construct a global image of the world and to understand as many elements of their experiences as possible. A worldview can in fact be perceived of as a map that people use to orient and explain, and from which they evaluate and act, and put forward prognosis and visions of the future.
>
> (Note 2006:83)

Chalk's (2006) assertion that there is but one worldview because, after all, there is just one world to view, has its own appeal. Yet despite all the nuances, similarities, and differences about worldviews, in reality there are only two basic worldviews, which revolve around one issue, God, and the simple question asked in Genesis 3:1 of the Holy Bible (New Living Translation, NLT), "'Did God really say you must not eat the fruit from any of the trees in the garden?" (ThM 2009). The notion of whether God has spoken or *not* determines the level of human autonomy, and appears to be the underlying principle that separates two dominant worldviews that shape our reality: on the one hand, the conceptual Western worldview, and, on the other hand, the numerous indigenous worldviews, which emerged as a result of people's close relationship with the sociological, physical, and spiritual environment (McKenzie and Morrissette 2003). We use the word "conceptual" advisedly to qualify the Western worldview because there are countries that share the Western worldview but are not necessarily geographically located in the West.

Cosmovisions have their luster and limitations. For the "positivist" school of the Western worldview, truth is an absolute certainty; "it is never just one of several possible ways of viewing the world; it is the only possible truth. Anyone who does not share their way of thinking is obviously uninformed or misguided or out of touch with reality" (Olsen et al. 1992:245). The "constructivists" see truth as socially negotiated through interactions such as between observer and observed, which means truth is always under construction. Yet people resort to worldviews to explain, justify, and legitimate their unwholesome behaviors and actions (Olsen et al. 1992). For instance, two-thirds of the population in rural areas of northern Ghana attribute their development challenges to the will of God, which suggests that some people may be the victims of an incapacitating worldview (Al-hassan 2011). Rather than viewing nature as something to be dominated, which is what adherents of the Western worldview do, devotees of African indigenous worldviews strive to negotiate an existence with nature. Land, for instance, is not just an economic commodity; it is where our ancestors are buried, which also explains why land custodians may be reluctant to give out land to "strangers" to farm. If, for instance, a drought leading to poor harvest is attributed to disequilibrium within the forces of nature – which requires sacrifices to appease unseen forces instead of the introduction of soil enrichment procedures and the adoption of "improved seeds" – does that stifle or stimulate creativity and innovation?

But cosmovisions, not least teleological and anthropocentric cosmovisions, have their charms too. Reverence for and veneration of the ancestors to a large extent dictate the use of resources, such as the way land, water, plants, and animals are to be used, how decision-making processes happen, how problems are solved, and how rural people organize their lives and livelihoods (Millar 2018). This holistic worldview engenders an ethical responsibility toward nature and all forms of life. Land is not only seen as a living organism, but it is also where the ancestral spirits domicile. This holistic worldview begets a systems approach to everyday decision-making where the interests of the ancestors, the living, and the yet-to-be-born are taken into strictest consideration. The intergenerational dimension of the concept of sustainability is therefore at the heart of indigenous people's adaptation strategies, which are organized to meet current needs without compromising the ability of future – and in our case including past – generations to meet their own needs and aspirations (Millar 2018).

Zhang et al. (2018:37) reiterates the need for systems thinking in problem-solving:

> The rapidly evolving environment in which we live requires responses based on careful analysis of alternative intervention options, especially when multiple and simultaneous challenges emerge. Decisions that do not consider the complex dynamics underlying the true causes of a problem risk unintended consequences or side effects.

We are at the cusp of a new revolution: Until recently, the conceptual Western worldview, which denies the existence of the metaphysical world, is beginning to accept the idea of multiple realities as against one reality. In their Manifesto for a Post-Materialist Science, Beauregard et al. (2014:273) argue that human self-perception has evolved through the successive shocks of three major scientific upheavals. "The first, due to Copernicus, reset our physical place in the universe. The second, due to Darwin, revised our biological place. *And the third, now in progress, will ultimately redefine our psychic/spiritual place in the cosmos*" (original emphasis). Thus, the framework for consensus and compromise has been established. There is now the need to prepare pathways for "western science" to accept truths from other worldviews, which could result in a fusion of the horizons of human wisdom (Subramanian and Pisupati 2010).

The African worldview and cosmovisions

The African worldview does not give time for theoretical idiosyncrasies because theory and reality are fused, even if this deepens the ontological debate. The African's worldview is a religious worldview because the two concepts – religion and worldview – are so intertwined that they are inseparable. "Africans are very spiritual people . . . [and], like people everywhere, need beliefs to explain and give meaning to the world they live in" (Richmond and Gestrin 1998). In writing about the Akan people of Ghana, Pobee (1979) makes similar observation:

> "The first thing that strikes one about Akan society's worldview is the fact that *homo Akanus* has a religious ontology" (Pobee 1979:44). In grounding the teleological and anthropocentric worldview of the African – that God is the Originator of the cosmos, and the Human is the center of this ontology – Opoku (1978:13) states "African traditional

> religion . . . represents our forefather's effort to explain the universe and the place of [hu]man in it in their own way . . .". African ontology is therefore a religious ontology. Likewise, the African cosmology is a religious cosmology. However, whereas "cosmological analysis [seeks] to discover what is true for this world, questions of the origin and structure of the universe – its creation or everlastingness – ontological analysis [attempts] to discover relations and distinctions which would be valid in any world".
>
> (Chalk 2006:68–69)

Diagrammatically, as depicted in Figure 5.2, three interlocking circles representing the natural world, the human world, and the spiritual world are used to graphically represent the interconnected relationships that characterize the African's cosmovision. It is a cosmovision in which the natural world provides the habitat for the spirits and humans and sends messages from the spiritual world to the human world. The spiritual world provides guidance, bestows blessings to, or exacts punishment on the human world. Humans are therefore duty bound to relate to the natural and the spiritual world (Millar 2004).

Since Africans live in a "holistic ecology involving both seen and unseen entities" (Millar 2005:137), interactions between the three worlds give rise to seven different constellations or ways of knowing. The seventh constellation, which represents an interaction between the natural, human (social), and

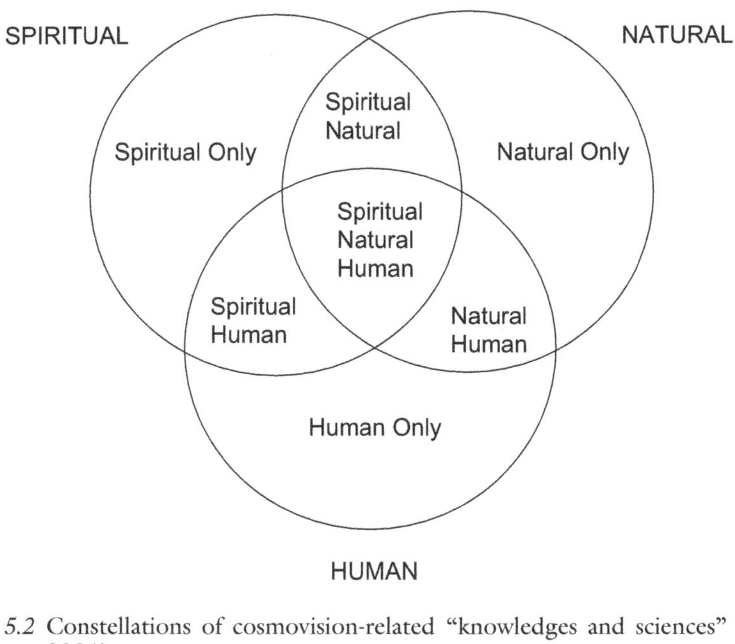

Figure 5.2 Constellations of cosmovision-related "knowledges and sciences" (Millar 2004).

spiritual worlds, creates balance. These constellations of knowing implicate a hierarchy of identities in the universe. Chalk (2006:22–26 cited in Mbiti 1969) identified five discrete categories of these identities that are consistently mentioned in African cultural and religious practices:

- God as the AllFather, the curator of the heavens and the earth and everything in between and beyond
- The spirits and other divinities made up of superhuman beings and ancestors
- [Hu]man, including human beings alive and those not yet born
- Flora and fauna or the remainders of biological life
- Phenomena and objects without biological life

The universe is also replete with vital forces (energy), and God is the source and ultimate controller of these vital forces. The spirits have access to some of this energy. Special human beings, such as Seers, Soothsayers, and Rain Guardians, have the knowledge and skills to tap into, operate, and use these vital forces, mostly for good but potentially for evil. In order to appease the divinities, rituals are performed, and sacrifices are made for occasions such as fertility of humans, crops, and animals; rainmaking, storm breaking, planting, harvesting, and thanksgiving (Millar 2004).

Though each of the three worlds stands independently, it is through the human world that we experience the physical and spiritual worlds. This anthropocentric view aligns with Aristotle's thinking that "the world would have existed if [hu]man[s] were not born into it, but it would not have been known to exist" (Jancar 1966:34; emphasis in original text).

The interlinkages of the three subworlds are critical to setting the stage for any knowledge development process – including here critical sustainability science or emancipatory science. All knowledges and sciences are incomplete if they do not satisfy all seven constellations, as shown in Box 5.1. The seventh constellation is the perfect state, which strives to be in balance or harmony with itself.

Box 5.1 The Three Circles Depicting the African Worldviews and Cosmovisions

Social/human world, spiritual world, and natural world – the interaction of the three worlds implies the following constellations of knowledges and sciences:

1 Knowledges and science resulting from social interactions only
2 Combination between the social and natural
3 Combination between the social and spiritual

4 Knowledges and science resulting from natural interactions only
5 Combination of the natural and spiritual
6 Knowledges and sciences resulting from spiritual only
7 Combination of social, spiritual, and natural

It cannot be emphasized enough that these constellations of knowing highlight the heterogeneity and complexities of African Sciences and therefore engender different bodies of knowledges and critical sustainability science options for the emancipation of knowledge. An understanding of these constellations sets the stage for appreciating and celebrating the ontologies of the African as manifest in critical sustainability sciences. Sustainability, in whatever form, is guaranteed by harmony, reverence, and the spiritual satisfaction of the African. The "first cause" or the ultimate goal is to live a fulfilled life, or simply, to "live well", which takes into consideration the spiritual, social, and material worlds at all levels of society. According to Delgado (2007), the spiritual world supports the integrity of human beings and reaffirms the principles and values that allow harmony, love, and complementarity. This ontological position makes a compelling case for broadening the actor/stakeholder categories in generating sciences and the desired emancipation necessary for sustainable knowledge generation and use.

So, what ontological discourses potentially nurture critical sustainability sciences? Can the Dagara concept of "Vurr" provide life-size examples, showcasing how each element in their cosmovision is expressed in ongoing processes of living, enabling the concepts to leap from the comfort of their abstract zone into the reader's bloodstream? For a world that is still struggling to come to terms with the pluriversality of knowledge, as against its presumed universality, how can the ontology of the Dagara of Southwest Burkina Faso and Dagara of Northwest Ghana – through the concept of *Vurr* – enable us to extend the "value chain" of knowledge? Can it provide a glimpse into the different ways of knowing and perhaps a glimmer of hope for once to challenge scientists to move out of the materialistic–positivistic comfort zone, push further the horizon of science, and extend the frontiers of knowing, and in so doing, advance the sustainability science agenda?

Dagara worldview and ontological thought

Dagara worldview and constellations of knowing

Dagara worldview is a typical indigenous worldview, which encompasses the spiritual, the natural, and the human worlds, as depicted in Figure 5.2. These worlds overlap and continually interact to give meaning to the universe and everyday decision-making among the Dagara. However, it is through the human world that the natural and the spiritual worlds are experienced. Different interactions between these worlds produce different constellations of knowing that have practical application.

An interaction between the human and the physical world motivates Dagara, who are typically smallholder farmers, to regard the earth as a living organism, which deserves to be treated with the highest consideration. Farming methods are therefore informed by the desire to at worst not hurt the soil and its organisms or degrade the physical environment, and at best enhance their value. This partly explains why Dagara smallholder farmers practice mixed-farming – keeping of cattle, sheep, goats, and poultry alongside crop production, and spread their risks by adopting mixed-cropping – combining cereals with legumes, root tubers, and vegetables (Millar 2018). Mixed farming and mixed-cropping, including crop rotation, "protects the integrity of the soil, and checks pests and diseases" (Bormann and Kellert 1991:67).

With regard to an interaction between the natural and the spiritual worlds, Dagara see certain diseases and adverse events as communication from the spiritual world through the natural world. This view is not dissimilar to Sackey (2006) and Azongo et al. (2015). In their study, Azongo et al. (2015:2) confirmed, "diseases are caused by invisible forces such as the gods and the ancestors". Snakebite on the farm or a destructive rainstorm will first elicit consultation with the ancestors through the medium of divination to see if the equilibrium has been breached before other measures are taken.

As far as the interaction between the human and the spiritual worlds is concerned, Dagara have a reverential relationship with the ancestors as intercessors through whom they gain access to the AllFather. This aligns with Azongo's (2014:12) assertion that "the Ancestors are rest-stops, not the destination". An interaction between all three worlds creates harmony in which the human being achieves inner peace, is at peace with neighbor and the natural environment, and align to her/his natural cause with the AllFather.

"Vurr" and "Dagara" ontological thought

The Dagara concept "Vurr" (Energy) – and its ramifications – is a philosophical belief that emphasizes the very importance of the notion of life being cyclical – *cyclical linkages between the living, the dead, and the yet unborn*, as depicted in Figure 5.3. Life is both cyclical and relational, that is, a relationship between the dead, the living and the yet unborn, and interaction between these three worlds generates or implies different constellations of sustainability.

A synthesis of our findings indicates that the following beliefs are prominent among the Dagara. "Vurr" is the precedence of life, life itself, and life as it exists in the future. Epistemologies beyond "Vurr" do not exist, hence "Vurr" is the essence of life, as the following illustrate. "Vurr" is an energy that gives opportunity for the existence of "Nyo-Vurr"; the nose of "Vurr" which is life, being, breath, fecundity, and the ability to reproduce, procreate, and re-create.

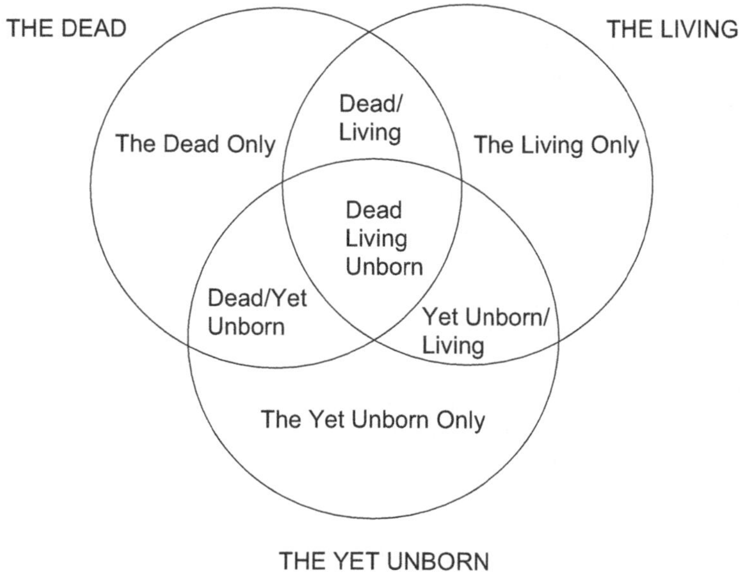

Figure 5.3 Onthological ramifications of "*Vurr*" and "sustianabilities".
Source: Authors' construction, 2021.

As the essence of life, other ramifications of the "Vurr" include the yet unborn: "Pug Vurr" (the womb), "Ma-Vurr" (the mother's life), "Bipila-Vurr" (the baby's energy or the baby's spirit). As far as the living are concerned, "Vurr" can manifest as "Vur-re" (breathing, has life); "Vur-ru" (has energy); "Nyo-Vurr Wogo" (long life); "Nyo-Vurr Song" (good life or living well, living a fulfilled life); and its antithesis "Nyo-Vurr Faar" (bad life). Then, as life grinds, inexorably, toward the end – "Che-Vurr" (shorten life/cut life); "Nyo-Vurr Baar" (finished energy/death), it is the end of a new beginning. Death is not an end; it is life in a different form, which means life and death alternate and occur in perpetual transition, thus "Nyo-Vurr Tegr" (substitution of spiritual life).

Life transitions into death, into the life of spirits, some of which could be bad, "Vurr-Nyuur" (Presence of Bad Spirits). But at that point, life, or death, assumes different abilities such as "Nyo-Vurr Ayi" (ability to survive in mysterious circumstances). And, just as a compass is aligned to the perpetual north, so is life aligned to its ultimate maker, "Nyo-Vurr Sob" (the AllFather/the chief of the gods).

It is important to note that the dead occupy a pride of place among the Dagara as uniquely placed between the living and the AllFather to intercede on behalf of the living. This is the *raison d'être* of Dagara elaborate rites and rituals in honor of the dead. They believe the dead can bestow countless blessings on the living, but the dead can also exact punishment for wrongdoing. Fear of the wrath of the ancestors is a strong restraining force against any form of harm to humans and degradation of the natural environment. Since Dagara believe life is cyclical and does not end at death, a comprehensive and cyclical understanding of sustainability is at the center of their ontology. This is reflected in a popular African adage to "treat the world well; it was not given to you by your Ancestors; it was lent to you by your children". This *manifestation is the perfect state, which strives for absolute sustainability.*

Above all, the Dagara ontology of the "Vurr" takes on a philosophical meaning as in "Nyo-Vurr Bedaar" (life cannot be bought). And this is where critical sustainability science comes in. Since life cannot be bought, it cannot and should not also be destroyed except, of course, by the one who created it, The AllFather. Human intellect and the indomitable human spirit are failing, and the resilience of nature is being overstretched, but Dagara still believe life will go on forever, as in "Nyo-Vurr Kubangbaar" (life without an end/eternal life). Thus, Dagara are among the people in the world who honestly believe there is hope for our world, the kind of hope that has been the crucial survival trait that has sustained our species since time immemorial.

But beyond the living, the dead, and the yet unborn, "Vurr" also manifests itself in everyday mundane behaviors and actions such as "Vurr-Nyor" or "Vurr- Tobe" (stubborn/holes in the ear), "Vurr-Nyee" (stupid/foolish), "Vurr Nour" (speechless/holes in the mouth), or "Vurr-Sang" (weave a mat). "Vurr" also depicts human character traits, both wholesome and unwholesome, for example, "Eire-me Nyo-Vurr" (extreme bother/disturbance); "Mmaa-me Nyo-Vurr" (extreme anger); "Vurr-tii" (persistence, perseverance); but also what is worth aspiring, "Nyo-Vurr Tomme" (life ambition).

So, the Dagara ontology of "Vurr", as it spans the world of the living, the dead, and the yet unborn, enables us to appreciate the cyclical nature of life through the lens of the Dagara of northwest Ghana and southwest Burkina Faso. Those who believe there is life after death behave differently on earth from those who do not believe there is life after death. But the question we should be asking ourselves now is not whether what the Dagara believe is true or false, but to see what practical uses we can derive from Dagara ontological positions since different beliefs about the world lead to different conclusions, behaviors, and actions that have significant implications for sustainability science. For instance, Dagara smallholder farmers, until recently, used compost and farmyard manure to fertilize their crops. The idea is to feed the soil, which then in turn feeds the crop, because their cosmovision enjoins them to treat the earth as a living organism. This sharply contrasts with chemical-based agricultural practices that focus on feeding the crop at the expense of the soil, which is informed by a materialistic worldview that fosters human domination of the earth. In the same vein, those who believe in life after death are more likely to hold the conviction that the way they treat others here on earth would determine how they are treated in the Hereafter.

What we distil from the discussion, so far, is that Dagara ontology is a spiritual ontology, anthropocentric and teleological. Their ontology is also respectful of different forms of life. The concept of "Vurr" has the potential to add a sliver of influence on our ability and willingness to leave behind a world that is worthy of our children and the children coming after them. And just in case, the Dagara have got it all wrong in their worldview, they err on the side of caution. So, in their quest to generate new knowledge, in what directions will sustainability scientists be inspired by this holistic cosmovision of the Dagara made manifest through the ontology of "Vurr"? Just as the earth rotates on its axis, ontology evolves on its praxis. What opportunities exist for building bridges across cosmovisions, cultures, ideologies, identities, histories, space, and time to strengthen sustainability science?

Dagara ontology and implications for critical sustainability sciences

Critical Sustainability Science (CSS) needs to be conscious of the people and the cultures in order to attain the goals so desired. The variables in CSS, such as human systems and environment systems, even when addressed in an integrated manner, will not attain sustainability of goals. Real sustainability and transformational sustainability can only be achieved when we move toward Culturally Centered Critical Sustainability Science (CCSS) or Transcendental Sustainability Science, as depicted in Figure 5.4.

A transcendental philosophical approach to research should underscore the dictates of conducting CCSS. Transdisciplinarity would then provide the basic methods, tools, and techniques for conducting research and development in an emancipatory manner – getting all stakeholders (including rural ontologically rich, but material resource–poor farmers) included and involved so as to own the processes in a collegial manner. The ontological unpinning of the Dagara phenomenology of "Vurr" makes it imperative to put spirituality at the center of human and environmental systems. Spirituality is *sine qua non* for CCSS; without a holistic analysis of actors and actor systems – both physical and metaphysical – marginalization and/or total exclusions do occur in objective setting and action targeting. This has dire implications for system sustainabilities.

Conclusion

Although Critical Sustainability Science (CSS) is a much more meaningful improvement on conventional sciences, it has succeeded in guaranteeing only a part (not a whole), a degree of sustainability that would not stand the test in time. It might lead to or be a potential for development conflicts of neglect. But the major contribution of Critical Sustainability Sciences is that it has pointed the way to more critical ontological debates. Even if we do not have to shift the goalpost, there is certainly the need to widen the frontiers of the sustainability science debates.

Figure 5.4 Cultural/spiritual context of Critical Sustainability Science from a transdisciplinary perspective.

Source: Authors' construction, 2021.

Cosmovision provides the compost for the sprouting of lifeworlds, but worldviews are in a state of flux. With the recent explosion of information available to people around the world and the advent of "the global village", the conflict between science and religion will magnify and intensify, and the exchange of ideas across the globe will cause people to question what they believe in light of new alternatives. This is because changes in a society are a function of, or a cause of, change in cosmovisions. The questions then are as follows: What is the future of belief? How wide can we keep extending the ontological debate? Will there be unity in a diversity of cosmovisions or will there be the unity of diverse cosmovisions into a global cosmovision? What does that portend for sustainability, and the future of critical sustainability sciences? Whichever way you look at it, the ontology of the "Vurr" of the Dagara in northwest Ghana and southwest Burkina Faso illuminates one pathway to culturally centered critical sustainability science linked to African spirituality and cosmovision.

Co-creation of sciences and coevolution of knowledges

Rightfully, we have argued that Critical Sustainability Science (CSS) (or emancipatory science for that matter) has a vital niche in contemporary science-based discourses. From an ontological point of view, we have argued for a modification/adaptation into Culturally based Critical Sustainability Science (CCSS) to basically deal with cultural difference and spiritual ramifications of peoples' knowledges. From a holistic science perspective, we thus recommend, going forward, a dialogue that will give rise to the *co-creation of sciences* (integrating the different expressions of sciences). We foresee a respectful dialogue under equal conditions, leaving open the ontological grounds

- to influence and be influenced,
- to strengthen and revitalize the marginalized sciences,
- to build a synergy and opportunities for mutual learning,
- to deal with the contradictions and exclusions of orthodox science, and
- to foster the emergence of completely new ways of knowing, or new levels of mutual respect and interaction between the two worldviews.

Such emerging co-created bodies of sciences would lead to plurality of dynamic knowledge bases that would be *co-evolving* in their utilization and in harmony with nature, in a transformational sustainable manner.

References

Al-hassan, S., 2011. The savannah accelerated development authority and poverty reduction in Northern Ghana: What are the people's expectations? In: *Center for Continuing Education and Inter-disciplinary Research (CCEIR) 5th Harmattan School Conference Proceedings*, February 2011, University for Development Studies, Tamale.

Apusigah, A. A., and Millar, D., eds., 2010. Endogenous knowledge and African development: Issues, challenges and options. *Proceedings of the First Africa Regional Conference on Endogenous Development*, p. 36. Bolgatanga, Ghana: CAPTURED Africa.

Azongo, T. B., 2014. *The Role of Divination in Health Seeking Behaviour Practices in the Talensi-Nabdam District of Northern Ghana*. Atlanta, GA: Scholar's Press.

Azongo, T. B., Yidana, A., and Der, E. M., 2015. Soothsaying as health seeking behaviour: Implications for medical treatment of diseases perceived to be supernatural. *Asian Journal of Humanities and Social Studies*, 3(1). Asian Online Journals (www.ajouronline.com).

Beauregard, M., et al., 2014. Manifesto for a post-materialist science. *Guest Editorial Explore*, 10(5), 272–274. Elsevier. www.sciencedirect.com/science/article/abs/pii/S1550830714001165 [Accessed 13 Feb 2022].

Black, M., 1952. The identity of indiscernible. *Mind*, 61(242), 153–164.

Bormann, F. H., and Kellert, S. R., eds., 1991. *Ecology, Economics, Ethics: The Broken Circle*. Binghamton.

Brundtland, G. H., and Mansour, K., 2010. *World Commission on Environment and Development (WCED), 1987*. Bonn, Germany: Our Common Future.

Chalk, J. P., 2006. *Genesis 1–11 and the African Worldview: Conflict or Conformity?* Thesis (PhD), Pretoria: University of South Africa.

Champagne, D., 2015. Centering Indigenous nations within Indigenous methodologies. *Wicazo Sa Review*, 30(1), 57–81.

Cornwall, A., and Jewkes, R., 1995. What is participatory research? *Social Science Medicine*, 41, 1667–1676.

Creswell, J. W., and Creswell, J. D., 2005. *Mixed Methods Research: Developments, Debates, and Dilemma*. Oakland, CA: Berrett-Koehler Publishers.

Delgado, F., 2007. Ancestral guidance in Africa's every day life. COMPAS. *Endogenous Development Magazine*. No. 1, December 2007.

Dewitt, R., 2004. *Worldviews: An Introduction to the History and Philosophy of Science*. Oxford: Blackwell Publishers.

Fann, D., Neuhauser, L., and Gibbs, P., eds., 2018. *Transdisciplinary Theory, Practice and Education: The Art of Collaborative Research and Collective Learning*. Berlin: Springer.

Filho, L. W., 2018. *Handbook of Sustainability Science and Research*. Berlin: Springer.

Gehman, R. J., 1999. *Who Are the Living-Dead?* Nairobi: Evangel Publishing House.

Honderich, T., 2005. *Ontology. The Oxford Companion of Philosophy*. Oxford: Oxford University Press.

Hunt, F., and Thornsbury, S., 2014. Facilitating transdisciplinary research in an evolving approach to science. *Open Journal of Social Sciences*, 2(4).

Imasogie, O., 1983. *Guidelines for Christian Theology in Africa*. Ibadan, Nigeria: University Press Limited.

Jancar, B., 1966. *The Philosophy of Aristotle*. In: Barrett, D. B. (Ed). (1982). *World Christian Encyclopedia*. Nairobi: Oxford University Press.

Kabisch, N., Qureshi, S., and Haase, D., 2015. Human – environment interactions in urban green spaces – A systematic review of contemporary issues and prospects for future research. *Environmental Impact Assessment Review*, 50, 25–34.

Kates, R. W., 2011. What kind of science is sustainability science? *Proceedings of the National Academy of Sciences*, 1087(49).

Kirk, J., and Miller, M. L., 1986. *Understanding the Validity of Qualitative Research*. Girard, KS: Little Blue Book.

Laan, D., and Vander, A., 1997. The ontology of impossible worlds. *Notre Dame Journal of Formal Logic*, 38(4), 597–620.

Lang, D. J., and Wiek, A., 2012. Transdisciplinary research and sustainability science: Practice, principles, and challenges. *Sustainability Science*, 7, 25–43.

Laszlo, E., 2006. A holistic worldview for a planetary civilization. In: Haverkort, B., and Reijntjes, C., eds. *Moving Worldviews: Reshaping Sciences, Policies and Practices for Endogenous Sustainable Development* (pp. 95–98). Leusden: COMPAS Series on Worldviews and Sciences 4.

Levin, S. A., and Clark, W. C, 2010. *Toward a Science of Sustainability*. Center for International Development. Working Papers 196. Cambridge, MA: Kennedy School of Government. Harvard University.

Lewis, C. S., 1952. *Mere Christianity*. www.samizdat.qc.ca/vc/pdfs/MereChristianity_CSL.pdf (Accessed 16.8.2016).

Lütteken, A., and Hagedorn, K., 1999. Concepts and issues of sustainability in countries in transition-an institutional concept of sustainability as a basis for the network. In: *Central and Eastern European Sustainable Agriculture Network First Workshop Proceedings* (vol. 61). Rome: REU Technical Series, 26–36.

Massari, M., and Pellegrino, V., 2019. Emancipatory social science today: Presentazione. *Quaderni di teoria sociale*, 1, 11–18.

Maxwell, J. A., 1992. Understanding the validity of qualitative research. *Harvard Education Review*, 62(3).

Mbiti, J. S., 1969. *African Religions and Philosophy*. Garden City, NJ: Anchor Books.

McKenzie, B., and Morrissette, V., 2003. Social work practice with Canadians of Aboriginal background: Guidelines for respectful social work. In: Al-Krenawi, A., and Graham, J. R., eds. *Multicultural Social Work in Canada: Working with Diverse*

Ethno-Racial Communities. Don Mills, Ontario, Canada: Oxford University Press, 251–282.

Millar, D., 2004. *Interfacing Two Knowledge Systems: Local Knowledge and Science in Africa*. Paper for the Compas panel in the conference: Bridging Scales and Epistemologies: March 2004. Tamale: Centre for Cosmovision and Indigenous Knowledge.

Millar, D., 2005. Endogenous development: Some issues of concern. *Ghana Journal of Development Studies*, 2(1).

Millar, D., 2018. *Farmers Experimentation: An Alternative Logic*. Accra, Ghana: AIT Consult.

Millar, D., and Derbile, E. K., 2018. *Operational Methodologies for Endogenous Development Research*. Accra, Ghana: AIT Consult.

Nicolescu, B., 2002. *Manifesto of Transdisciplinarity*. Translated from French by Karen-Claire Voss. New York, NY: State University of New York.

Note, N., 2006. Reflections about worldviews, the western worldview and intercultural polylogues. In: Haverkort, B., and Reijntjes, C., eds. *Moving Worldviews: Reshaping Sciences, Policies and Practices for Endogenous Development*. COMPAS Series on Worldviews and Sciences 4, Tamale: COMPAS, 83–94.

Olsen, M. E., Lodwick, D. G., and Dunlap, R. E., 1992. *Viewing the World Ecologically*. Boulder, CO: Westview Press. Volume: 13 (issue: 4), page(s): 245–246. Issue published: August 1, 1993. https://doi.org/10.1177/027046769301300491.

Opoku, K. A., 1978. *West African Traditional Religion*. Accra, Ghana: FEP International Private Limited.

Parrinder, E. G., 1969. *Africa's Three Religions*. London: Sheldon Press.

Patton, M. Q., 2002. *Two Decades of Development in Qualitative Enquiry: A Personal Experience Perspective*. Sagepb.com.

Pobee, J. S., 1979. *Toward an African Theology*. Nashville, TN: Abingdon.

Richmond, Y., and Gestrin, P., 1998. *Into Africa: Intercultural Insights*. Yarmouth, ME: Intercultural Press.

Rigolot, C., 2020. Transdisciplinarity as a discipline and a way of being: Complementarities and creative tensions. *Humanities and Social Sciences Communications*, 7, 100.

Rist, S., Bottazzi, P., and Jacobi, J., eds., 2019. *Rethinking Sustainability Science: Critical, Intercultural and Emancipatory Perspectives*. Temuco (unpublished report).

Sackey, B., 2006. *New Directions in Gender and Religion: The Changing Status of Women in African Independent Churches*. Accra: Rowman & Littlefield Publication.

Smart, N., 1983. *Worldviews: Crosscultural Explorations of Human Beliefs*. New York: Charles Scribner's Sons.

Subramanian, S. M., and Pisupati, B., eds., 2010. *Traditional knowledge in policy and practice: Approaches to development and human well-being*. pp. 272-293.

ThM, P. D., 2009. *Understanding the Importance of Worldviews*. Christian Apologetics Project. http://capro.info/understanding-the-importance-of-worldviews/ (Accessed 21.3.2018).

Wright, E. O., 2007. *A Framework for an Emancipatory Social Science*. Madison: University of Wisconsin.

Zhang, Z., Li, J., Zhu, P., Zhao, H., and Liu, G., 2018. *Modeling Multi-Turn Conversation with Deep Utterance Aggregation*. arXiv preprint arXiv:1806.09102.

6 Contributions of the notion of cosmosophy to the formulation of critical sustainability sciences

Jéssica Sepúlveda Pizarro

Cosmosophy a necessity for critical sustainability sciences?

In this chapter, we have proposed to investigate the notion of *cosmosophy*, as we understand that its contribution is vital for a reconceptualization of the term sustainability from a critical, emancipatory, and intercultural perspective; in fact, its relevance is such that it becomes a central dimension that articulates and gives meaning to the entire field of critical sustainability sciences. By way of context, it is striking that the more critical sustainability proposals that have emerged and that have dismantled the developmentalist and capitalist vision of the term still lack a cosmosophical perspective that allows understanding the phenomenon of life and the human being more comprehensively.[1]

Undoubtedly, this deficiency denotes the monocultural thinking present in the dominating dualist or materialist thinking on sustainability, which prevents it from being open to other ways of inhabiting, and hence feeling and thinking, the world. Thus, incorporating the cosmosophical dimension to critical sustainability sciences implies a program of decolonization, not only for non-Western cultures, but also for the West itself. The latter has forgotten those "pre-modern" cosmologies inspired by the intuition of an *anima mundi*. The "pre-modern" cosmologies were undermined by modern rationality, cutting the human being from its vital connection with the cosmos. Today, more than ever, the words of Emerson (2008) resound, written in the essay *Nature* (published in 1836), where he invites us to resume an original relationship with the universe that was lost:

> Our age is retrospective. It builds the sepulchres of the fathers. It writes biographies, histories, and criticism. The foregoing generations beheld

1 An interesting document to review is the *Manifiesto por la vida por una ética para la sustentabilidad* (2002), which, from a critical perspective, incorporates the principle of an "Ethics of being and the time of sustainability" where it recognizes the rhythms that mark natural and social history, taking a step toward the valuation of the cyclical time of renewal. In spite of this, it fails to establish a cosmological vision of sustainability.

DOI: 10.4324/9781003043577-6

God and nature face to face; we, through their eyes. Why should we not enjoy an original relation with the universe?

(Emerson 1836:4)

Emerson's call for an original relationship with the universe suggests the existential mutual influence between the human being and the universe, and that it goes beyond an understanding about the "functioning" of the cosmos. In fact, the terminology *cosmosophy* in its etymology is a notion composed of two Greek words: kósmos (κόσμος) and sophia (Σοφία), so its meaning alludes to a *wisdom of the cosmos*. This meaning puts the accent on the word wisdom, which unlike the word cosmology (*logos* about the cosmos) places us in the realm of transformative praxis. Undoubtedly, the term has resonances with that of *ecosophy*, the wisdom of the *oikos* (home), which is currently taking on significant academic weight through authors such as Arne Naess, Felix Guattari, Josef Estermann, and Raimundo Panikkar. Each one deals with different approaches with the term ecosophy. The last two authors stand out for nurturing their conceptualization with a fruitful intercultural dialogue that allows them to incorporate cosmic visions. Estermann (2013) suggests, based on an intercultural hermeneutic, an approach of Andean ecosophy that gathers the wisdom of the Andean people. It represents an alternative proposal for cosmic coexistence, where the human being is "arawina, that is, 'caretaker' or 'guardian' of the pacha and its cosmic order" Estermann (2013:6, own translation). Panikkar's (2008) notion of ecosophy is the fruit of intercultural dialogue between three religious traditions: Christianity, Hinduism, and Buddhism, adding, in addition, the interdisciplinary harmonization of theology, philosophy, and chemistry.

Undoubtedly, the life experiences of these authors become a reference for the formulation of critical sustainability sciences, where their ecosophical proposals cannot be understood without interculturality. In fact, as will be indicated later in the text, the fathers of environmental ethics themselves are influenced by non-Western cultural and religious traditions.

For our case, the cosmosophical approach is inspired by Raimon Panikkar's (2008) definition of ecosophy, which highlights in its formulation the subjective genitive of the term, that is, the wisdom *of* the Earth, taking a step toward the deanthropomorphization of the very notion of wisdom:

The wisdom of the Earth itself, of our habitat, of our dwelling place, which reveals itself to us once we are open to understand it, to surrender to the spell of what it is revealing to us. It is the wisdom of the Earth, not human expertise.

(Panikkar 2008, own translation)

The question that arises is about the pertinence to use both notions, *cosmos* (cosmosophy) and *oikos* (ecosophy). The two of them refer to the notion of dwelling. Rather than entering the discussion about the pros and cons of using them, we believe that both notions provide us with horizons of intelligibility that nourish critical sustainability sciences, yet in both notions, the ontological, anthropological, and ethical assumptions that are at the basis of our current crisis of civilization are removed. The choice of cosmosophy is, therefore, not due to an exercise of exclusion of the proposal of ecosophy. We rather want to highlight the vital experience with the cosmos that is so present in many people who inhabit the different continents, and who are, despite of their presence and relevance, tragically ignored by the hegemonic culture of the West.

The notion of cosmosophy is clearly inspiring and, at the same time, transgressive to the very foundations of our modern civilization. To highlight the scope of what has been said, in our opinion, the notion removes three major assumptions. First, it suggests that the cosmos has wisdom and that we as human beings are grounded in a cosmos that is seen as a living organism, a bearer of morality, and, therefore, also as source of ethical normativity in the conduct of humans. Second, and concomitant with the previous one, the animist vision removes that narrow vision of understanding the cosmos as a material entity and object of manipulation. In such animistic views, the cosmos is seen as being inhabited by beings of biological and spiritual nature. Moreover, the demarcation line between the material and the spiritual is not as clear as the modern dualistic approach; on the contrary, the earth is seen as material, sacred, and revered alike. Third and finally, cosmosophy situates the place of the human being in the cosmos, meaning we have to engage into a profound anthropological discussion, which, according to the philosopher Fornet-Betancourt (2007:14), means dealing with the following question:

> Shouldn't we promote an anthropological turn whereby the human being says goodbye to the protagonism of the activist seeing himself as the "sustainer" of the earth and recover its natural organicity as a "sustained being"?
> (Fornet-Betancourt 2007:14, own translation)

Clearly, this anthropological–cosmosophical vision is not foreign to that of indigenous people who self-define their identity in relation to the cosmos. Choquehuanca Céspedes and Rist (2024) in Chapter 3 speak of the *Cosmoser*, which is understood as an identity that transcends the narrowness of the modern *subject*, and that unfolds its process of self-realization in a cosmological becoming with all the beings of nature.

Having made this somewhat extensive introduction, is necessary in order to realize the deep significance of the incorporation of cosmosophy into the field of critical sustainability sciences, and to become aware that its scope is much greater than just the inquiry into the functioning of the cosmos, for the meditation of our readers, we will expand on the following topics:

- The wisdom of a living cosmos
- Cosmogenesis: the wisdom of liberation in a coevolutionary becoming

- A sacred and reverential cosmos: the cosmic responsibility of the human being
- A relational cosmos: inter-independence, beyond substantiality

The wisdom of a living cosmos

The intuition that the Earth is a living organism has been present in many indigenous nations and people and even today, as we will see subsequently, it is once again resurfacing. Panikkar (1999:165) summarizes this intuition as follows:

> Earth is alive. She is the mother. The union between heaven and earth generates all creatures: she gives them life and sustains that life. Life is not only humans' privilege, yet humans participate in the life of the universe. Human beings, precisely because they live, have been called a microcosm. The model is the macrocosm, not the other way around, and that macrocosm is a living being.
>
> (Panikkar 1999:165, own translation)

This intuition of the earth is accompanied by its attributes of fertility and giver of life: it is the *terra mater*, a symbol attested by scholars of religions such as M. Eliade.[2] An example of this is given by the Mapuche people, in whose lives this experience plays a central role (Curivil 2020:44):

> We belong to the land because we live in harmony with it, *Mapu* means land and *che* people or person. We belong to the land because the land shelters us every day and we say that we emerge from the land because our conception of land, according to our ancestors, is that we are born from the land and Mapuche means that.[3]
>
> (Curivil 2020:44, own translation)

It is interesting to note that the symbol of the *terra mater* is not alien to the original meaning of the word nature in Western thought: its etymological origin comes from the Latin word *natura* and the Greek word *physis*. Both are related to the act of being born. In fact, if we dig deeper into the original meaning, the word *physis* for the Greeks was an all-encompassing concept of reality that contemplated three elements: "a) *process*, becoming or development; b) the starting point or beginning of the process, *arche*; c) the consummation or fulfillment of a process, that is, the structure of reality, *cosmos*" (Calvo Martínez 2000:32, own translation). That is to say, the term *physis*

2 On *terra mater* studies in M. Eliade, see Chapter III, "The Sacredness of Nature and Cosmic Religion", in his book *The Sacred and the Profane* (2017).

3 The preceding quote is from Tránsito Segundo Lipan, interviewed by the Curivil (2020:44, own translation).

contemplated the entire universe, seen as nature (the structure of reality) and affected by movement (principle and process). Therefore, the conception of nature as an inert entity is a great mutilation of its original meaning, which has brought, as a consequence, the negative mutation of the symbol of *terra mater* to *terra nullius*, which means denying its regenerative capacity and the rights of the original inhabitants, since they are considered as "lands without people, empty, barren and underutilized" (Shiva 2001:69, own translation), justifying within a colonial and patriarchal culture its usurpation and exploitation.

Prior to this tragic outcome, what is certain is that in the Greek, medieval, and Renaissance worlds, a cosmological model loaded with an animistic imprint prevailed. This model is rooted in ancient Greece, with the cosmology of Plato's *Timaeus* having a great influence on Western thought. From this cosmological model, we are interested in highlighting three ideas that serve as elements of understanding of our notion of cosmosophy: the wisdom of the cosmos related to an eternal order, the animistic vision of the cosmos, and the relationship between cosmology and anthropology.

We have seen that the notion of physis in the Greeks implicitly carries the notion of cosmos, as the structure of reality, but this "structure" is not reduced to a reified and mechanical vision; on the contrary, its meaning is that of a "beauty resulting from an order" for Brague (2008:39–38, own translation). The same author specifies that the term cosmos appeared thematically in ancient Greece and that its etymology is still unclear, but its sense of order ("good order") and beauty ("adornment") already appears in the *Iliad*. He also adds that the term is found in various pre-Socratic authors, Pythagoras being the first to call *cosmos* the gathering of all things because of an order.

The interesting thing about the original meaning of the Greek cosmos is that it has relevant implications at the anthropological and cosmological levels, since it shows an order that precedes human activity and, in turn, influences it; therefore, Heraclitus of Ephesus states in his fragment 1 and more concretely in fragment 51 the following:

> This order of the world, the same for all, was not made by God nor by any man, but was always, is and will be; fire always alive, lit according to measures and extinguished according to measures.
>
> (quoted in Bernabé 2010:135, own translation)

In the preceding quotation, reference is made to a self-sufficient totality whose order is established in the idea of "measure". Heraclitus' idea of a cosmic order is reaffirmed in fragment 1 with the presence of a unique and common *Logos*, which is very different from the *logoi* of people who are unable to grasp the true reality. Therefore, for Heraclitus, it is the sage who grasps the *Logos* and gains access to true reality and true knowledge. Thus, the idea of cosmic order is associated with a wisdom, a *Logos*, which has relevance in the life of human beings if they can grasp it.

Concomitant to Heraclitus' idea of an autarchic cosmos, Plato (2010) in his work *Timaeus* elaborates his discourse on nature, making explicit his animistic theory of the world, which had great influence in the West. Plato there indicates: "[30b] We must say that this world, which is truly a living being endowed with soul and intelligence [30c] was generated by the rational foresight of god" (Plato 2010:203, own translation). Therefore, the cosmos is a living being that possesses a body, constructed of the four elements, and a soul, endowed with reason. The soul is the intelligent cause that produces order and is responsible for subduing the errant cause that produces disorder in the world. Both principles are present in Plato's cosmological becoming, understanding order as better than disorder, thus guiding a teleological movement toward order in his living universe.

Another interesting aspect to highlight, in the animistic cosmology of *Timaeus*, is the relationship between macrocosm and microcosm, where the human being is a cosmic being. This relationship is configured by the similarity of elements with which the soul of the world was created with the immortal soul of the human being,[4] determining the relationship of correspondence between the macrocosm and microcosm. However, Plato recognizes that this relationship of correspondence is affected in the becoming of the human being by the action of the errant cause, which is that principle that tends to disorder and disharmony, for example, the diseases of humans. Undoubtedly, this approach has much more resonance with the cosmologies of indigenous and oriental people, in which diseases are conceived as the product of transgressive actions to the laws of the universe.

In short, Plato's cosmology refers us to a "teleological and mythical-theological image of the cosmos" (Gómez-Heras 2010:106, own translation), charged with an animistic imprint, since Plato's universe is an animated whole. However, his metaphysical conception will lead him to give priority to the animist element over the corporeal, since matter has an ontological precariousness. In this sense, the normativity of nature is given by an ordering intelligence (transcendental ideas) that impels to the excellence of things.

In opposition to Plato's vision, his disciple Aristotle (2008), who, not following the idealistic steps of his master, proposed an immanent teleological interpretation of nature, pointing by it to those identities "that possess the principle of movement in themselves and by themselves" (Aristotle 2008:215, own translation). Without deepening the great differences between these two great thinkers, we can infer that in the classical Greek world, an animistic vision of the cosmos and nature, whose structure of a cosmic order was relevant to human life, has prevailed. For Brague (2008), the Greek cosmological model referred to the idea of order, and its influence will be such that – together with

4 One should remember Plato's tripartite theory of the soul: intellect, irascible, and concupiscible, where the latter are mortal and are affected by the errant cause, that is, disorder and disharmony.

the Abrahamic model – it will dominate the thought of the ancient and medieval[5] worlds. It is this animistic cosmological model that is present in the approach of the ***anima mundi***, which had influence, with diverse interpretations, until the Renaissance. The further implications of this concept will not be explained in this work. The important thing is that this model was exterminated with the birth of modern science, in which nature became an inert object and the human being lost his vital connection with it: the modern autonomous and sovereign subject was born, but with it, it lost its identity as a microcosm.[6]

Cosmogenesis: the wisdom of liberation in a coevolutionary becoming

Science, especially physics–mathematics, has been gradually conceiving in its discoveries a new animated image of the physical world. For Hathaway and Boff (2014:37–38), this new understanding of the cosmos is reminiscent of an earlier cosmology, still common among many indigenous people, where the universe is a living organism, but which, unlike traditional cosmologies, the universe is conceived of as evolving and not as a static or eternal entity.

This change of conception assumes the temporalization and evolution of nature within the framework of a morphogenetic explanatory theory that is opposed to any mechanistic and closed conception of nature. This temporalization of nature burst with force in the philosophical and scientific thought of the 19th century and its main task "consisted in giving an interpretation of natural dynamics and organization that took into account the reality of time; in situating forms and laws in an *open universe*" (González 2004:190, own translation).

This resonates with the work of the biologist Rupert Sheldrake (1994), who saw, in the same century, the introduction of the concept of *field* as a principle of self-organization of physical reality and the concept of universal *energy* as a unifying principle of reality. As a result, both concepts, field and energy, have come to constitute the basis of the conception of physical reality in contemporary scientific thought. In addition, there is the recognition

5 Undoubtedly, there were currents of thought that also questioned this cosmological model, such as the Gnostics and the atomists. These currents of thought will not be entirely rejected because processes of syncretism will occur, as, for example, the gnosis present in the revealed religions (Brague 2008).

6 It is important to point out that the triumph of the mechanical vision to the detriment of the animistic and organic vision of reality took place in a cultural panorama, characterized by the influence of the Hermetic tradition that had a resurgence mainly in the 16th century. In addition, as a counterpart to the mechanistic conception of nature that gestated in modern science, the movement of Romanticism, which developed, in the mid-19th century, as a critique of the rationalist thought of the Enlightenment, played an important role. In its influence, an appreciation of the organic, even mystical conception of Nature can be appreciated; despite this, causality prevailed as the dominant, mechanistic, and conception of nature.

of indeterminism and chaos in complex processes of reality, and, with it, the affirmation of a degree of spontaneity and freedom inherent in nature. This reanimation of the physical world has led this part of science to repropose teleological explanations, which were formerly invalidated by the mechanical worldview of Newton's physics. In this sense, theories that explain intentional features of living organisms are re-emerging, as it is also in the case of James Lovelock's Gaia hypothesis. For Sheldrake (1994), the reanimation of the physical world is evidenced in the following characteristics, assigned to nature by contemporary science:

> Now nature is again considered to be self-organizing. Instead of the soul of the universe and the other types of soul it contained, the basis of this self-organization is now attributed to the universal gravitational field and the other types of fields it contains. Indeterminism, spontaneity, and creativity have re-emerged in the natural world. Immanent purposes or ends are now modelled as attractors. And beneath it all, akin to an atomic underworld, is the inscrutable realm of dark matter.
>
> (Sheldrake 1994:107, own translation)

Undoubtedly, the new animistic image of contemporary science is not exempt from controversy, as has been Lovelock's Gaia hypothesis itself; in fact, many of the scientific explanations for a living planet are based on theories of complex systems of self-organization and freedom (Lovelock 2000). Beyond the divergences raised by an animism that brings us the problem of intentionality and consciousness in the cosmos, it is true that there is a certain convergence related to the dismissal of the dualistic metaphysical explanations between soul and body, which is very much present in the intuition of the *anima mundi*. Thus, in Sheldrake's thought, the soul of the universe becomes a morphic field of self-organization, and the intentionality toward an end becomes an attractor. The latter remains a mystery to the human intellect since the universe is in evolution.

The question arises whether in this open universe, it is possible to conceive the presence of a cosmological wisdom, that is, a cosmosophy with relevance in human life, as it was thought in the Greek cosmological model and in many traditional cosmologies still in force. For Hathaway and Boff (2014), the answer is affirmative, but their approach to such wisdom is through the ancient Chinese word *Tao*, as they point out that:

> The Tao can be understood as a principle of order that constitutes the common foundation of creation: it is the fluid structure of the universe, which cannot be described, but only proved. The Tao is the wisdom that resides at the very heart of the cosmos, and which contains the essence of its purpose and direction.
>
> (Hathaway and Boff 2014:340, own translation)

For these authors, cosmic wisdom manifests itself as an "attractor", which they call the hidden wisdom of the *Tao*, the driving force behind the unfolding of reality. This unfolding denotes a finality, which is not predefined, because it evolves. In this context, it is important to point out that when referring to evolution, for many authors it does not necessarily manifest in linear time. This is the case of Panikkar (2012:88) who, in his conception of time and space, gravitates the human and cosmic experience in the temporal density of the present; hence he points out, "we are no longer *viatores* (travellars). We have become *comprehensors* (complete, perfect, omnicomprensius)" (Panikkar 2012:88, own translation).

Thus, the realization of the human being occures through aligning to the wisdom of the *Tao* and "finding our unique contribution to the writing of cosmic history" (Hathaway and Boff 2014:344, own translation). This description of cosmic wisdom as *Tao* does not cease to be somewhat cryptic, even more so for the modern subject, who rather than being affected by life, is more interested in "producing" it. However, in the light of contemporary science, we can glimpse dynamics that account for the cosmic history of evolution and that can be translated into a wisdom in which the human being, together with other living beings, participates in a common destiny. A suitable framework for Hathaway and Boff to interpret this wisdom is the cosmogenic principle that Brian Swimme and Thomas Berry (1992 cited in Hathaway and Boff 2014) outline in their book of poems *The Universe Story*. According to the universe story, there are three key characteristics that mark the cosmic history of evolution: differentiation, autopoiesis, and communion. The existence of a structured cosmos depends on the manifestation and balance of these three characteristics (Berry and Swimme cited in Hathaway and Boff 2014):

> If differentiation did not exist, the universe would collapse into a homogeneous blob; if subjectivity [or autopoiesis] did not exist, the universe would collapse, become an inert, dead extension; if communion did not exist, the universe would collapse into ontic singularities.
>
> (Berry and Swimme cited in Hathaway and Boff 2014:344, own translation)

Undoubtedly, this cosmogenic principle is based on relational ontologies and on the paradigm of inter-independence,[7] which are the basis for a critical and comprehensive notion of sustainability, since specialization helps ecosystems to be more interdependent, as pointed out by the science of conservation biology. Hence, the biocultural diversity is good for the health of ecosystems and societies. In fact, taken to the political and ethical plane, this principle is

7 Inter-independence is a term used by R. Panikkar and will be dealt with in the fourth section.

found in Shiva's (2006:142) approach to an "Earth Democracy", according to which living societies, living cultures, ecosystems, and living organisms are characterized by the three characteristics of the cosmogenic principle: (1) diversity, (2) self-organization, self-regulation, and self-renewal, and (3) reciprocity among systems. The relationality of these three characteristics is explained as follows:

> Our diversity makes mutuality and a culture of mutual concession possible. Mutuality makes self-organization possible. Profoundly autonomous, self-organized and, at the same time, deeply interconnected human beings (with the Earth, with all species and among themselves) are creating the conditions for their future survival. An Earth Democracy is thus being reborn, even immersed as we are in the violence and war that surrounds us.
> (Shiva 2006:142, own translation)

This cosmogenic principle leads us not only to rethinking the political and ethical community, but also to review the very notion of liberation, generally restricted to the human. In this way, cosmic wisdom implies the unfolding of the life of all beings that inhabit the universe, and that, therefore, the realization of each human being, carries with it the destiny of the entire community of life in the cosmos; hence, each human being is related to a cosmic responsibility that weighs on the action of each human being, and, therefore, this action cannot be confined to a restrictive time and space. In the current age of the necrocene, marked by systematic ecocides and genocides, it becomes more evident and necessary to recover this wisdom of liberation to repair the damage inflicted on our peers and the planet. Therefore, wisdom implies the experience of liberation, understood from a cosmic perspective. In this context, in the words of Hathaway and Boff (2014), individuals and human societies are liberated to the extent that they

- become more diverse and complex, respecting and highlighting differences;
- deepen the aspect of internalization and awareness, fostering creative processes of self-organization, and
- strengthen their bonds of community and interdependence, including communion with the general community of life on Earth.

(Hathaway and Boff 2014:23, own translation)

The process of cosmogenesis does not help to understand that the notion of wisdom of liberation from a cosmic perspective carries with it the idea of coevolution; therefore, such wisdom is accompanied by the ideas and images of a community of interdependence, where all beings participate in a great collaboration that is life. In this community, not only beings of material nature are participating, but also spiritual beings, yet they are part and apparently also fulfill cosmological functions. The relevance of this is further detailed in the following section.

A sacred and reverential cosmos: the cosmic responsibility of the human being

We have seen that contemporary science has initiated in its discoveries a process of reanimation of nature that abandons the mechanistic parameters to explain life. Today, it is known that nature possesses degrees of freedom and self-organization, as explained in the previous section. This paradigm shift makes us think of nature without or with a divine dimension. If the latter would be true, it means to think in the divine, not so much in the traditional image of transcendence, but to discover it, in the immanence of life. Thus, the challenge is to abandon the dualistic and substantialist visions that make the cosmos and the divine two separate realities.

Such a discussion is undoubtedly alien to most religious traditions, as Eliade (2017) makes explicit in his studies of the histories of religions. He shows that the phenomenon of the desacralization of nature is a recent phenomenon, proper to human beings and modern science. Contrary to this vision, Eliade notes that the *homos religiosus* nature is never only "natural" because it is always loaded with a religious value that transcends its materiality, and that makes it sacred as it is a divine creation. In the case of cosmic religions, the sacred is revealed in the structures of the world, and in its diaphanousness, it unveils the multiple aspects of the sacred (Eliade 2017):

> The sky reveals directly, "naturally", the infinite distance, the transcendence of the God. The Earth, likewise, is "transparent": it presents itself as mother and universal wet nurse. The cosmic rhythms reveal order, harmony, permanence, fecundity. The cosmos is at once a real, living and sacred organism: it discovers at once the modalities of being and sacredness. Ontophany and hierophany meet.
>
> (Eliade 2017:87, own translation)

Eliade's strong historical data on this vision of the sacredness of nature, present in many cultures, is striking in current times in which the moral foundations for environmental/ecological ethics are being sought. It is not unremarkable, as Baird Callicot (2017:67–68) points out, the existence of the great influence of Eastern philosophy exerted on Western environmental philosophers. The spiritual richness of Hinduism has been a source of inspiration in environmentalists such as R.W. Emerson, H. Thoreau, and A. Naess, among others.

This background brings us back to the original discussion of a non-dualistic vision between the divine and nature. It is clear that a sacred mountain, as many people venerate, corresponds to a vision that transcends matter, but does not dispense with it either. That is to say, there is a relationship that unites matter with a reality that transcends it. This transcendent reality has been called by scholars of the religious phenomenon as mystery, power, sacred, suprareality, to mention some denominations, and that, as such, are characterized by their "in-objective" nature that manifests itself as a symbol that carries with it the dialectic between power and form; hence a sacred mountain is a symbol charged with reality. The same Greek word *symballein*, from which symbol comes, means to unite these two orders of reality. Therefore, eventually, and from a dialectical viewpoint on the sacred and the profane, they do not represent two realities, since the sacred, which is power and potency, is the reality par excellence, which ontologically founds the world, and which is therefore far from that profane reality that is volatized in a subjective relativism.

The understanding of the sacred and symbols leads us to an anthropological discussion since such experiences occur in human consciousness. It is then necessary to ask ourselves about an anthropology that shelters this type of experiences, that is, that allude to a transcendent order of reality. For Panikkar (2007:41), the challenge for the modern conception of the human being, marked by the body–soul dualism, means to return trinitarian anthropologies, present in many cultures, still today, and which refer to the triad: body, soul, and spirit. Moreover, the spirit is not an exclusive feature of the human being; but it is the element that allows the state of consciousness of the sacred to become manifest. Therefore, the sacred can never be manipulated by the whims of the human being.

On the other hand, and in order to avoid confusion, which is typical of dualistic ontologies in which we might think that the spiritual is a separate "part" of this anthropological triad, Hathaway and Boff (2014) indicate that, from a holistic vision, spirituality accounts for a way of being of the person that allows the experience of depth of the human being, understood as the ability to go beyond the perception of the mere appearances; it means to stop seeing the appearances as things only, but understanding them as symbols of a reality that transcends the appearances. For these authors, spirituality is a moment of consciousness in which we not only experience the value of things, but through which "we become aware of the totality and of ourselves, as part of that totality" (Hathaway and Boff 2014:380, own translation). They further argue that this experience is present in many religious traditions that experience non-duality, and that binds and relinks them "one to another and to all beings, making them a cosmos instead of chaos, a symphony instead of a cacophony" (Hathaway and Boff 2014:380, own translation).

It is from this experience of the human being's "re-ligation" with the cosmos, with the totality and the source of life, that we can understand the intercultural symbol of *Terra Mater*, whose symbolic force is predisposing the human being to the veneration of the sacredness of earth. As we have already indicated, this symbol is very much present in the Mapuche people, where the earth is not only a physical place, but also encompasses an immaterial dimension, and this is indeed so, because of their spirituality that enables them to perceive the earth as sacred, a temple to be venerated, as emphasized by Curivil (2007:40, own translation): "according to the Mapuche religious mentality, nature is 'the temple' par excellence".

Such a vision undoubtedly conflicts with the imposition of development projects in indigenous territories. An example of this is the resistance of the Mapuche Williche communities against the imposition of private businesses and the state of Chile for building a hydroelectric plant on the Pilmaikén River. For the extractivists, the building place is only a space that offers "natural resources", while for indigenous communities, it is a "sacred ceremonial space where the ngen mapu kintuante, protective spirit of the crops, of the lawen, the Mapuche medicine, dwells" (Manque 2016:53, own translation).

Unfortunately, this clash of visions is framed in a historical context of dispossession where the Chilean justice administration, both in the past and at present, has systematically and improperly criminalized indigenous movements in the southern macro-region of Chile, with harmful practices of racism, which expresses itself in symbolic, judicial, and physical violence.[8]

Another fundamental part to recognize in the sacralized vision of earth is the understanding of the human being as a ritual being since the sacred vision of the earth predisposes him/her to a reverential attitude toward the earth. This attitude has a much greater scope than the simple respect toward the earth and to assume a responsibility of "friendly decisions with the environment"; on the contrary, in the rite, the human being reactivates the primordial cosmogonic moment and therefore assumes responsibilities in the cosmic plane, starting to recollaborate with the Gods in the restoration of the balances of the world. This restoration of equilibrium is what allows the cosmos to be an order, a dwelling place, and thus not to precipitate into chaos. Thus, for Byung-Chul Han (2020:12, own translation) "Rituals can be defined as *symbolic techniques of home installation*", thereby making the world habitable, quite contrary to what happens in neoliberal societies, in which "The permanent pressure to produce leads to a loss of home" (Byung-Chul Han 2020:18–19, own translation). The installation of home means that rituals generate a community of resonance which is capable of harmony, of a common rhythm:

> Rituals create axes of resonance that are socioculturally established, along which one can experience vertical (with the gods, with the cosmos, with time and with eternity), horizontal (in the social community) and diagonal (referring to things) resonance relationships.
>
> (Rosa quoted Byung-Chul Han 2020:22–23, own translation)

8 To be informed about the systematic action of the Chilean Justice in the Walmapu, I recommend the following reading where reference is made to the prisons of ethnicity: "Understanding justice and injustices in Wallmapu through intercultural studies as proposals of con-vivir" (Le Bonniec and Payas 2021, own translation).

We see that the community of resonance is not restricted to the human, since rituals are characterized by this triple relationship, where the symbols used will always refer to the transcendental. Moreover, this vision of a resonance community already settles us in relational ontology, and as explained by Byung-Chul Han (2020:23), resonance is not an echo of the self, but it is harmony; therefore, the dimension of the different is inherent to it. In tune with this approach, Hathaway and Boff (2014) emphasize that spirituality "creates space for the logic of coexistence, cordiality and veneration before the unique reality – the otherness – of each being, and before the communion with all things and with God". (Hathaway and Boff (2014:382, own translation). This idea of a community of resonance, where harmony and difference are relevant, allows us to move on to the next theme.

A relational cosmos: inter-independence, beyond substantiality

We have seen that the cosmogenic principle developed by Berry and Swimme (1992 cited in Hathaway and Boff 2014) has three characteristics also applied to the ethical–political plane in the proposal of the Earth Democracy of Shiva (2006): differentiation, autopoiesis, and communion. These three characteristics place us in the emerging paradigm of inter-independence, where the relationality of life is postulated, converging in it processes of individuation and relationality. Moreover, we have seen that the sacred dimension of the earth and of life places us in a profound experience of re-ligation and coexistence that is the basis of the postulate of relationality.

But what does this ontological relationality mean? This idea is anchored in a *vital experience and not in a logical argumentation*. Moreover, we can observe that the discourses of relational ontologies are posed in normative terms they aim at a radical *metanoia* in order to advance toward them and overcome the dualistic ontologies that have had dire consequences in our present time of necrocene.

The importance of interculturality, a fundamental pillar of critical sustainability sciences, helps us to transcend our horizons of intelligibility, limited to our cultural tradition where dualistic ontologies predominate. An example of this is provided by Estermann (2013:6), a Swiss-Andean philosopher, who explains that for the people of the Andes, the principle of relationality is present, yet there is an equivalence between "relationship" and "life"; they, therefore, often say "without relationship there is no life". Therefore, interculturality is forged as an ethical imperative in current times, and if complemented with the relational cosmology of contemporary science, it can help us to advance in this radical *metanoia*, or to "an ontological reconversion", in the words of Escobar (2014), who also states that this path is not easy, since entering the field of relationality implies "transforming our daily practices and spaces according to such an inspiration" Escobar (2014:59, own translation).

A first difficulty posed by a dualistic ontology is to think of relational ontologies in terms of pluralism. In fact, for Escobar (2014), a key concept in this regard is the notion of pluriverse, pointing to the argument that "there are many ontologies or worlds that, although ineluctably interrelated, maintain their difference as worlds" (Escobar 2014:58, own translation). Without any doubt, this assertion is very complicated for our modern epistemology, and even for a field as important as ethics because a quantitative interpretation of the pluriverse with a foundation of an atomistic reality necessarily leads us to an ontological and moral relativism.

Therefore, relational ontologies carry implicitly the tense relationship between the one and the multiple, but not from a logical argumentation, but, above all, from the vital experience in the territories in which human beings meet themselves in nature. If we think, in the same experience of inter-independence, which is proper to relational ontologies, the concept embraces both, the conscience of diversity and unity. This leads us to ask, whether a new epistemology is required to understand, from experience, this relationship of unity and diversity, as an interplay of polarities and not as two exclusive entities?[9]

Moreover, many questions arise regarding the approach of relational ontologies, some of them are of the logical and metaphysical order; this is clear, since in these ontologies the concept of substance is being questioned in all its radicality. Escobar (2014) points out "that all things in the world are made of entities that do not pre-exist the relations that constitute them" (Escobar 2014:59, own translation), that is, the relation is primary in the constitution of beings. The same argument referring to the ontological emphasis of the *relation* is found in one of the principles that postulates the movement of "deep ecology" of Naess (2007:9), which makes him reject the image of *man-in-environment* in favor of a relational image, of *field-total* as he calls it, where he "considers organisms as knots in the biosphere network or field of intrinsic relations". This idea that he will emphasize in his mature thinking referred to his proposed concept of ecosophy (Naess 2018) as follows:

> Relationism has ecosophical value because it facilitates the undermining of the belief in organisms or people as something that can be isolated from their environments. To speak of interaction between organisms

9 This experience, which implies the relationship between the one and the multiple, Panikkar (1990) bases it on the experience of a-duality (Hindu *advaita*, neither one nor two), the trinitarian experience of Christianity (the three persons of the trinity, neither one nor three) and the experience of harmony, which is capable of unifying without phagocytizing plurality. In these experiences, the author tries to overcome monism and dualism. In the case of monism, reality is one and plurality is secondary. This vision, in the political plane, we can see it when the figure of a state subjugates diversities, even for a "common good". On the other hand, when faced with the argument of a monism that unites spirit and matter, and thus overcomes a dualistic ontology that has fatally conceived them as separate realities, Panikkar refutes that they are not one, since they are different, but neither will he say that they are two, since they are interrelated. For a deeper understanding of Panikkar's thought on this matter, I recommend his book *Sobre el diálogo intercultural* (1990:28–34).

> and the environment gives rise to erroneous associations, for an organism is interaction. Organisms and environments are not two things – if a mouse were placed in an absolute vacuum, it would no longer be a mouse. Organisms presuppose the environment.
>
> (Naess 2018:102–103, own translation)

Panikkar (1996) bases the idea that reality is relationship on three religious traditions: the trinity of Christianity, the *advaita* of Hinduism, and the intuition of the *pratīyasamutpāda* of Buddhism. In all three traditions, a relational thinking prevails where reality is neither reduced to one (monism) nor to two (dualism); entities are not substance "in themselves" but relationships. In order to go deeper into this approach, which leads the author to the radical affirmation that reality is relation, we will see his arguments in relation to Buddhism. Etymologically, the Sanskrit word *pratīyasamutpāda* seems to indicate for the author the following:

> A reciprocally conditioned origination by virtue of a cosmic functionalism. Moreover, the word indicates that it is a single network of connections between the various elements, that is, a universal concatenation, rather than a substantialist causality or an ontological causality.
>
> (Panikkar 1996:120, own translation)

In this sense, the intuition of *pratīyasamutpāda* is the intuition of pure contingency, understood in its pure positivity: mutual dependence, radical interdependence, without any other external support, and, therefore, disconnected from any foundation. The experience of contingency is radical and at the same time devastating, since it is discovered that nothing has ontological consistency (including the subject that discovers it) and that all eagerness to substantialize being is useless, because there is no such being: "Every phenomenon is empty (śunya), it is only in relation and dependence (pratīyasamutpāda): pure dynamism" (Panikkar 2007:233, own translation). He emphasizes at this point the radical relativity of each being: each being is precisely what it is, in intimate and constitutive relation with other beings. In other words, beings "are" in their relation of inter-independence with other beings, that is, "things 'are' insofar as they cease to be in order to give themselves to others; they are nothing but mutual constitutive relations; there are no monads without any openness" (Panikkar 1996:244–245, own translation).

This calls us to a *relational thinking*. For Panikkar (1996), the difficulty of our reason is found in its thinking substantively; that is, substance has been considered the basis of "relation". Relation is not a "thing" that depends on previously existing "others"; but on the contrary, things are nothing but relation, their "is" is conditioned to the relation with others. Relations are always polyvalent and polyhedral: a mother "is" not only in relation to her child, but she is also a wife, a citizen, a colleague, etc.; hence, a mother "is nothing but radical relativity: in itself it is nothing" (Panikkar 1996:237, own translation).

And in this sense, relational polyvalence allows us to discover that no "being" is totally explained by a limited number of relations since radical relativity "is the constitutive openness of the whole universe in all its relations" (Panikkar 1996:238, own translation).

Panikkar (1996) postulates that another relevant aspect to consider in the primary vision of relationality is that the concept recovers the word *harmony*, lost in modern thought due to its substantialist conception of reality. The primary vision of relationality does not account for a numerical reality that can be reduced to unity, nor to duality. Any quantification of reality destroys uniqueness. For Panikkar (1996), this uniqueness is harmony and the introduction of *love* as the ultimate level of reality is what enables this vision. Applied to the world of science, we agree with the ecologist A. Leopold's (2005) critique who is pointing to science blindness to perceive harmony and its practice of dismembering reality through specialization. For the author, harmony is not only the field of poets (Leopold 2005):

> There are men who have been assigned the task of examining the construction of the plants, animals and soils that constitute the instruments of the great orchestra. They are called professors. Each of them selects an instrument, and then spends his whole life taking it apart and describing its strings and sounding boards. That process of dismemberment is called research. The place of dismemberment is called a university.
>
> (Leopold 2005:130–131, own translation)

Along the preceding sections, we have reviewed some authors who propose relational ontologies, and we can already observe in them the diversity of interdisciplinary and intercultural sources that enrich their approaches. However, there is still a key element of interest for us to study: if relational ontologies refer to an *experience*, we should ask ourselves about the nature of this experience, always understood in a pluriverse. This is not a phenomenological inquiry that would give us an account of a certain morphology of the experience of inter-independence, since we think, like Panikkar (1996), that this new paradigm is just emerging, so that our epistemological resources are still limited for its observation. Nevertheless, we would like to reflect on these experiences, in relation to two themes: the first, referring to the personal relationship of human beings with nature, and the special place of language, and second, the question on how the traditional vision of community is altered through these experiences.

With respect to the first point, personal relationship and language, relational ontologies are not based on the dualistic, subject/object view between human beings and nature. Following Martin Buber (2006), we could speak of an I–Thou relationship, but this is certainly more evident among human beings where it is mediated by language. In the words of Buber (2006), in the coexistence with nature, "the relationship floats in the shadows, and is beneath language" (Buber 2006:14, own translation). This obviously challenges us to understand this relationship, even more so when we hear stories in which human beings dialogue with nature, as the Mapuche poet Elicura Chihuailaf (2014) tells us:

> Sitting on my grandmother's knee, I heard the first stories of trees and stones in dialogue with each other, with animals and with people. Nothing more, she told me, you must learn to interpret their signs and to perceive their sounds, which are often hidden in the wind.
>
> (Elicura Chihuailaf 2014:17, own translation)

In the case of the Mapuche people, their language, ***Mapuzungun***, means the speaking of the earth, so the earth speaks for and through the human being; hence ***Mapuzungun*** is an onomatopoeic (imitative) language. This intimate and *sui generis* relationship with nature can be seen in the case of Don Lorenzo Aillipan, who, at the age of 8–9 years, was initiated through a dream as Man–Bird–Mapuche or Üñumche (üñum = bird, *che* = human being) and who as such can communicate with birds in a ritual context as explained in the work de Aillapan and Rozzi (2004:421, own translation): "Before doing so, Aillapan asks permission on each occasion to the birds and Mother Nature to speak of them and for them". His winged poems not only show the brotherhood with birds, but also the possession of an ecological knowledge that can even be a contribution to science itself.

An example of this is the winged poem about the hummingbird (***pinda*** in *Mapuzungun* and *sephanoides sephaniodes* in its scientific name) where an intriguing fact for scientists is announced, and which refers to the dormancy or torpor throughout the winter of the hummingbird. This fact is also attested by locals, who have found some species sleeping in tree concavities during the winter, so Aillapan in Aillapan and Rozzi, (2004) writes:

He sleeps or dozes off in April.
And it is awakened or revived in October, when there are
Many flowers.
Its nest is made of cotton bangs
And on a gold weighing scale weighing two "tomins[10]", he and
her nest.
Subtle as the little birds that on the margins of the
hours of prayer
put in the illuminators.
Pin pin pin pin pin pin pin d aaaa pin pin pin pin pin pin d aaaa!
Müpümüpütuy ka ülkantuy.

(Aillapan and Rozzi 2004:425, own translation)

10 One "tomin" is equal to 0.5953 grams of gold, in an ancient measure of weights of Spanish colonizers.

Undoubtedly, the interrelationship between nature and language that *Mapuzungun* presents is a problematic aspect from a Western point of view, yet in this view language is only an instrument of human thought. However, in the history of the West, there have also been voices that approach this *sui generis* relationship between language and nature. An example of this is the transcendentalist philosopher Emerson (2008), already mentioned at the start of this chapter, who explains such a more comprehensive relationship in his essay *Nature*. He points out that nature is the vehicle of language and thought, that is, every word expressing an intellectual or moral fact, taken to its root, results in the borrowing of a material appearance, thus "*Right* means *straight*; *wrong* means *crooked*; *Spirit* means in the first place, *wind*" Emerson (2008:54, own translation). This relationship between nature, and language is relevant to Emerson, yet he sustains that when the language of man distances itself from visible things and is not accompanied by a natural image, the corruption of language appears, and with it, the corruption of man. The intrinsic relationship of the language of nature with human language and the corruption of the word is also expressed by the Mapuche poet Elicura Chihuailaf (2014):

> The language of nature is a clear, transparent whole. Because it is part of it, so is – in essentials – also the language of human beings. What has happened then to the Word? What happened to the blue melody of understanding, the wisdom that gives it life?
>
> (Elicura Chihuailaf 2014:42, own translation)

This *sui generis* personal relationship with the beings of nature, expressed in language, undoubtedly shakes the experience we have with the territories we inhabit, and with it, the traditional notion of community, understood from modern thought, only inhabited by human beings. Aldo Leopold (2005) in his *Ética de la Tierra* defines the extension of the borders of the community "to include soils, water, plants and animals; said in a collective way, the earth" (Leopold 2005:135, own translation). However, in our opinion, where he best expresses this idea is in one of his essays where he registers those experiences of extension of the borders of community (Leopold 2005):

> Books or no books, it is a patent fact, to my dog and to me, that at dawn I am the sole owner of all the acres I can walk on. Not only do boundaries disappear, but so does the idea of being limited. Dawn knows expansions, ignored by deeds or maps, and solitude, which supposedly, no longer exists in my county, extends everywhere, as far as the dew reaches.
>
> (Leopold 2005:73, own translation)

This idea of not being limited disappears when the sun comes in:

> I can already feel the sun. The chorus of birds has run out of breath. The roars of a tractor warn me that my neighbor is already in action. The world

> has shrunk to the small dimensions known to county clerks. We're going home, and to breakfast.
>
> (Leopold 2005:76, own translation)

In the experience of dawn, Leopold recounts the spectacle of the birds that begin with their proclamations about the trees they own. His perceptive insight is of such magnitude that he registers not only the symbiotic relationship between the birds and the trees, but also highlights the influence of this experience on his identity, where self-limits disappear: "the idea of being limited".

Will this experience have to do with the identity of Choquehuanca Céspedes' and Rist's (2024) "Cosmoser" or Arne Naess's "Ecological I"? We have no answer in this pluriverse, but it is clear that in such experiences, the "limited" vision of the dualistic ontologies that cut off the human being from nature has been overcome.

This experience of separation is reflected in the cruelty of the biologism of the meat industry and the taxonomic bias of science: more than 85% of existing studies on the planet's biodiversity concentrates on vertebrate animals, at the detriment of invertebrates, which represent 97% of the species recorded in biosphere. (Rozzi 2019:92). The neglect of science regarding the invertebrates is clearly believed to be evident because of their morphological distance from humans, unlike mammals.

It seems that the ecological knowledge provided by the ancestral knowledge, such as that of don Lorenzo Aillipan (in Aillapan and Rozzi 2004), is at the basis of the human being's capacity to extend the frontiers of community. However, it is important to note that this experience also has a political facet, as pointed out by Escobar (2014):

> Thus, the concept of community, initially centered on humans, expands to include non-humans (which can range from animals to mountains to spirits, all depending on the specific territories). Consequently, the terrain of politics opens to non-humans. What impact does it have for the modern conception of politics that it is not restricted to humans?
>
> (Escobar 2014:103, own translation)

Undoubtedly, the question posed by Escobar (2014) is relevant in current times, where new legal categories, such as ecocide, land as a subject of rights, among others, come to confirm the challenge implied by the new conception of community. However, in this journey, interculturality is more necessary than ever, as it helps us to vitalize relevant notions about our dwelling place, and thus avoids being petrified by Western languages and legal instruments, concepts, and legalities. Undoubtedly, for a cosmos to be habitable, mythical, and symbolic language is necessary.

Synthesis and future steps to move toward critical sustainability sciences

In this chapter, with the meditation of the concept of cosmosophy, we have advanced from the narrow vision of the cosmos, as seen from a dualistic ontology. The four topics discussed here inspire us with novel ideas to be considered in critical sustainability sciences, and which are largely absent in traditional sustainability discourses. I outline the most important of them:

- Critical sustainability sciences must regenerate a sustainability concept that is based on thoughts rooted in an organic and animistic vision of reality: the earth is alive, as we are taught by the indigenous people, who constantly dialogue with it. This vision must be nourished by scientific and ancestral knowledge; hence, the dialogue of knowledge, interdisciplinarity, and transdisciplinarity are fundamental. However, critical sustainability sciences must contribute to knowledge that promotes synthesis, and this can only be achieved through mythical and symbolic thinking.
- This means that the experience of inhabiting a living earth is only an experience for those who inhabit a myth and symbol, for example, such as the *terra mater*, and this means going beyond the conceptual contribution, provided by the sciences on the interdependence of ecosystems. This is undoubtedly a challenge for critical sustainability sciences, but it is necessary to assume it. If we do so, we can even contribute to the integration of the human being, since it is not only an intellectual being that also carries *mythopoetic* capacity, creator of myths, and before the current scenario of disintegration of many of the myths of modernity, it is necessary to contribute to the emergence of new myths that make the cosmos again our habitable dwelling and that drives us to take care of it, heal it, and venerate it.
- This chapter focuses on the ideas of R. Panikkar, A. Escobar, M. Eliade, and V. Shiva, among others. We outlined the existence of an intrinsic relationship between the novelty of their positions and the experiences of lives that emerge out of interculturality and openness to otherness. Undoubtedly, interculturality helps us not to absolutize our own view and ensure that we can rethink the concept of sustainability, as an exercise of critical reflection necessary for critical sustainability sciences, beyond the ontological imperialism of the dominating narratives of sustainability. Moreover, interculturality establishes a terrain of hope. In a technical-scientific society that has abolished the utopian capacity of human beings, interculturality helps us to recover and construct new myths of co-ownership among the different beings that inhabit the Earth. Intercultural praxis restores that ontological latency, which gives hope for a mobilizing agent for change.
- Critical sustainability sciences must incorporate a cosmosophical perspective, and thus deepen the reciprocal influence of the relationship between nature and the human being. Recognizing that the cosmos possesses a wisdom of which the human being is a part, it is vital to enter the depths of this affirmation that takes us out of a leap of anthropocentrism.[11] Indigenous

11 The intercultural critique of modern anthropocentrism is addressed by Sepúlveda (2021) in the chapter "Sustainability and Interculturality: Moving Towards a New Way of Inhabiting the Earth".

peoples have much to teach us in this regard, they know that cosmic wisdom guides the daily lives of the inhabitants, even in the moral sphere. It is a challenge to dare to advance in this knowledge.

- Assuming the cosmosophical perspective in critical sustainability sciences implies the conceptual revision that still revolves around anthropocentrism. An example of this is the term liberation that we have reviewed and that now deserves a reinterpretation from a process of coevolutionary cosmogenesis.
- Critical sustainability sciences must incorporate and substantiate the sacred visions of nature. At present, there is recognition of these visions among indigenous and non-Western peoples, but their existential and ontological value is relativized by seeing them only as "cultural" manifestations. It is no coincidence that many of today's environmental movements advocate the sacredness of the earth and that many environmental thinkers draw their thinking from diverse spiritual sources.
- It is relevant that critical sustainability sciences deepen in relational ontologies and thus in the new paradigm that is emerging: inter-independence. It anchors us in the vital spaces of our existence, so that entering this experience implies removing the ontological, epistemological, ethical, and anthropological assumptions, which are rooted in dualistic ontologies. As emphasized by A. Escobar, relational ontologies imply an "ecological reconversion", which could be at the heart of critical sustainability sciences and contribute, through this, to forge new vital practices in the personal, community, and political spheres.

In summary, we would like to emphasize that critical sustainability sciences should stimulate anthropological reflection, since, as Fornet-Betancourt mentions, we must undoubtedly move from a "sustaining-being" to a "being sustained" by all the natural organicity of life. To emphasize and conclude this chapter dedicated to cosmosophy, I leave the words of R. Panikkar (1999:162–163, own translation):

> To participate in the unfolding of life, to attend the cosmic exhibition of all the forces of the universe, to be witnesses of the unfolding of time . . . being aware of the systole and diastole of the world, feeling the assimilation and de-assimilation of the cosmos on the macro- and microcosmic scale, being sensitive to the stars and the atoms, being a mirror of the universe and reflecting it without distorting it, suffering also in one's own flesh the disorders of the world, being oneself the laboratory where medicines and antibodies are created, without ignoring the forces of evil, nor the tendencies of history, but also not allowing ourselves to be suffocated by them, dominating each one of us these demons in our lives, understanding the song of the birds, the sounds of the forest and even all human noises as part of the vitality of the reality that expands, living, breathing in and breathing out, no longer to go somewhere else (and never arrive), but simply to be, to live, to exist in all planes of existence at

the same time: the tempestuous explosion of the adventure of Being. . . .
This is transhistorical existence.

Acknowledgments

I would like to thank Stephan Rist for translating the original Spanish version of my chapter into English. I would also like to acknowledge the support received in this research by my university, Universidad Católica de Temuco, Chile, and the Project Centro Internacional de Cabo de Hornos – Cape Horn International Center (CHIC) with project number ANID/BASAL FB210018.

References

Aillapan, L., and Rozzi, R., 2004. Una etno-ornitología mapuche contemporánea: poemas alados de los bosques nativos de Chile. *Ornitología Neotropical*, 15, 419–434.
Aristotle, 2008. *Metafísica. Introducción.* Traducción y notas de Tomás Calvo Martínez. Madrid: Gredos.
Bernabé, A., 2010. *Fragmentos Presocráticos de Tales a Demócrito.* Introducción, traducción y notas de Alberto Bernabé. Madrid: Alianza.
Brague, R., 2008. *La Sabiduría del mundo.* Historia de la experiencia humana del universo. Madrid: Encuentro.
Buber, M., 2006. *Yo y Tú y otros ensayos.* Buenos Aires: Lilmod.
Byung-Chul, H., 2020. *La desaparición de los rituales.* Barcelona: Herder.
Callicot, J., 2017. *Cosmovisiones de la Tierra. Un estudio multicultural de éticas ecológicas desde la cuenca del Mediterráneo hasta el desierto australiano.* Madrid: Plaza y Valdés.
Calvo Martínez, T., 2000. La noción de Physis en los orígenes de la filosofía griega. *Revista de filosofía*, 21, 21–38.
Chihuailaf, E., 2014. *Recado confidencial a los chilenos.* Santiago, Chile: LOM Ediciones.
Choquehuanca Céspedes, D., and Rist, St., 2024. A culture that understands that everything is interrelated, that nothing is divided, and nothing is outside. In: Rist, S., Jacobi, J., and Bottazzi, P., eds. *Critical Sustainability Sciences: Intercultural and Emancipatory Perspectives.* London: Routledge.
Curivil, R., 2007. *La Fuerza de la religión de la tierra. Una herencia de nuestros antepasados.* Santiago, Chile: Ediciones Universidad Católica Silva Henríquez.
Curivil, R., 2020. Cultura mapuche: Un antiguo ideal de persona para una nueva historia. *Utopía y praxis latinoamericana*, 25(88), 41–54.
Eliade, M., 2017. *Lo sagrado y lo profano.* Barcelona: Paidós.
Emerson, R., 1836. *Nature.* Metcalf, Torry, and Ballou: Cambridge Press.
Emerson, R., 2008. *Naturaleza y otros escritos de juventud.* Madrid: Biblioteca Nueva, S.L.
Escobar, A., 2014. *Sentipensar con la tierra. Nuevas lecturas sobre desarrollo, territorio y diferencia.* Medellín: UNAULA.
Estermann, J., 2013. Ecosofía andina: Un paradigma alternativo de convivencia cósmica y de Vivir Bien. *Revista FAIA*, 2(9–10), 21 págs.
Fornet-Betancourt, R., 2007. ¿Es la sostenibilidad una perspectiva interculturalmente sostenible? Elementos para la crítica de un concepto bien intencionado, pero insuficiente. *Pasos*, 129, 8–14.

Gómez-Heras, J., 2010. *En armonía con la naturaleza. Reconstrucción medioambiental de la filosofía.* Madrid: Biblioteca Nueva.
González, J., 2004. *Teorías de la vida.* España: Síntesis.
Hathaway, M., and Boff, L., 2014. *El tao de la liberación. Una ecología de la transformación.* Madrid: Trotta.
Le Bonniec, F., and Payas, G., 2021. Entender la justicia y las injusticias en Wallmapu a través los estudios interculturales como propuestas de con-vivir. In: Samaniego (ed.). *Estudios interculturales desde el sur. Procesos, debates y propuestas.* Santiago de Chile: Ariadna, 37–64.
Leopold, A., 2005. *Una Ética de la Tierra.* Madrid: Catarata.
Lovelock, J., 2000. *Gaia a New Look at Life on Earth.* Oxford: Oxford University Press.
Manifiesto por la vida por una ética para la sustentabilidad, 2002. *Ambiente & Sociedade,* 10, 1–14.
Manque, C., 2016. Pilmaikén: la fuerza espiritual del kintuante frente a las hidroeléctricas. In: Colectivo Editorial Mapuexpress, eds. *Resistencia Mapuche al extractivismo.* Santiago de Chile: Quimantú, 47–57.
Naess, A., 2007. Los movimientos de la ecología superficial y la ecología profunda: un resumen. *Revista Ambiente y Desarrollo de CIPMA,* 23(I), 98–101.
Naess, A., 2018. *Ecología, comunidad y estilo de vida. Esbozos de una ecosofía.* Buenos Aires: Prometeo.
Panikkar, R., 1990. *Sobre el diálogo intercultural.* Salamanca: San Esteban.
Panikkar, R., 1996. *El silencio del Buddha. Una introducción al ateísmo religioso.* España: Siruela.
Panikkar, R., 1999. *La intuición cosmoteándrica. Las tres dimensiones de la realidad.* Madrid: Trotta.
Panikkar, R., 2007. *De la Mística. Experiencia plena de la Vida.* Barcelona: Herder.
Panikkar, R., 2008. *Ecosofía.* www.grupotortuga.com/Ecosofia (Accessed 7.6.2022).
Panikkar, R., 2012. *El Ritme de L'Ésser.* Les Gifford Lectures. Col·lecció: Opera Omnia Raimon Panikkar. Vol. X/TOM I. Barcelona: Fragmenta.
Plato, 2010. *Timeo.* Edición bilingüe de José María Zamora. Notas y anexos de Luc Brisson. Madrid: Abada Editores.
Rozzi, R., 2019. ¡Chovinismo Taxonómico, No Más! Antídotos de Hume, Darwin y la Ética Biocultural. *Environmental Ethics,* 41(9998), 73–112.
Sepúlveda, J., 2021. Sustentabilidad e interculturalidad: Avanzar hacia un nuevo modo de habitar la tierra. In: Samaniego, ed. *Estudios interculturales desde el sur. Procesos, debates y propuestas.* Santiago de Chile: Ariadna, 65–181.
Sheldrake, R., 1994. *El Renacimiento de la Naturaleza. La nueva imagen de la ciencia y de Dios.* Barcelona: Paidós Ibérica.
Shiva, V., 2001. *Biopiratería. El saqueo de la naturaleza y del conocimiento.* Barcelona: Icaria.
Shiva, V., 2006. *Manifiesto para una Democracia de la Tierra. Justicia, sostenibilidad y paz.* Barcelona: Paidós.

7 Toward a "nature alliance"

Why sustainability must be rethought in terms of relationality

Beat Dietschy

Introduction

There are many reasons to rethink "sustainability". In particular, a critical revision is called for regarding the political use of the double term "sustainable development", which comes from the discourses of more than 70 years of development agendas. It has a heavy legacy to bear. Heralded by President Truman (1949:44) with the promise to "embark on a bold new program to make the benefits of our scientific advances and industrial progress available for the improvement and growth of underdeveloped areas", this epoch was proclaimed as a transition from the era of colonialism to the era of "development". Despite the passing of the "old imperialism" (53), colonial structures lived on. Poorly concealed as development discourse, poverty, inequality, and dependency persisted in the Global South in a modernized and refined form. The racially underpinned "mission of 'civilizing the uncivilized' was replaced by the worldwide establishment of social institutions along European-Anglo-Saxon lines", as Aram Ziai (2006:41) concludes in his study:

> The myriads of possibilities of human society are reduced to the replication of Western European industrial capitalism, the formation of which is transfigured as a necessary stage in human history. The discourse of "development" is thoroughly Eurocentric (40).[1]

This refers to more than perceiving others through one's own glasses. The hegemonic discourse of development not only confirms the worldview of European modernity, but in any case it generates a normative knowledge of development that decides "about the wrongness of other people's ways of life and the way of necessary correction" (31). With the concept of development, as set forth in Truman's speech, a large part of humanity became "underdeveloped" in one fell swoop. For Gustavo Esteva (2010:1–2), the "invention

1 This text includes various citations from publications that are written in Spanish, German, French, or Portuguese. If the reference of these citations is given in those languages, the citations were translated by the author of this chapter.

DOI: 10.4324/9781003043577-7

of underdevelopment" for two-thirds of the people in the world is simultaneously accompanied by the devaluation and subjugation of their way of life, which is linked to their colonization and economization:

> Disvalue transmogrifies skills into lacks, commons into resources, men and women into commodified labour, tradition into burden, wisdom into ignorance, autonomy into dependency. (15)

Even if this development-critical position is summarized very briefly, it nevertheless draws attention to how closely "development" is linked to the planned reshaping of social relations and the global implementation of capitalist structures taking place within the framework of reshaping the order of knowledge under Western hegemony. In this sense, it can even be said that "from the start, development's hidden agenda was nothing else than the Westernization of the world" (Sachs 2010b:XVIII).

"Sustainable development" seems – at first glance – to break with these neocolonial geopolitics. It was launched toward the end of the "lost development decade" of the 1980s in response to criticism of the ecological consequences of the Western development model and was explicitly propagated by the Brundtland Commission, which included representatives of industrialized and developing countries, with a view to "our common future" (UN 1987). For development-critical voices like Esteva (2010:13), however, "in its mainstream interpretation, sustainable development has been explicitly conceived as a strategy for sustaining 'development', not for supporting the flourishing and enduring of an infinitely diverse natural and social life". In any case, "development" finally "emerges rejuvenated from this liaison" (Sachs 2010c:28) with environment.

The Agenda for Sustainable Development adopted by the UN in 2015, again, presents a paradoxical picture: to the multiple crises, which are basically to be seen as a crisis of the globalized Western development model, it certainly does intend to provide, for the first time, a truly global response, which, unlike the previous Millennium Development Goals, calls all nations to account, and includes rather ambitious sustainability targets in 17 key fields of action. However, the "great transformation" calling for achieving the goals by 2030 seems to be limited to aspects of a socio-ecological modernization of the existing development model. In any case, "the debate is meaningfully silent" about a fundamental change in class and gender relations, structural racism, or systemic extractivism, but, above all, about fundamental mechanisms of capitalist socialization such as the transformation of property relations (Brand and Wissen 2017:28–33). "It lacks . . . a perspective of emancipation" (37).

Against this background, it is even more important for critical sustainability research to explore ways and means to bring to light the deeper reasons for the prevailing "mal development" and to elaborate "concrete utopias" (Bloch) for overcoming it. To this end, I will discuss first decolonial and intercultural approaches and then bring into play Ernst Bloch's concept of "Naturallianz". They can all contribute to a multiverse of convivial practices of liberation or emancipation.

Decolonial critique of development

"Our entire global civilization faces an integral crisis of multiple dimensions, which is rooted in the world's power structure" (Park 2015:85). This is what the South Korean theologian Park Seong-won wrote at the time when the 2030 Agenda was adopted in New York. And he continues:

> The current civilization is unsustainable as the foundation, and the order of life of all living beings – in the biological, ecological, sociological, geo-political and religious-cultural dimensions – is rapidly being destroyed.
> (85–86)

With his statement, the Presbyterian theologian Park stands in a long tradition of church statements about questions of economic and ecological justice. Thus, at the Assembly of the World Alliance of Reformed Churches in Accra (2004), it was recognized and confessed that the massive threat to life on earth and to the dignity of many people was the product of an unjust economic system and that its justification should be judged theologically as idolatry:

> The policy of unlimited growth among industrialized countries and the drive for profit of transnational corporations have plundered the earth and severely damaged the environment.
> (WCRC 2014:8, cf. 10)

At the 10th Assembly of the World Council of Churches held in 2013 in Busan (Korea) Park called for a "radical transformation" of the life-destroying civilization model that prevails today (Dietschy 2014:65).

Others rather refer to an unsustainable and colonialist model of development. But nothing fundamentally different is meant by this, as philosopher Raúl Fornet-Betancourt (2017b:38–39) points out: the term "development", practically transformed into a synonym for progress and economic growth, has "become world . . . because for us, in the 21st century 'development' is the name behind which stands the hegemonic civilization of the present". The answer to the question of whether it is colonialist is clear for Fornet-Betancourt: "Yes, this concept of development is colonialist, and as such it represents a concept from which violence emanates in several respects, from social and political to epistemological violence" (37–38).

In Latin America, two schools of thought have made it their task to recognize these connections, both of which draw conclusions for a liberating change of their life contexts: intercultural philosophy and decolonial thinking, which I will discuss in more detail subsequently. Both are as multifaceted as the worldwide development critique. This is also because their questions arise from specific experiences and sociocultural contexts in which social movements "counter post-colonial social forms with a multiplicity of differently decolonized life worlds" (Rist 2015:111). Core themes about which research is conducted are nevertheless identifiable. For example, there is an international network that examines the connections between European modernity, capitalism, and coloniality from different perspectives. The starting point for this line of research[2] was the work by the Peruvian sociologist Aníbal Quijano on the "coloniality of power". In an article written in 1989, he says:

> With the conquest of the societies and the cultures which inhabit what today is called Latin America, began the constitution of a new world order, culminating, five hundred years later, in a global power covering the whole planet.
>
> (Quijano 2007:168)

This means the hegemonic power order of a globalized capitalist society that exists today has as its prerequisite a coloniality of power that was first formed in the "New World".[3] America became space and time to bring forth a new model of power. It was not discovered as it was – in all its multiplicity of cultures – but reinvented.[4] This is, as Enrique Dussel (1995:9–10) has expressed, a process of invention of the Other, in the course of which Europe invents itself as a center of power. As Dussel states:

> Modernity appears when Europe organizes the initial world-system and places itself at the center of world history . . . against a periphery, equally constitutive of modernity.

Dussel calls the forgetting of this peripheral origin of modernity by great European thinkers "the Eurocentric fallacy". This conceals not only the alterity of the Other, but also the fact that only "by controlling, conquering, and violating the Other, Europe defined itself as discoverer, conquistador, and colonizer of an alterity likewise constitutive of modernity"(12).[5] While Dussel, in the book cited here, primarily traces the constitutional process of modern

2 It has become known as "Modernity/Coloniality and Decoloniality" (MCD). In addition to Quijano, this circle includes, among others, Arturo Escobar, Walter Mignolo, Enrique Dussel, Maria Lugones, Catherine Walsh, Edgardo Lander, Nelson Maldonado-Torres, and Santiago Castro-Gómez.

3 The binomial expression "modernity/coloniality" underlines that the history of colonial domination of the West cannot be understood as an intra-Western process.

4 A statement from the Nasa people in Colombia's Cauca Valley (quoted in Escobar 2020:166, fn 3) says: "One day in 1535, the Conquista came. One night we went to sleep as Nasa. We woke up as Indians".

5 "Europe never discovered (des-cubierto) this Other as Other but covered over (encubierto) the Other as part of the Same: i.e. Europe" (12).

subjectivity between 1492 and 1636, that is, from an "I conquer" to the formation of Descartes' solipsistic ego cogito ruling over the outside world (18 and 48), Quijano is more concerned with elaborating the social-structural factors[6] that led to the conclusion that modernity is inconceivable without coloniality and represents a profoundly contradictory social formation.

With his analyses, Quijano "enabled a decolonial critique of modernity that was not grounded on modernity's own narrative but on the thinking and experiencing of coloniality" (Icaza and Vázquez 2017:50). This change of sides is fundamental for decolonial thinking: "Decolonizing ourselves means seeing reality with our own eyes and no longer with the eyes of the West", said Choquehuanca Céspedes (2020) in his inaugural speech as vice president of Bolivia. Liberation is not possible without epistemological rupture or decoupling from coloniality and ultimately unfolds from the "exteriority" of modernity, into which it has banished the non-modern, "non-civilized", and "non-developed". Liberation, finally, requires a critical sifting through the sites, modes, and situations of continuing coloniality. Therefore, decolonial thinking resists definitive definitions of decoloniality. Shared conviction, however, is that coloniality is constitutive of modernity.

Maria Lugones, for example, went beyond Quijano's design of a coloniality of power in that she not only engendered it, but addressed the "dark/hidden side of the Western-style modern-colonial gender system" (Lugones 2008:53) and highlighted the violence and power that non-white women are exposed to. She thus takes up their critique that hegemonic feminism overlooks the intersectionality of racism, class domination, sexism, and gender (Lugones 2016:13). "It is only when we perceive gender and race as intermeshed or fused that we actually see women of color" (18).

Catherine Walsh goes one step further by extending Lugones' intersectional alliance-building to nature. She examines "coloniality of nature",[7] which for her is one of life/existence in general: "A pattern of power that occurs at the intersections of the cultural, ontological, existential, epistemic, territorial, cosmological and socio-spiritual" and that ultimately enforces a single dichotomous worldview of human (i.e., white, educated male) domination over nature (Walsh 2015:103). Since the Incas and the Spanish colonial–imperial domination has been installing hierarchies and divisions, based on the "ideas" of race, gender, sexuality, and nature, where previously gender relations were more "dynamic, fluid, open and non-hierarchical" (106).

6 There are two categories in particular that Quijano considers crucial for the formation of colonial power relations: on the one hand, the basic racializing distinction between conquerors and conquered, or European–non-European. On the other hand, "all the forms of labour and control of labour were not only simultaneously performed in America, but they were also articulated around the axis of capital and the global market" (Quijano and Ennis 2000:550).

7 By this, Walsh (2006:40) understands the dichotomy of nature and society established with "modernity/coloniality" and the associated ruptures between the "biophysical, human, and spiritual worlds".

Walsh (101), a scholar and activist in Ecuador, traces the intertwining of "life, nature, and gender" in Andean cosmology in order to strengthen decolonial thought and action by exposing its "otherwise". The same is true for the Colombian anthropologist Arturo Escobar. For him, it is central to consider the "interwoven character of the discursive, material, social and cultural dimensions of the social relations between human beings and nature" (Escobar 2005:21). Only in this way, he argues, can the otherness of cultures be understood, the basis of which is a "relational ontology".[8] However, these peoples and their culture are made invisible and destroyed by the mechanisms and modes of operation of the dominant "one-world-world".[9] In the sense of the "sociology of absences" of Boaventura de Sousa Santos, Escobar is therefore concerned with making the diversity of negated worlds visible through a "political ontology".

Escobar (2011:68) notes that seemingly capitalist modernity has declared war on every ecosystem on the planet and underlines that few places, since the mid-1990s, have become as exemplary of this destruction as Colombia's Pacific coast. He therefore adds to the many fields of the "coloniality of modernity" that of the "coloniality of nature". He makes a point of exploring the other half of the picture as well, namely, the ecological and cultural practices of Afro-Colombian communities, that "have had an intrinsic notion of sustainability" (69). It differs from the usual Western conception of sustainability. Escobar (2020:46) speaks of an "earth-form of life" that does not separate human and non-human life: "every living being exists, because all others exist" (Escobar 2016:27). Nasa leader Marcus Yule expressed this non-anthropocentric worldview with the words: "Somos la continuidad de la tierra, miremos desde el corazón de la tierra" ("we are the extension of the earth, let us think from the earth's heart", 27).

In 2004, the Nasa published a declaration entitled "The Liberation of Mother Earth". It deals with both the restitution of stolen land and the defense and promotion of life: "We must learn to live with the earth, not merely from her, consuming her. We free ourselves with the earth" (Escobar 2020:58). For Escobar, the example from the Cauca Valley shows: "The proliferation of struggles in defense of territory and cultural difference suggests that what emerges from such struggles are entire relational worlds" (70).

The struggles and conflicts over land and territories have permeated the entire history of Latin America, since the Conquista, and have been closely

8 With this term, Escobar (2020:71) wants to underline the peculiarity of "epistemologies of the South" (Boaventura de Sousa Santos): The mangrove world, for instance, "is enacted minute by minute, day by day, through an infinite set of practices carried out by all kinds of beings and life forms". It is this "dense network of interrelations" what he calls a "relational ontology", a world in which "*nothing preexists the relations that constitute it.* Said otherwise, things and beings are their relations, they do not exist prior to them" (71–72).

9 Escobar uses this term from John Law to express the colonial and exclusionary nature of capitalist, rationalist, and patriarchal modernity (Escobar 2016:15).

linked to the plunder of the continent's natural resources. In the era of international development policies, the concept of "development" has also become the object of protest and criticism from peasant, indigenous, and Afro-Latin American populations affected by displacement, land grabbing, or environmental destruction through mining and agro-industrial monocultures. The post-2000 global commodity boom, which exploded the number of extractive megaprojects in Latin America and gave rise to "neo-extractivism" as a new development model (cf. Svampa 2019:3),[10] led to an intensification of struggles over nature and habitats and an increased coalescence of peasant, indigenous, environmental, and even popular feminist movements and forms of resistance. Maristella Svampa with regard to these socio-ecological struggles speaks of an "eco-territorial turn" (40), which opens up perspectives for agroecological and ecofeminist alternatives, as well as for indigenous communities that insist on decolonization. Similarly, Alimonda (2011:36–37) has complemented the concept of the "decolonial turn" with a "natural-colonial turn" that underlines the importance of indigenous struggles for land, territory, and also the associated cosmocentric worldview like the respect for Pachamama (Mother Earth).

In a context in which "development" has once again become a global meta-discourse under the label of the sustainability, social movements around territoriality and culture in Latin America are delinking from development and renewing a critique of development that had already been formulated since the late 1980s. What had initially appeared as a "post-development" discourse is now more strongly invested with indigenous and decolonial connotations. Escobar, who himself has contributed significantly to building bridges between decolonial and post-development approaches, as well as between research and grassroots movements, points out that it is precisely the public debates about "Buen Vivir" that have promoted a paradigm shift in some countries and disciplines: no longer only alternative development, but alternatives *to* development (and to extractivism) have been sought since then.

What does this mean for "sustainable development"? In which category does it belong? What matters, is not the term alone, but its use in geopolitical terms. In this sense, Quijano has located concepts, such as development and progress, in the long history of the global hegemony of Eurocentrism. They have, by being linked to a linear course of history and a social dichotomy, established Western hegemony:

> Thus, all non-Europeans could be considered as pre-European and at the same time displaced on a certain historical chain from the primitive to

10 She calls this strategy a "development illusion" (15). It perpetuates the myth of inexhaustible resource wealth and goes back to the colonial utopia of Latin America as an "El Dorado" (cf. Dietschy 1993).

> the civilized, from the rational to the irrational, from the traditional to the modern, from the magic-mythic to the scientific.
>
> (Quijano and Ennis 2000:556)

In the face of this still effective narrative, which makes coloniality disappear from modernity and non-European cultures appear as retarded, indigenous peoples like Nasa ask themselves: "This being so, should we get on board the progress train? As errand boys, as bosses?" (cited in Escobar 2020:48). Thus, they don't allow themselves to be relegated from the simultaneity of modernity to a non-simultaneity, but insist, like the Zapatistas: "We are from times past, yes, but we are something new".[11] This means that after 500 years of oppression, marginalization, and genocide, they do not seek inclusion in this world of oppressors, but rather a coexistence based on self-determination and work for "a world where many worlds fit". For Escobar (2020:63), it is clear that these indigenous perspectives are hardly compatible with the concepts of sustainable development or the "green economy", since, like the manifesto of the "Liberation of Mother Earth", they aim at something different than "development" and actually stand for a new way of being-in-the-world. "The lucid knowledge of this indigenous people from northern Cauca imbues us with the idea of a civilizational change, from the Man-form (that of anthropocentric modernity) to the Earth-form of life (relationality and biocentrism)" (9).[12]

It is important to note that the primary intention of decolonial critique is not to denounce or change the content of development discourses. Rather, it consists in getting to the bottom of their rules: it examines "their power effects, its world-making processes . . . or the erasure of other worlds, . . . the anthropocentric consumption of the earth" (Icaza and Vázquez 2017:58–59). It is essentially a critique of the epistemic foundations of European modernity and creates conditions for a grammar of decoloniality that makes another being and knowing possible. A first step toward such a grammar consists in " Learning to Unlearn" (Tlostanova and Mignolo 2012). It is thus a matter of liberation or "decoupling" from the imposed worldview that underlies the concept of "development".

Resistance and justified criticism are evoked by any worldview that demands subjugation,[13] especially when it makes people lose local livelihood and cultural autonomy. This is also true, as Ariel Salleh (2019:42) points out, when "top-down plans for 'sustainable development' are underwritten

11 From a speech by Subcomandante Marcos on March 7, 2001, at the march "El color de la tierra" (Dietschy 2007:3).

12 From a decolonial perspective, modern thinking is anthropocentric from the ground up, because it is "based on the othering of nature and the formation of the human as an individual sovereign self" (Icaza and Vázquez 2017:58). A sustainability narrative following the logic of development belongs to this genealogy (55).

13 "Our struggle is against all types of subjugation and against the colonial, patriarchal, single thought, wherever it comes from" (Chapter 3 in this book).

by extractivisms". However, the rule of "development experts" who claim a monopoly of definition has to be even more denounced, if linked to a twofold hierarchized worldview – in gender as well as in nature relations. "Sustainable development" therefore can only be an option in decolonial discourse if it is a "strong" notion of sustainability, biocentric rather than anthropocentric, and based on development critique from indigenous, eco-feminist, or eco-territorial perspectives. This presupposes an epistemic break with occidental universalism and the "modern project of One World", which "aims to transform the many existing worlds into one" (Escobar 2014:134). "A strong notion of sustainability", Escobar argues, "must be epistemically decolonizing, economically and socially liberating, and depatriarchalizing; it must also be able to design alternatives *to* 'development'" (65). If this is the case, it could give rise to a dialogue between different perspectives:

> Decolonizing sustainable development means recovering human relationships with nature and with life, to reconnect with spirituality and ancestral memory; in this way, social progress is given in terms of a genuine dialogue of knowledge, through an intercultural, alternative and insurgent project, which goes against the modern logic, based on power relations, characteristic of global coloniality.
>
> (Arteta Melgarejo et al. 2021:296)

Intercultural philosophy's perspective

These statements are also shared in Latin America by intercultural philosophy, as represented by Raúl Fornet-Betancourt, to which I refer in the following. It has emerged in dialogue with Dussel's philosophy of liberation and other currents of Latin American critical thought such as liberation theology and feminist theories (cf. Fornet-Betancourt 2015b:159). The starting point and "locus enuntiandi" for intercultural practice is equally "the concrete historical situation in which the 'made invisible' find themselves" (Fornet-Betancourt 2012:129), and the transformation of the dominant tradition this philosophy strives for to free thinking "from its imprisonment in conceptual conceptions, dominated by the West" (Fornet-Betancourt 2015b:166).

Fornet-Betancourt (2012:138) agrees with the "decolonial turn", but advocates supplementing it with an additional "intercultural turn". What is needed, he says, is a critical understanding of interculturality that also questions "internalized Eurocentrism", one's own ways of thinking and academic institutions, in order "not to idolize a new concept, but to constantly bring it forth anew" (139).[14] He recalls José Martí's dictum that the colony lives on in the republics, because only its forms have been changed, not its

14 Walsh (2014:52) states: "interculturality does not exist by itself; it is not a given fact, but something in permanent journey, insurgency and construction".

spirit (139). To practice critical interculturality means "to shake the foundations of the dominant world order so that other possible foundations may emerge" (125–126).

In this sense, for Fornet-Betancourt, as for Catherine Walsh (2014:53), "interculturalizing and decolonizing go hand in hand". Both are "bets, proposals and projects of life, of re-existing, feeling-thinking and living in the cracks of the dominant order and the modern/colonial matrix" (47). That both approaches complement each other is also evident with regard to the critique of the "hegemonic development model and the associated idea of progress" (Fornet-Betancourt 2015b:169), which the latter describes as one of the tasks of Latin American intercultural philosophy. It is no less radical than that of the decolonial critique: "'Development' as a civilization of progress

- represents . . . the historical process within the framework of which a planetary neo-colonization of mankind is taking place on an unprecedented scale.
- means . . . further the penetration and domination of nature, people, and cultures by a system of laws with a logic of exteriority.
- means . . . the construction of a time progressively assembled according to the idea of the new that will soon appear . . . with the consequence of the devaluation of the "products that have become old"

(Fornet-Betancourt 2017b:40–42)

Sustainability is part of this problem area, "because the devastation of hegemonic civilization is so great that it can no longer be a question of repairing what exists. Rather, what is required is a re-foundation of all conditions" (Fornet-Betancourt 2017b:43). Accordingly, the project of "sustainable development" is not seen by Fornet-Betancourt (2007:121) as a solution, but rather as an indicator of the non-sustainability, caused by our model of civilization. He sees one origin of the problem in the threefold grievance that man's self-confidence has experienced on the part of Western-modern science: the cosmological (Copernicus), no longer being in the center of the universe, the biological (Darwin), no longer being a species removed from the animal world, and the psychological (Freud), not even being master in one's own house (123). These devaluations would be compensated with the economically and the technologically–scientifically justified "promise of a continuous and limitless increase in the power of man" (125).[15] The prevailing concept of "sustainable development" follows the "logic of development and progress" (129) that emerged from all of this.

However, Fornet-Betancourt rethinks sustainability interculturally. According to him, this must begin with the past and the damage done by this civilizing

15 Recall Bacon's (1902:Aph. 129) project to enlarge through science and technology "the power and empire of mankind in general over the universe".

mission, namely, "with the question of justice and the right with which we invade the space and time of other peoples and flood the entire earth with products and artefacts that basically no one needs" (130–131). This is not for the sake of a romantic restoration of traditional ways of life, but to achieve a process of decolonization that allows "a blossoming of the diversity of life in its many paths" (132). For only the "relationship among real worlds in the spaces and times liberated by the movements of decolonization opens the possibility for the intercultural revision of the concept of sustainability" (132).

Relation means dependence, recognition of relationality as a constitutive element of reality, and also the abandonment of Western fantasies of omnipotence by a subject that imagines itself as sovereign (cf. 134–135).[16] As already mentioned, this implies that the "European Sonderweg" that has led to the global domination of the Western capitalist model must be revised: it cannot be seen simply as a product of an intra-European history but must also be seen in relation to the other worlds that made it possible. The European narrative of development as progress sees only itself as producing world history. This "universal history", however, as can be seen from its unilinear thinking about progress, stands for a universality that is not truly universal, since it must abstract from and negate multiple histories and their cultural contexts in order to become solely dominant. For Fornet-Betancourt (2012:130), this abstract and monocultural rationality is not the only one that can claim universality; rather, he sees in interculturality "a movement of increasing participatory processes of universalization". These processes are participatory and interacting with "the different" by not overwhelming or appropriating the other, but rather by developing *with* it and thus recognizing diversity. True universality can be found only "in a common horizon of co-dignity able to affirm the normative conviction of a dialogued intercultural encounter" (Fornet-Betancourt 2010:13).

Interculturality is distinguished from a mere management of the multiculturalism of coexisting cultures by a culture of interactive practice, capable of transforming coexistence into conviviality. The interculturalization of social relations turns the "there are many strangers among us" of multiculturalism into a "we are different" that makes diversity the basis of participatory belonging (cf. Fornet-Betancourt 2012:128). It implies, as in decolonial thought, "a liberation of intercultural relations from the prison of coloniality" and also "the freedom of all peoples to choose between various cultural orientations" (Quijano 2007:178). This includes not only a radical critique of "the dictatorship of a *logos* that seeks domination over this world" (Fornet-Betancourt 2010:15), but beyond that, an ethical and spiritual dimension:

> The will to do justice to the others and their realities, is understood concretely as the will to share the world and history. Intercultural dialogue

16 This also includes the abandonment of that activist protagonism that sees humans as sustainers and "bearers of the earth" instead of beings carried by it (cf. Chapter 6 in this book).

> is not simply a dialogue of knowledges, but aims at . . . a reconfiguration of the world from the knowledges of all.
>
> (Fornet-Betancourt 2012:134)

The spiritual dimension, as Leonardo Boff puts it, is to be seen in that we become aware of the whole and of ourselves as part of that whole. It is for the inhabitants of the Andes linked with what the ongoing coloniality of being has tried to destroy: "the (re)construction of a collective sense of belonging and being that makes alliance with those who came before" (Walsh 2006:39–40).

In the same sense, Esteva (2014:38–39) describes listening to one another as the key to intercultural dialogue. According to him, the indigenous understanding is characterized by the fact that, in contrast to the Western one, it does not place speaking above listening and requires a "*listening* to the cosmos" ("cosmo*audición*"), even a "cosmoviviality" ("cosmovivencia").

For an intercultural and interepistemic dialogue, which is facing strong asymmetries of power under the given historical circumstances, this means that especially the dominant Western cosmovision and epistemology must be questioned and subjected to a fundamental revision. A main criticism is that "the science recognized as such today" (Fornet-Betancourt 2014a:19) makes a monopoly claim, which reminds us that "epistemological monism has always been a companion of empire" (2017a:26). This is compounded by the industrialized control of the production and transmission of scientific knowledge, which is accompanied by the globalization of methods and knowledge interests (2014a:14). Fornet-Betancourt criticizes not only the power-serving "configuration and institutionalization of technoscience" (2017a:7) and the "narrowing and impoverishment" (2014a:20) of the thus accessible world, but also the "expulsion of spirituality from the processes of human knowing" (19). In view of this constant "epistemological *coup d'état* that we suffer" (20), he calls for responding with a spiritual uprising, that is, "with a rising of the spiritual knowledge of humanity, able to break the prevailing epistemic totalitarianism" (20). Hence, we need promoting the recognition of the diversity of forms of cognition and knowledge (see also Fornet-Betancourt and Senent de Frutos 2017).

Seen in this light, the project of intercultural philosophy seeks ways to transform science and philosophy that lead to more justice in interepistemic relationships. This normative claim points to the fact that interculturality, like decoloniality, is to be understood in a (self)critical sense and is fundamentally different from corporate and governmental policies that adorn themselves with a purely functional interculturality. Walsh underscores this by quoting Luis Macas, one of the founders of Ecuador's indigenous umbrella organization CONAIE:

> Experience and history have made us realize that we live a colonized interculturality, seen and created from a Western and colonial logic. We

> have, therefore, the great task of decolonizing interculturality, to dispose Eurocentrism, to de-monopolize life in view of our resistances and our projects.
>
> (Walsh 2014:75)

It is precisely this compass that intercultural philosophy follows when it calls for a spiritual (up)rising: "the 'rising' that we speak about implies an ethical rebellion in favor of the dignity and the rights of those that are humiliated by the hegemonic system" (Fornet-Betancourt 2017c:15). Therefore, interculturalizing is to be understood as a transformation process on an individual, as well as on a collective and systemic level, which presupposes "unlearning" and is based on an "epistemic disobedience".[17] This unlearning is directed against the normative power of big monopolies, which through their functional logic demand conformity, instead of conviviality, as Marianne Gronemeyer (2015:51) has formulated in the tradition of Ivan Illich:

> "Thou shalt agree with me and trust my evidence", says natural science. "Thou shalt desire to prevail over your neighbor", says economy. "Thou shalt have machines work in your place, have them serve you and care for you", says technology. "Above all, thou shalt not disturb things", says bureaucracy.

In contrast, intercultural philosophy seeks, especially in the societies of the globalized world, "to initiate a transition from 'coexistence' to 'convivence' or 'conviviality'" (Fornet-Betancourt 2021:134). In doing so, it is not only based on the traditional wisdom of indigenous peoples, for whom the search for balance, reciprocity, and complementarity between people as well as dialogical relationships with all forms of life play a key role (see Delgado and Rist 2016:338–344 and Choquehuanca Céspedes 2020). Fornet-Betancourt (2021:139) also asserts – with Ortega y Gasset – that there is basically no personal experience of life for human beings "except in the form of an original conviviality. Life, therefore, is 'living together'". From this follows a kind of pre-political gratitude toward the Other(s): "Even before the contractual recognition of the rights to one's difference and consequently, one's rights as a citizen in the public sphere . . . we owe the other unconditional acceptance as a fellow-living being" (141–142).

17 Walsh and Mignolo refer to the decolonial learning processes initiated by the Intercultural University of Indigenous Nationalities and Peoples Amawtay Wasi in Ecuador (cf. Mignolo 2011:64, note 17, and Tlostanova and Mignolo 2012:225–236). Among the places of similar unruly and self-determined learning are also the "Universidad de la Tierra", co-founded by Esteva in Oaxaca or the "Unitierra-CIDECI" in Chiapas, close to the Zapatistas.

Bloch's rescuing subaltern traditions

The schools of thought mentioned so far have, as Fals Borda (1988:103) recognized early on, been "fed by the knowledge of life experience" ("conocimiento vivencial") formed in the struggle for alternatives to the "monopoly of the positivist paradigm" and in "defense of life and cooperation with nature" in Latin America.

If we search European alternatives to the dominant thinking regarding "progress", Bloch's concept of "nature alliance" can be interesting. At first glance, this may be disconcerting because the German philosopher seems to share some of the main premises of Eurocentric modernity. Is it not the "dreams of a better life", encyclopedically dealt with in *The Principle of Hope*, that fuel the euphoria of development and the hunger for growth? And did he not celebrate the beginning of modernity as birth time of "novum" and Columbus' expedition as an "exodus out of the old into the new" (Bloch 1995:751)? He even describes his geographical discoveries with the hoped-for "Eldorado-Eden" as the horizon of all utopian intention, calling it "a golden space" still to be expected and a symbol for the "spatial possibility of a new heaven and a new earth" (794). Columbus is the "leading figure of border crossing", who, unlike Odysseus or Faust, embodies a historically real figure (1204). Ultimately, the "plus ultra", the crossing of all onto-epistemic boundaries following Bacons' project, can not only be considered the myth of origin of Western capitalist expansion dynamics and technoscientific modernity, but it is also a leitmotif of Bloch's "ontology of not-yet-being".

However, taking the definition of Occidentalism proposed by Fernando Coronil (1996:57) as a basis, it becomes clear that Bloch's philosophy, as much as it stems from the Western tradition of thought, contradicts the Occidentalist worldview in main points. It subverts the dualisms of culture and nature, civilized and barbaric, or modern and pre-modern, with which European modernity established its colonial supremacy and enforced asymmetrical and hierarchical relations between peoples, societies, forms of knowledge, and genders. Above all, it challenges the Cartesian dichotomy of mind and matter, through which a patriarchal and anthropocentric relationship of dominant subjectivity and controllable external world could be established. In Bloch's judgment, this reifying and instrumentalizing of the world has paved the way for nature becoming calculable and robbed of its qualities and intrinsic value. This led to a situation, in which humankind, with its techniques of domination, behaves toward nature "like an army of occupation in enemy territory" (Bloch 1995:696).

Bloch recognized and described the coloniality of power in the capitalist relationship to nature. Likewise, with regard to human history, he highlighted the aporia of the "straightforwardly European" concept of progress and rejected the "mere succession of progress", "that European, if not predominantly German, time axis" (Bloch 1970:125), which relegates the non-European and nature to prehistory. Instead of the homogeneous linear time of "before and after" (141), he pleads for a "multiverse" in history that creates space for the "multiverse of cultures" (128f). This allows grasping the unfolding processes of humanity in their polyphony and non-contemporaneous polyrhythms – without "Europeanizing rape" (146) and standardization.

In dealing with the multiverse of times, Bloch insists in recognizing what he calls an "***unaccomplished actuality***" (151) or "future in the past" (152), that calls for realization (Bloch 1970:153, 151). He was fond of quoting the German peasant song that recalls their defeat in the Peasants' War of 1525: "Defeated we go home, our grandchildren will carry on a better fight" (151). In this sense, for him there is "tradition as revolution of the deceased, in the midst of revolution as tradition of the emerging" (153). Tradition, stated the Peruvian Marxist José-Carlos Mariátegui (1970:117–120), who can be described as a pioneer of liberation philosophy and decolonial theory, is not opposed to revolution; on the contrary, it is "alive and in motion" when it is renewed, he emphasized in 1927 regarding the Indian majority in his country, which continues to be oppressed since the conquest.

Bloch, Mariátegui, as well as Fornet-Betancourt (2010:18), consider the "cultural memories" that bear witness to the history of the oppressed highly important. They contain a different reference to time than those of the victors' history and, with their "active latency", they give a new orientation to what is happening today. Moreover, they seek to rehabilitate or reconstruct what has been eliminated by the dominant rationality as irrationality or myth (25). This rescue of a tradition is not to be confused with traditionalism or even an "ancestor cult in philosophy" (Bloch 1995:204). Rather, it is a matter of "saving critique" (Benjamin) that blasts open the petrified past in order to release the unrealized enclosed therein – "a subversive intervention into the traditional philosophical universe" (Rehmann 2018:25) that seeks to transform canonized or reified forms of thought, and indeed all that which has gone bad.

Not everything that is or was, is worth saving. Criteria are needed that ultimately have emancipatory significance. If the question is "how interculturality can be realized in today's world as a method for transforming existing forms of thinking and acting" (Fornet-Betancourt 2007:8), then this includes the question whether this transformation also achieves the opening of Western habits of thinking for "the wisdom that expresses itself in the many life worlds and cultures of humanity" (Fornet-Betancourt 2014b:93). If the heritage of pre-capitalist ways of thinking and being is at issue, then for Bloch (1990:200) it has to be "disruptive in the capitalist economy", so that as "non-contemporaneous" it does not simply work toward "capitalist contemporaneity" (206), instead of reminding us of the "tradition of the yet unarrived" (Bloch 1970:153) future the humiliated and insulted, the colonized and oppressed peoples hope for.

As far as the relationship with nature is concerned, Fornet-Betancourt (2018:323–324) poses the question, whether "an intercultural dialogue that proposes the recovery of experiential knowledge as alternative wisdom present in many cultures" will succeed in liberating us from the dictates of technoscientific reason? He understands the latter as the end point of the formation of a "domineering and destructive" type of knowledge established by European Modernity. It is based on a subject–object dualism and an aggressive anthropocentrism, in which "vital 'communication' between nature or the cosmos and humans does not exist. Nature is not spoken to, it is only exploited, processed and researched" (327).

Nevertheless, more dialogical forms of relating to the world can also be found in the Western tradition of thought. For Bloch, this includes in particular the Stoic concept of the "phantasia katalêptikê": according to Zeno, a mostly sensual perception evokes an impression in the soul, which, through an act of assent by reason, generates an accurate "phantasia", that is, a presentation representing what is actual (cf. Sedley 2002:146). According to Bloch, this is a mutual grasping ("katalêpsis") in which "object and subject meet helpfully, embrace each other reciprocally" (Bloch 1975:256). In the Stoic sage, katalêpsis becomes systematic knowledge, as he strives to recognize the connection and interdependence of the parts (sympátheia) of the world and to practically follow the nature thus grasped (cf. Dietschy 2021:79). The ethics that emerge from this aim is a life "in harmony with nature" (Gr. ὁμολογουμένως τῇ φύσει ζῆν [homologouménōs tê phýsei zên]). This means living in concordance (homología) with the Logos as the divine principle of nature or the cosmos, but also with the particular human nature that is part of the universal nature.

For Bloch, there are also intercultural aspects that play an important role in his uncovering of veiled approaches toward a different relationship with nature and thus proposes a new type of materialism. When referring to Aristotle's conceptualization of the relationship between matter and form, which is fundamental for Western thought, he does not follow the interpretation given by Christian scholasticism. Due to dogmatic reasons, the latter attributed the dominant position to the spiritual principle of form and defined matter and the corporeal world, primarily by means of their passiveness and deficiency. Bloch associates this approach with what he calls "Aristotelian Right". And in contrast to this, he points to another, more hidden, line of thought "that leads from Aristotle not to Thomas and the spirit of the beyond, but to Giordano Bruno and the "*flourishing* all-matter" (Bloch 1972:481).

Consequently, this is what Bloch calls "Aristotelian Left". He locates this viewpoint in Islamic philosophers such as Avicenna (980–1037) and Averroës (1126–1198), as well as the Jewish poet and philosopher Avicebron (1021–1070), and the Christian pantheist Bruno (1548–1600): "They simultaneously rescued and transformed the Greek light", Bloch (497) asserts, which illustrates his own method of a transformative reception of cultural heritage.

Through this long-neglected tradition, matter is valorized by being connected with active, generative forms such that a self-generating emergent collaboration between form and matter is produced. "Thus, the concept of a separate form must dissolve" and, with Avicenna "transfer a part of its effect-actuality [Wirk-Wirklichkeit] to matter" (501).

Moreover, Averroës is speaking of an "eductio formarum ex materia", that is, "the forms emerge from matter itself" (501). It is no coincidence that Averroës brings in the Latin translation of a passage from Aristotle the expression of natura naturans – naturing nature or creative nature – into play for the first time. Finally, in Bruno, who is the "cosmologist of an infinite pregnancy of form" (519), creative matter, which is at the same time "inseminating and impregnated universal life" (494), takes the place of a world-transcending Creator-God. "No wonder", Bloch remarks (502), "the Islamic orthodoxy condemned Avicenna and Averroës" and burned their works, "just as the Christian Inquisition later burned Giordano Bruno in the flesh".

Bloch's vitalist concept of matter, which is as far removed from the dead clod of mechanistic materialism, as it is from the Cartesian dualism of mind and matter, owes much to the "Aristotelian left", but also to its continuation in Spinoza, Schelling, and Marx. The latter two are decisive for Bloch, insofar as they allow him to comprehend social relationships to nature as an interactive process of production in which both, the human being and nature, participate actively and transform each other. This is linked with an epistemological change of perception which is no longer – as in previous materialism – "conceived only in the form of the *object, or of contemplation*, not as *sensuous human activity, practice*, not subjectively" (Marx 1845:3).

Recalling Schelling's recourse to the concept of natura naturans as an "original productivity of nature", Bloch (1995:688) emphasizes that being active and producing cannot be restricted to the human subject. That allows him to conceive of the human–nature relationship in terms of co-productivity and even envisage the possibility of a dialectical connection between producing "subjects" in society and nature. All this is connected with a remodeling of the dualistic conception of reality. The subject of modernity's self-conscious reason, which constitutes its objects, is thus to be dethroned, for it separates what nature has united. Schelling's concern is that such a rift between man and the world be eliminated: "Contact and reciprocal action must be possible between the two, for only so does man become man" (Schelling 1988:11).

That Marx is not far from these thoughts is clearly shown by his Paris manuscripts: "The worker can create nothing without *nature*", he says there. "It is the material on which his labour is realised, in which it is active, from which and by means of which it produces" (Marx 1844:273). Even more explicitly he states: "The *human* aspect of nature exists only for *social* man; for only then does nature exist for him as a *bond* with man" (298). This sociogenesis of a different relation to nature can only be accomplished through the historical movement of people acting consciously and in solidarity. Through "the *positive* transcendence of *private property* as *human self-estrangement*", it is "as fully developed naturalism, equals humanism, and as fully developed humanism equals naturalism" (296). This chiasmus is brought by Bloch into the formula "naturalization of man and humanization of nature" (Bloch 1995:131), which formulates the basic idea leading to the alliance with nature (see Dietschy 2021).[18]

Bloch's "nature alliance"

Bloch's concept of "nature alliance", together with the concept of an "alliance technology", is developed in the chapter "Will and Nature, the Technical Utopias" of *The Principle of Hope*. However, it is important to note that he is not concerned with a technology-based utopia, but with human liberation, thanks to an alliance based on "the social exchange process with nature itself" (Bloch 1995:286). This perspective became apparent to him in the course of the 1920s through his in-depth engagement with Marx. As chapter 19 of *The Principle Hope* shows,[19] it leads him to a Marxism that understands itself as a Philosophy of Praxis, that is, as a "*science of occurrence and change*" (284), which comprehends social and natural reality in its unenclosed potentiality of turnability and capability-of-becoming-other (234). This means, if they are presented in an objectivist manner as reified power, forcing "petrified relations . . . to dance by singing their own tune to them" (Marx 1843:178).

The "Archimedean point" of this *historical* materialism consists in the double "relation of people to people and to nature" (286), so that one can speak of an "ontology of relation".[20] For Bloch (2000:235), "Marxism" stands for "the re-transformation of reifications into process-like mutable relations". In this sense, he also calls for a "Marxism of technology" being capable of re-transforming what is reified in the bourgeois relationship to nature; it would make possible the "release and mediation of the creations slumbering in the womb of nature" (Bloch 1995:695).

Connected with this is the criticism of the "vulgar-materialistic" narrow-mindedness in science and technology and the attempted rescue, not only knowledge banished by modernity, but also of the qualities of nature, lost with it. Bloch criticizes the increasing de-qualification, quantification, and abstractness that prevail in the course of the assertion of modern rationality: a rationality of calculus dominates, "alienated not only from human beings but also from things, one indifferent to their content" (666). This is not so much an attack on the mathematization of natural sciences as on the fact that their

18 It is not surprising that Bloch rephrases the substantival Marxian chiasmus of "naturalism" and "humanism" into "naturalizing" and "humanizing". Hanspeter Dürr points out that in the language of recent physics, activity words such as "bind", "act", or "bubble" take the place of nouns: "Basically, there is only relationship, connection, *religio*, *connectedness*, processuality, but all this without a reference to a substantial something" (Dürr and Panikkar 2008:24).

19 "Changing the World or Marx's Eleven Theses on Feuerbach" (Bloch 1995:249–286).

20 Etienne Balibar (2013:32, 117) sees this as the hallmark of Marx's philosophy, which examines the "transindividual" results of the multiple interactions between individuals.

"abstract laws are different from the substratum of content with which these laws maintain no relation" (667). Bloch does not deny that the "bourgeois-abstract calculation" finds an equivalent in nature itself (694). However, he points out, in agreement with Horkheimer and Adorno (2002:3), that it disregards "anything which does not conform to the standard of calculability and utility".

From this point of view, regaining the *quality of nature* means "attempting to abolish the separation of ethical, aesthetic, sensual qualities from the economically reduced and politically administered object-nature" (Danner 1981:I). It means searching for signs of another relation to nature amidst the "abstractness (alienness) of the bourgeois material relation" (Bloch 1995:665). Bloch even finds them in the progenitor of modern science: Bacon's "'House of Solomon' cannot do without Solomon, that is, without natural wisdom" (657) – it contains reference to its counterpart, the nature.

Moreover, Schelling's "dynamic-qualitative philosophy of nature" is a testimony to "this other side of nature" (Bloch 1959:804), albeit only as a placeholder for the socially not yet accessible reality. In this respect, art can also perform an important heuristic function by pointing through allegorical formation of figures to "real ciphers" of nature. Then, "something may speak to us in the impression of the seemingly mute, as we feel addressed by it" (Bloch 1975:214).

For Bloch (221), aesthetic practice forms "the bridge between man and nature" that needs to be crossed. It anticipates and promotes a possible alliance building through which "the creative forces of a frozen nature will be released again" (Bloch 1995:690). This would, of course, presuppose a peaceful interaction between humans and non-human nature and a "technology without violation" (691), contrasting with the violence of modern appropriation of nature. The latter resembles with its cunning technology a "tamed, guarded colony" (670), but not an "affinitive inhabitation of nature" (671). The technical accident up to and including the technical catastrophe, and the equally inevitable economic crises stem, according to Bloch, "ultimately from a poorly mediated, abstract relationship of human beings to the material substratum of their behaviour" (694). They follow a logic of economic value that is alien and external to the core of human being, as well to the natural substrate.

Nature has therefore "not become good friends with its caning master" (694). This would require an "alliance technique" that does not arise from the perfection of technological solutions, but from a liberation in the interhuman relationship capable to transforming the relationship with nature, "so that man no longer needs to stand in nature as in enemy territory, with the technical accident as a constant threat" (Bloch 1975:251). When speaking of a "befriending" in the nature relationship, Bloch (1995:695) does not mean a "philanthropy for maltreated metals". Solidarity with nature is not a mere repair shop for the destruction of environment but aims at a practice that has left behind colonial relations, violation, and overexploitation. It obviously "presupposes interhuman concretizing, i.e., social revolution; until this occurs, there will not even be any steps, let alone a door, to the possible alliance with nature" (696).

Nevertheless, this does not make this alliance unattainable. Like the aforementioned examples from a different culture's worldview, Bloch (1990:358–359) relies on the multilinearity of historical courses and their specific interconnectedness in time: "There were times – and they could return at a highly rational level – when human beings had a more *cosmomorphic* relationship than that of cunning, than that of mere 'mastery', 'exploitation' of natural forces". The fact that he insists on world compatibility and concludes that in order to ultimately "lift society and nature on to their hinges" (Bloch 1995:248), according to which history must be founded in the still "open totum" of a *cosmology* has generated much criticism toward him. For he understood this explicitly as "the problem area of man and of his work with the possible subject of nature" (Bloch 1995:1175–1176). To critics in the German Democratic Republic who accused him of a "Marxist-disguised religious doctrine of world redemption", he replied that there was "no new Marxist anthropology without a new Marxist cosmology" (Bloch 1970:145).

It is true that "nature alliance" and "alliance technology" refer primarily to "our surrounding nature, in which we can intervene and with which we are in constant metabolism" (Bloch 1975:251). For a practice that is wary of being "only in relation to human society" (252), however, questions also arise about the relationship to the "extra-human as a whole", as they arise not only by the ecologically surrounding but also by the "vaulting" cosmic sphere of nature (177). Fornet-Betancourt (2010:19, 29), in his critique of the impoverishment of cultural horizons through an overly anthropocentric humanism, has (with Panikkar, see Chapter 6) called not only for an "anthropological turn", but also for a recovery of cosmological questions. This is also Bloch's concern. For a long time, he – along with Henri Lefèbvre, almost as the only one from the spectrum of 20th-century Marxisms – has consistently held the view that "the two sides belong together: what we have made historically and what we have not made . . . but what we have included and can include" (Bloch 2000:447). In other words, only a practice that considers the non-human conditions of human life "can lead the matter pending in the historical process: the naturalization of man, the humanization of nature, out of real possibility into reality" (Bloch 1995:247).

This formula, used very often by Bloch, expresses that he is concerned neither with a mere human utopia nor with a nature utopia, but with the indissoluble mutual interweaving of both aspects. *Humanization* of nature does not aim at domestication or instrumentalization of the non-human cosmos of life, but at the restoration of the "metabolism between man and earth" that the capitalist mode of production destroys. It means to establish this exchange recognizing the inherent dynamics of natural processes and "in a form adequate to the full development of the human race" (Marx 1982:638). *Naturalization* of the human being means that the "*development of the richness of human nature*" (Marx 1863:497) or "*human* wealth" according to

Bloch (1995:1375) is to be expanded and merged with wealth "of *nature* as a whole". In this way, an adequacy could be achieved that would make the earth a home and dwelling place for man: human "inhabitation of nature" (671) and the earth's habitability are mutually dependent. This requires definitely quitting the colonial "standpoint of the exploiter and animal tamer" (695) and regarding nature adopting "an explicit technology of alliance which integrates into nature rather than expropriating it" (Zimmermann n.d.:fn. 6).

Bloch's proposal of an alliance-building which aims at re-embedding human practice in the web of nature is certainly not sufficient to achieve the transformations needed. If contextualized with other approaches such as that of intercultural philosophy and decolonial thought and the corresponding practices, it can however contribute to reformulate "sustainability" in terms of relationality. The following points speak in favor of this:

1 The shared conviction of the aforementioned positions is without doubt "that the idea of 'development as progress' needs to be deconstructed to open a way for cultural alternatives that nurture and respect life on Earth" (Kothari et al. 2019:XVII).
2 Whereas the West has succeeded in imposing its idea of "One World", which is supported by its own norm of scientificity, alter-globalization movements plead for a "pluriversality" based on the multiplicity of "ways of worlding" (XXXIV). According to Bloch (1990:358), "precisely in the dialectical method the world thus operates pluralistically, but all the more so as a multiverse versus Unum nondum inventum", that is, as "a *demand* on the Unitas and towards it" (357).
3 From an intercultural and historical perspective, "humanity can make different futures possible" (Fornet-Betancourt 2000:155). Conceivable are, says similarly Sachs (2010c:35), "efforts to elucidate the much broader range of futures open to societies which limit their levels of material output in order to cherish whatever ideals emerge from their cultural heritage".
4 The pluriversal perspective expressed in the Zapatista slogan "a world where many worlds fit" contains a concrete universal claim. It differs from the abstract claims to universality of the West, thanks to its origin in local histories of the oppressed, from which autonomous and communal futures emerge. At the same time, pluriversality refers to the "coexistence of the human and the non-human in profound interrelation" (Escobar 2020:63). Thus, indigenous and peasant communities, as well as environmental movements, seek "to ground human activities in the rhythms and frames of nature, respecting the interconnected materiality of all that lives" (Kothari et al. 2019:XIX). Especially "feminist activists advocate for the simultaneous decolonization and depatriarchalization of society through the interweaving of the community, the environment and the spiritual" (Lozano Lerma 2019:230).
5 The fact that these movements and intellectual currents are mostly critical of abstract Western rationality and globalism does not mean that they

close themselves off to any universalism becoming provincialist. Where they meet without claiming to replace the others, a "universal" can emerge that is "rich with all that is particular, rich with all particulars, the deepening and coexistence of all particulars" (Aimé Césaire, cited in Grosfoguel 2012:95). While Escobar insists more on non-universal, relational, and place-based knowledge, the intercultural philosophy of Fornet-Betancourt (2010:13) advocates more strongly for an "enlargement of the world" and a "temporarily pluralistic universality" through "dialogued intercultural encounter".

6 The power imbalance that has been established in development discourse between the "experts" and those whose development is being talked about also returns in the even more complex field of sustainability between Western and non-Western epistemologies, especially when it comes to popular experiential knowledge or traditional wisdom. The interepistemic dialogue between cultures of knowledge must therefore free itself from the monopoly position of the dominant scientific canon and "break through the barriers of these monocultural structures" (Fornet-Betancourt 2000:151).

7 This is all the more urgent because, in the course of the assertion of the coloniality of modernity, an "epistemology of the hunter" (Panikkar) with an aggressive anthropocentrism has been established, which has increasingly transformed the human cognitive faculty into an instrument for taking possession of the world's objects (cf. Fornet-Betancourt 2000:114–215). Unsustainability, ecocide, and epistemocide – that is, the eradication of other cognitive cultures (Dietschy 2014:67–68) – are undoubtedly partly caused by "dualistic concepts, particular those that divide nature and culture, humans and non-humans, the individual and the communal, mind and body" (Escobar 2016:28).

8 Bloch's plea for a life in alliance with nature, despite the ongoing war against it, can be seen as a way out of this civilizational impasse, as a path that makes colonizers co-inhabitants of the earth to which they belong with other beings (cf. Sepúlveda Pizarro 2021:168). For Fornet-Betancourt (2007:136), this path is about developing "an idea of sustainability reimagined from cosmological and anthropological experiences of interculturality in the sense of a 'deep ethics'" and "designing approaches beyond sustainability to support ethics, i.e., a practice of dwelling on the earth and living another way of life". He calls it an "ethics of embrace" because it leads us to "treat our *environment* as a proximity celebrated through embrace".

9 "Embrace" can also be seen as a symbol of relationality that is the hallmark of non-dualistic worldviews and ways of life we can see in a wide variety of manifestations all over the world. These include ancient forms of mutual aid, as well as those of collaborative action in cyberspace, indigenous and peasant communities that maintain their spiritual and material relationships to the land or to the Pachamama, and emerging "ecoterritorial"

movements that, according to Maristella Svampa (2020:46), are the result of a confluence of indigenous communitarian and autonomous narratives with an environmentalist orientation and popular feminisms. Thus, "Buen Vivir" ("Living Well") is often brought into play as a response to the colonial development doctrine ("desarrollismo") and as an image for the double conviviality – between people and with nature. However, the diverse indigenous expressions (Sumak Kawsay, Suma Qamaña, Teko Kawi, Ñandereko, etc.) are to be distinguished from Western concepts of good living. They rather point at Living Well, together in a (cosmically) comprehensive set of relationships.

10 The inclusion of nature's rights in the new Ecuadorian Constitution of 2008 and in Bolivia's "Law of Mother Earth" in 2012 is part of the changes that were already announced by the emergence of "Buenos vivires" discourses. For Walsh (2015:113–118), the explicit "recognition of Nature as the subject of rights" – including "Nature's right to reparation and restoration" – represents an important break with colonial and anthropocentric understanding of rights, although it has not yet brought about a fundamental change in Western-style constitutional logic, nor a departure from extractivism. Acosta (2015:13) sees this as a first step toward liberating nature from its lawlessness, which at the same time allows for the creation of a kind of ecological "meta-citizenship".

11 One could also interpret Bloch's hypothetical "nature-subject" similarly, namely, as an attempt to underline the autonomy and dignity of non-human nature as a complement to the dignity of human nature. In any case, "nature alliance" is to be understood as the anticipation of a relationship with nature that transforms latent or manifest enmity into partnership and into a collaborative relation. The prerequisites for this alliance, however, are, on the one hand, the liberation of social actors from oppression and structural violence (cf. Rist 2015:117), and, on the other hand, the unfolding of the "co-productivity of nature" (Bloch 1995:686). This needs the practice of a "plural, dialogical and convivial reason" (Fornet-Betancourt 2015a:41).

Acknowledgments

I would like to thank Noa Zanolli and Stephan Rist for their help in revising the English version of the chapter and for their valuable comments on the content.

References

Acosta, A., 2015. El Buen Vivir o sumak kawsay – Una oportunidad para construir otros mundos. In: Stöckli, B., ed. *Questioning Development, Repenser le développement. Repensar el desarollo. Entwicklung neu denken.* Bern and Lucerne: dialogue4change, 9–16, https://repository.globethics.net/handle/20.500.12424/228857.

Alimonda, H., ed. 2011. *La naturaleza colonizada. Ecología política y minería en América Latina.* Buenos Aires: CLACSO.

Arteta Melgarejo, Ximena Carolina, et al., 2021. Colonialidad de la naturaleza: Aspectos decoloniales para el debate sobre el Desarrollo Sostenible. *Revista de Ciencias Humanas, Teoría Social y Pensamiento Crítico,* 14, 288–300.

Bacon, F., 1902. *Novum Organon.* New York: Collier.

Balibar, E., 2013. *La philosophie de Marx.* Paris: La découverte.

Bloch, E., 1959. *Das Prinzip Hoffnung,* Bd. 1–3. Frankfurt/M: Suhrkamp.

Bloch, E., 1970. *Tübinger Einleitung in die Philosophie.* Frankfurt/M: Suhrkamp.

Bloch, E., 1972. *Das Materialismusproblem, seine Geschichte und Substanz.* Frankfurt/M: Suhrkamp.

Bloch, E., 1975. *Experimentum Mundi. Frage, Kategorien des Herausbringens, Praxis.* Frankfurt/M: Suhrkamp.

Bloch, E., 1990. *Heritage of Our Times.* Translated by Neville Plaice and Stephen Plaice. Berkeley: University of California.

Bloch, E., 1995. *The Principle of Hope.* Vols. I–III. Translated by Neville Plaice and Stephen Plaice, and Paul Knight. Cambridge, MA: MIT Press.

Bloch, E., 2000. *Logos der Materie. Eine Logik im Werden. Aus dem Nachlass 1923–1949.* Edited by G. Cunico. Frankfurt/M: Suhrkamp.

Brand, U., and Wissen, M., 2017. *Imperiale Lebensweise. Zur Ausbeutung von Mensch und Natur im globalen Kapitalismus.* München: oekom.

Choquehuanca Céspedes, D., 2020. *Speech of Assumption of the Vice Presidency of the Plurinational State of Bolivia.* La Paz: Vicepresidency of the Plurinational State of Bolivia.

Coronil, F., 1996. Beyond Occidentalism: Toward nonimperial geohistorical categories. *Cultural Anthropology,* 11(1), 51–87.

Daxner, M., 1981. Vorwort. In: Daxner, M. et al., eds. *Andere Ansichten der Natur. Arbeitskreis Naturqualität.* Münster: SDZ.

Delgado, F., and Rist, S., eds. 2016. *Ciencias, diálogo de saberes y transdisciplinariedad. Aportes teórico-metodológicos para la sustentabilidad alimentaria y del desarrollo.* La Paz, Bolivia: CLACSO, AGRUCO-UMSS, CDE University of Berne.

Dietschy, B., 1993. Das Niemandsland *oder* die Entdecker-Utopie Europas. In: Greive, W., ed. *Alexander von Humboldt. Die andere Entdeckung Amerikas.* Loccum: Loccumer Protokolle 10/92, 179–194.

Dietschy, B., 2007. Ohne Ungleichzeitigkeit keine Zukunft. Indigene Versuche, eine andere mögliche Welt zu bauen. *Concordia. Internationale Zeitschrift für Philosophie,* 71, 3–14.

Dietschy, B., 2014. Von der Monokultur zum Multiversum der Erkenntnis. In: Fornet-Betancourt, R., ed. *Gerechtigkeit, Erkenntnis und Spiritualität,* Aachen: Wissenschaftsverlag Mainz, 65–78.

Dietschy, B., 2021. Vom Band zum Bund. Wie Bloch zur Naturallianz fand. *VorSchein. Jahrbuch der Ernst-Bloch-Assoziation,* 37, Nürnberg: Antogo, 77–97.

Dürr, H.-P., and Panikkar, R., 2008. *Liebe – Urquell des Kosmos. Ein Gespräch über Naturwissenschaft und Religion.* Freiburg: Herder.

Dussel, E., 1995. *The Invention of the Americas: Eclipse of "the Other" and the Myth of Modernity.* New York: The Continuum.

Escobar, A., 2005. Depois da Natureza. Passos para uma Ecologia Política Antiessencialista. In: Parreira, C., and Alimonda, H., eds. *Políticas Públicas Ambientáis Latinoamericanas.* Brasilia: Abaré, 17–64.

Escobar, A., 2011. Ecología política de la globalidad y la diferencia. In: Alimonda H. ed. *La naturaleza colonizada*. Buenos Aires: CLACSO, 61–92.

Escobar, A., 2014. *Sentipensar con la tierra. Nuevas lecturas sobre desarrollo, territorio y diferencia*. Medellín: CLACSO.

Escobar, A., 2016. Thinking-feeling with the Earth: Territorial struggles and the ontological dimension of the epistemologies of the South. *Revista de Antropología Iberoamericana*, 11(1), Madrid, 11–32.

Escobar, A., 2020. *Pluriversal Politics. The Real and the Possible*. Durham: Duke University Press.

Esteva, G., 2010. Development. In: Sachs, W., ed. *The Development Dictionary*, London: Zed, 1–23.

Esteva, G., 2014. Aprender a escuchar. ¿Cómo imaginar un mundo en que quepan muchos mundos? In: Fornet-Betancourt, R., ed. *Gerechtigkeit, Erkenntnis und Spiritualität*, Aachen: Wissenschaftsverlag Mainz, 35–44.

Fals Borda, O., 1988. Por un conocimiento vivencial. *Concordia, Internationale Zeitschrift für Philosophie*, 12, 97–105.

Fornet-Betancourt, R., 2000. *Interculturalidad y globalización. Ejercicios de crítica filosófica intercultural en el contexto de la globalización neoliberal*. Frankfurt/M: IKO.

Fornet-Betancourt, R., 2007. *Interkulturalität in der Auseinandersetzung*. London: IKO.

Fornet-Betancourt, R., 2010. Teoría y praxis de la filosofía intercultural. *Ricerca. Revista de pensament i anàlisi*, 10, 13–34.

Fornet-Betancourt, R., 2012. *Interculturalidad, Crítica y liberación*. Aachen: Wissenschaftsverlag Mainz.

Fornet-Betancourt, R., ed., 2014a. *Gerechtigkeit, Erkenntnis und Spiritualität, Dokumentation des XVII. Internationalen Seminars des Dialogprogramms Nord-Süd*. Aachen: Wissenschaftsverlag Mainz.

Fornet-Betancourt, R., 2014b, Thesen zur interkulturellen Transformation der Vernunft. *VorSchein*, 32, Nürnberg: Antogo, 85–98.

Fornet-Betancourt, R., 2015a. La philosophie interculturelle et les défis pour la rationalité philosophique traditionnelle ou eurocentrique. In: Stöckli, B., ed. *Questioning Development. Repenser le développement. Repensar el desarollo. Entwicklung neu denken*. Bern and Lucerne: dialogue4change, 37–41. https://repository.globethics.net/handle/20.500.12424/228857

Fornet-Betancourt, R., 2015b. *Zur Geschichte und Entwicklung der Interkulturellen Philosophie*. Aachen: Wissenschaftsverlag Mainz.

Fornet-Betancourt, R., 2017a. *Elementos para una crítica intercultural de la ciencia hegemónica*. Aachen: Wissenschaftsverlag Mainz.

Fornet-Betancourt, R., 2017b. Ist der Entwicklungsbegriff kolonialistisch? Elemente einer interkulturellen Kritik. In: Estermann, J., ed. *Das Unbehagen an der Entwicklung. Eine andere Entwicklung oder anders als Entwicklung? Dokumentation der Debatte zum Entwicklungsbegriff bei COMUNDO, 2015–2017*. Aachen: Wissenschaftsverlag Mainz, 37–46.

Fornet-Betancourt, R., 2017c. Introduction. In: Fornet-Betancourt, R., and Sement de Frutos, J. A., eds. *Spiritualities and Religions: Their contribution to Justice and Knowledge in the Global Society*. Aachen: Wissenschaftsverlag Mainz, 13–16.

Fornet-Betancourt, R., 2018. Del conocimiento teórico contemplativo al saber dominador y destructor. *Valenciana*, 21, 321–346.

Fornet-Betancourt, R., 2021. *Tradition, Dekolonialität, Konvivenz. Themenfelder zur Erprobung Interkultureller Philosophie*. Aachen: Wissenschaftsverlag Mainz.
Fornet-Betancourt, R., and Senent de Frutos, J. A., eds., 2017. *Spiritualities and Religions: Their Contribution to Justice and Knowledge in the Global Society*. Aachen: Wissenschaftsverlag Mainz.
Gronemeyer, M., 2015. Conviviality. In: Bollier, D., and Helfrich, S., eds. *Patterns of Commoning*. Amherst, MA: The Commons Strategies Group, 50–52.
Grosfoguel, R., 2012. Decolonizing Western universalisms: Decolonial pluriversalism from Aimé Césaire to the Zapatistas. *Transmodernity. Journal of Peripheral Cultural Production of the Luso-Hispanic World*, 1(3).
Horkheimer, M., and Adorno, T. W., 2002. *Dialectic of Enlightenment: Philosophical Fragments*. Stanford: Stanford University.
Icaza, R., and Vázquez, R., 2017. Notes on decolonizing development. In: Estermann, J., ed. *Das Unbehagen an der Entwicklung. Eine andere Entwicklung oder anders als Entwicklung? Dokumentation der Debatte zum Entwicklungsbegriff bei COMUNDO, 2015–2017*. Aachen: Wissenschaftsverlag Mainz, 47–62.
Kothari, A., et al., eds., 2019. *Pluriverse. A Post-Development Dictionary*. New Delhi: Tulika Books.
Lozano Lerma, B. R., 2019. Latin American and Caribbean feminisms. In: Kothari, A., et al., eds. *Pluriverse*, New Delhi: Tulika Books, 228–230.
Lugones, M., 2008. Colonialidad y género: hacia un feminismo descolonial. In: Mignolo, W., ed. *Género y descolonialidad*. Buenos Aires: Ediciones del signo, 13–54.
Lugones, M., 2016. The coloniality of gender. In: Harcourt, W., ed. *The Palgrave Handbook of Gender and Development. Critical Engagements in Feminist Theory and Practice*. London: Palgrave Macmillan, 13–33.
Mariátegui, J. C., 1970. *Peruanicemos al Perú. Obras completas, 11*. Lima: Biblioteca Amauta.
Marx, K., 1843. Contribution to the critique of Hegel's philosophy of law. Introduction, *Marx Engels Collected Works* (MECW)3. London: Lawrence & Wishart.
Marx, K., 1844. *Economic and Philosophic Manuscripts of 1844*. MECW, 3. London: Lawrence & Wishart.
Marx, K., 1845. *Theses on Feuerbach*. MECW, 5, London: Lawrence & Wishart.
Marx, K., 1863. *Theories of Surplus-Value (1861–63)*. www.marxists.org/archive/marx/works/1863/theories-surplus-value/
Marx, K., 1982. *Capital. A Critique of Political Economy, Vol. 1*. Harmondsworth: Penguin.
Mignolo, W., 2011. Epistemic disobedience and the decolonial option: A manifesto. *Transmodernity. Journal of Peripheral Cultural Production of the Luso-Hispanic World*, 1(2). https://doi.org/10.5070/T412011807
Park, S.-W., 2015. A new ecumenical journey – Towards life-enhancing civilization. In: Stöckli, B., ed. *Questioning Development, Repenser le développement. Repensar el desarollo. Entwicklung neu denken*. Bern and Lucerne: dialogue4change, 85–99. https://repository.globethics.net/handle/20.500.12424/228857
Quijano, A., 2007. Coloniality and modernity/rationality. *Cultural Studies*, 21(2), 168–178.
Quijano, A., and Ennis, M., 2000. Coloniality of power, Eurocentrism, and Latin America. *Nepantla: Views from South*, 1(3), 533–580. www.muse.jhu.edu/article/23906
Rehmann, J., 2018. Ernst Bloch als Philosoph der Praxis. *Das Argument*, 60(1), 9–30.

Rist, S., 2015. Klar, ohne Befreiung geht das nicht – Gedanken zur "grossen Transformation" in der Nachhaltigkeitsdebatte. In: Stöckli, B., ed. *Questioning Development. Repenser le développement. Repensar el desarollo. Entwicklung neu denken.* Bern and Lucerne: dialogue4change, 100–119. https://repository.globethics.net/handle/20.500.12424/228857

Sachs, W., ed., 2010a. *The Development Dictionary. A Guide to Knowledge as Power* (2nd ed.). London: Zed Books.

Sachs, W., 2010b. Introduction. In: Sachs, W., ed. *The Development Dictionary*, London: Zed Books, XV–XX.

Sachs, W., 2010c. Environment. In: Sachs, W., ed. *The Development Dictionary*, London: Zed Books, 24–37.

Salleh, A., 2019. Earth system governance. In: Kothari, A., ed. *Pluriverse*, New Delhi: Tulika Books, 40–43.

Schelling, F. W. J., 1988. *Ideas for a Philosophy of Nature as Introduction to the Study of This Science.* Cambridge: Cambridge University Press.

Sedley, D., 2002. Zeno's Definition of *phantasia katalêptikê.* In: Scaltsas, T. and Mason, A. S., eds. *The Philosophy of Zeno. Zeno of Citium and His Legacy.* Larnaca, 135–154.

Sepúlveda Pizarro, J., 2021. Sustentabilidad e interculturalidad: avanzar hacia un nuevo modo de habitar la tierra. In: Samaniego Sastre, M., ed. *Estudios interculturales desde el Sur. Procesos, debates y propuestas.* Santiago de Chile: Ariadna, 165–181.

Svampa, M., 2019. *Neo-Extractivism in Latin America. Socio-Environmental Conflicts, the Territorial Turn, and New Political Narratives.* Cambridge: Cambridge University Press.

Svampa, M., 2020. *Die Grenzen der Rohstoffausbeutung. Umweltkonflikte und ökoterritoriale Wende in Lateinamerika.* Bielefeld: Transcript.

Tlostanova, M., and Mignolo, W., 2012. *Learning to Unlearn. Decolonial Reflections from Eurasia and the Americas.* Columbus: Ohio State University.

Truman, H. S., 1949. *Inaugural Address.* www.bartleby.com/124/pres53.html

UN, 1987. *Report of the World Commission on Environment and Development: Our Common Future.* New York: World Commission on Environment and Development.

Walsh, C., 2006. Interculturalilad y colonialidad del poder. Un pensamiento y posicionamiento otro desde la diferencia colonial. In: Walsh et al., eds. *Interculturalidad, descolonización del Estado y del conocimiento.* Buenos Aires: Ediciones del signo, 21–70.

Walsh, C., 2014. Decolonialidad, Interculturalidad, Vida desde el Abya Yala-Andino. Notas pedagógicas y senti-pensantes. In: Borsani, M. E., and Quintero, P., eds. *Los desafíos decoloniales de nuestros días.* Neuquén: Educo, 47–78.

Walsh, C., 2015. Life, nature and gender otherwise: Feminist reflections and provocations from the Andes. In: Harcourt, W., and Nelson, I. L., eds. *Practising Feminist Political Ecologies. Moving Beyond the 'Green Economy'.* London: Zed, 101–130.

WCRC, 2014. *The Accra Confession.* Hannover: World Communion of Reformed Churches. http://wcrc.ch/accra/the-accra-confession.

Ziai, A., 2006. *Zwischen Global Governance und Post-Development. Entwicklungspolitik aus diskursanalytischer Perspektive.* Münster: Westfälisches Dampfboot.

Zimmermann, R., n.d. *The Philosophy of Ernst Bloch.* www.annette-schlemm.de/gast/bloch.htm

8 Society–labor–nature

The potential of conflict[1]

Nora Räthzel

Introduction

When we were researching the environmental policies of trade unionists in countries of the Global North and Global South,[2] we found that their policies were connected to the ways in which they perceived the labor–nature relationship (Räthzel 2021b; Räthzel and Uzzell 2019). Studying how other researchers have made sense of such perceptions, I found that the concept of work is underrepresented in theories of the nature–society relationship.

In what follows, I reflect how work as the mediator of human and non-human nature could be included into existing theories of the society–nature relationship. However, this is not a comprehensive literature review. I take the liberty to choose six scholars discussing the nature–society relationship which I find useful for my own work: Noël Castree, a human geographer; Jason Moore, a Marxist environmental historian; Donna Haraway, a feminist historian of consciousness; David Choquehuanca Céspedes, presenting an indigenous vision of the human–nature relationship; Ariel Salleh, a Marxist ecofeminist sociologist; and Frigga Haug, a Marxist-feminist sociologist. I do not claim that these scholars are "representative", but they are influential within and beyond their disciplines.

My thesis is that there is already a potential conflict at the human–nature interface (exacerbated by the capitalist mode of production) because, independent of the different modes of production in times and places, human work is characterized by its universality, by its potential to transform non-human natures on a global scale, irrespective of the specific cultures, modes of production through which it is shaped.

In the first section, I put several ways of thinking the society–nature relationship into dialogue with each other: nature as social, capital as producing nature and being produced by nature, humans and nature as mutually

1 Parts of this text have appeared in Räthzel (2021a).

2 David Uzzell, Ragnar Lundström, and myself within research projects funded by the Swedish research foundations FORTE and VR.

DOI: 10.4324/9781003043577-8

producing each other, Vivir Bien claiming an ontological harmony between humans and non-human natures, and reproductive work as a perspective of how to overcome the alienation of industrial work. In the second section, I discuss Marx' concept of work. The third section suggests a synthesis between the different perspectives and Frigga Haug's concept of gender relations as relations of production. I suggest that a broad theoretical framework is needed to understand the mutual interdependency of the production of life (a concept with which I replace "social reproduction") and the production of the means of life. In the fourth concluding section, I argue that within that theoretically framework, we can recognize the empirically existing conflicts and common interests between workers of all sectors across the globe.

The society–nature relationship

Nature as "inescapably social"

As Noël Castree maintains in his introduction to the book he edited with Bruce Braun, *Social Nature*, geography has a long tradition of investigating the society–nature interface (Castree and Braun 2001). They assemble authors who see nature as "inescapably social", arguing that at the time of publication this was a new, surging perspective.

The societal character of nature was perceived on three dimensions. First, it was "socially defined and determined", "often to serve specific, and usually dominant, social interests". This meant, Castree argues, that "the social and the natural were intertwined in ways that made their separation – in either thought or practice – impossible" (ibid.:3). Second, there is the "Knowledge of nature", which, as Castree writes, "is invariably inflected with the biases of the knower/s" and therefore there can be "only particular, socially constituted knowledges, in the plural" (ibid.:10). This is not a relativist statement in the sense that "we can know nothing", it means that our knowledge is never complete, can never reflect the world around us (nor who we are) in a one-to-one relationship. Third, seeing nature as social implies that seemingly natural events like storms, draughts, and floods have different effects on different societies and social groups. Those who have the least resources to protect themselves are hit hardest. A draught is only a draught for those whose lives depend on working the soil in a way that requires water. Another example is the climate crisis, which is created by richer but affects poorer countries much stronger. Thus, technological fixes do not solve the problems because they do not address the deeper issues of societal inequality. Relations of power determine the ways in which societies intervene in nature as when hazardous facilities are sited where people have the least resources to contest them. Climate Justice is the concept and practice aiming to address these inequalities.

These ways of understanding the society–nature relationship as necessarily social and determined by social power relations which have material effects, Castree argues, make it necessary to deconstruct given knowledges of nature.

That is, "showing them to be social products arising in particular contexts and serving specific social or ecological ends that ought to be questioned" (ibid.:12). He criticizes the ways in which scholars speak about society "impacting on", or "destroying" environments. Against such wording, critical geographers argue that the social and the natural cannot be physically disentangled. All there is, Erik Swyngedouw maintains, is 'socionature' (Swyngedouw 1999:443).

The question arises: If it is not possible to speak about "impacting on" nature, how can it be said that human actions are a decisive cause of the climate crisis? Castree argues that "physical opportunities and constraints of nature" (ibid.:12) are relative, that, for instance, the Amazon rainforest will have different physical attributes depending on who wants to use it. This seems trivial. The rainforest presents itself differently to communities that live within and through it than to those who see it as nothing but a source of timber. But does that mean that the physical characteristics of nature are "not fixed" (ibid.:13) or rather that out of a range of physical characteristics, different ones have different meanings for different actors, depending on how they want to use those characteristics and on the social power they have to pursue their interests?

Castree maintains, "We must live with this inability to know nature 'as it really is', while still remaining committed to the idea that some knowledges of, and practices on, nature are better or worse than others" (Castree 2001:16). What we learn from this brief examination is how difficult it is to find a language that represents the dialectics of the nature–society relationship. Castree's text is an example of this difficulty when he assures us that we cannot talk about humans "acting on" nature because nature and society are one, while writing a few pages later that "some . . . practices *on nature* are better or worse than others" (emphasis added). There is a dilemma between conceptually absorbing nature into society while still maintaining that society (humans) acts toward nature. The following scholars present new languages to overcome this dilemma.

Finding a new language: a Marxist approach

The environmental historian and geographer Jason Moore (2015) emphasizes the need for a language able to overcome the century-old division between "Nature" and "Society", which, as he maintains, is older than but central to capitalism. Descartes' separation of mind and body and his idea of a rational universe, Moore argues, "can be viewed as both symptomatic of, and contributing to, the seventeenth century's massive reorganization of power, capital, and nature" (ibid., introduction).

Like the geographers presented earlier, Moore criticizes conceptions seeing Society (or Capitalism) "acting on" nature because they fail to understand that "the species-specificity of humans is already co-produced within the web of life" (ibid.). Moore talks about the "web of life", the "flow of flows in which the rest of nature is always moving through us", "life making . . . that views the boundaries of the organic and inorganic as ever-shifting", about "world within worlds", and "webs within webs" (ibid., introduction). It becomes clear that trying to articulate how human and non-human lives co-produce each other is not an easy undertaking. Moore offers the term oikeios to express that capitalism needs to be seen as a producer as well as a product of nature. It originates from the Greek, meaning household:

> [T]he oikeios is a relation that includes humans, and one through which human organization evolves, adapts, and transforms. Human organization is at once product and producer of the oikeios: it is the shifting configuration of this relation that merits our attention.
>
> (Moore 2015:7)

This concept serves Moore to define the difference between his understanding of the society– nature relationship and that of others who have claimed that society and nature are one. Oikeios, he says, provides a bridge between such philosophical claims and his historical method, an investigation of the specific ways in which the nature–human relationship is constantly changing historically. What changes are the ways in which specific aspects of humanity, such as civilizations, "'fit' within nature" (introduction). It is, indeed, his historical method that differentiates Moore's work from other attempts to make sense of the society–nature relationship. His approach is useful because it allows him to pose new questions: "Instead of asking what capitalism does to nature, we may begin to ask how nature works for capitalism? If the former question implies separation, the latter implicates unification: capitalism-in-nature/nature-in-capitalism" (introduction).

These inverting processes constitute what Moore calls the "double internality", that is, nature and capitalism (the specification of the transhistorical notion of society) as always already inside each other.

Despite a language which at times seems to assume an equation of "human and extra-human work", Moore does mark the difference between capitalism-in-nature and nature-in-capitalism when he explains the double internality:

> [T] he first is capitalism's internalization of planetary life and processes, through which new life activity is continually brought into the orbit of capital and capitalist power. The second is the biosphere's internalization of capitalism, through which *human-initiated projects* and processes influence and shape the web of life.
>
> (Moore 2015:10; emphasis added)

This recognizes the "uneven process" of the double internality: while capitalism internalizes planetary life by changing it, the biosphere's internalization of capitalism is an effect of *human-initiated projects.* In the co-constitution of capitalism and the biosphere, it is capitalism that is responsible and needs to be taken to account for the ways in which it has transformed nature. The challenge for understanding the nature–society co-construction is to stick with the simultaneity *and* the unequalness of the processes. However, what constitutes the difference between capitalism's internalization of planetary life and the biosphere's internalization of capitalism is not further developed. We will get back to this question in the section about work.

Moore overcomes what he calls the dualism between nature and society by finding a spatial language for the way in which they process and proceed within each other, nature in society and society in nature. Donna Haraway, a feminist historian of science and technology focuses on the practices of human and non-human natures to conceptualize the society–nature relationship.

Finding a new language: a feminist approach

Like *Moore*, Haraway does not shy away from inventing a new language to help us understand that

> what used to be called nature has erupted into ordinary human affairs, and vice versa, in such a way and with such permanence as to change fundamentally means and prospects for going on, including going on at all.
> (Haraway 2016:40)

This could also be a description of how nature has erupted into Haraway's thinking. In the 1980s she became famous with her "Cyborg Manifesto" arguing that we need not fear but celebrate the fusion of human bodies and machines. Much of her following work dealt with the way in which nature was "invented" by scientists, for example, primatologists, reproducing narratives of gendered and racialized power relations by "discovering" them in the behaviors of animals. In her recent work, Haraway has redirected her gaze to investigate the relationship between humans and non-humans, including nature as an active agent. Numerous authors have thought of nature (and things in general) as active,[3] many of which Haraway discusses in her book. I discuss Haraway here as an example of finding a language that directs us toward the concept of work/labor.

Reflecting on the work of philosophers like Isabel Stengers and latest insights into microbiology, she suggests words that describe the process in which humans and non-humans act together to create a common world. One of these words is "sympoiesis":

> Sympoiesis is a simple word; it means "making-with". Nothing makes itself; nothing is really autopoietic or self-organizing. . . . That is the radical implication of sympoiesis. Sympoiesis is a word proper to complex, dynamic, responsive, situated, historical systems. It is a word for

3 Authors thinking of nature as active belong to quite different epistemological schools of thought. To name just a few, Latour (2000); Bloch (1973); Bennett (2010); Barad (2007); and Stengers 2010.

> worlding-with, in company. Sympoiesis enfolds autopoiesis and generatively unfurls and extends it.
>
> (ibid.:58)

Although Haraway has used similar words like Swyngedouw's socionature when speaking about natureculture (Haraway 2008), she goes beyond the idea of nature as being always already social. Sympoiesis, "working with," thinks of non-human nature as active in its relationship to humans, irrespective of the specificity of the processes (different in time and space) through which humans engage with non-human natures. This overcomes the problem of simply absorbing nature into society.

With humans and nature as actors working together, it becomes possible to understand their relationship more concretely by examining the different character of agency that humans (as part of nature) and non-human natures perform. Haraway quotes Stengers describing Gaia[4] as follows:

> [A]n intrusive event that undoes thinking as usual. She is what specifically questions the tales and refrains of modern history. There is only one real mystery at stake here: it is the answer we, meaning those who belong to this history, may be able to create as we face the consequences of what we have provoked.
>
> (Ibid.:44)

Not unlike Moore's conception of the double internality, this narrative describes Gaia as provoking an answer from humans while being itself an answer provoked by human activities. While Stenger emphasizes the violence within this unequal relationship, Haraway claims that humans and non-humans are also "necessary to each other's becoming" (ibid.:64) emphasizing their reciprocal interdependence on each other. However, humans can decide to create an alliance[5] with non-human nature or to work with/within nature irrespective of the consequences this has for their relationship. Humans can be constructive or destructive *intentionally* in their relations with non-human natures. Non-human natures cannot decide to create an alliance with humans

4 "What I am naming Gaia was in effect baptized thus by James Lovelock and Lynn Margulis at the start of the 1970s. They drew their lessons from research that contributed to bringing to light the dense set of relations that scientific disciplines were in the habit of dealing with separately – living things, oceans, the atmosphere, climate, more or less fertile soils" (Stengers 2015:44).

5 I take the notion of the labor–nature alliance from Ernst Bloch's Naturallianz. The concept appears first in his main work: Prinzip Hoffnung (Bloch 1973). He argues that an alliance with nature, a synergetic relationship between humans and non-human natures, requires to overcome the capitalist mode of production and thus the dominance of abstract over concrete labor. Beat Dietschy provides an excellent discussion of Bloch's Naturallianz in Chapter 7 of this volume.

or to keep away from them. Therefore, nature cannot be made responsible for the way it interacts with humans, whereas humans are responsible.

Stengers makes this difference clear when she criticizes the way in which Lovelock thinks about Gaia's activity today:

> [E]voking a vindictive Gaia is not just to attribute to her a memory but also an interpretation of what happens in terms of intentionality and responsibility. For the same reason, to speak of the "revenge" of Gaia, as James Lovelock does today, is to mobilize a type of psychology that doesn't seem relevant.
>
> (Stengers 2015:47)

That Gaia does not care about human beings implies that while its actions are provoked by human actions, its existence is not dependent on humans. One could argue that once humans appear on earth and transform it, nature needs humans to tend to its needs while tending to their own. Nevertheless, nature and humans do not "need each other" in the same way. While humans need non-human natures to survive, non-human natures can exist without humans.

Haraway's concepts, with which she represents the symbiotic relationship between human and non-human natures as being necessary for each other's becoming, avoid the problem of absorbing nature into society but they do not clarify sufficiently the difference between their respective interdependence and the different character of their actions.

Vivir Bien: insisting on harmony

The approaches presented so far have something in common; conflict is at the core of their argumentation: Moore's analysis of the capitalist mode of production defined by the exploitation and appropriation of humans and non-human natures; Haraway's sympoiesis that includes the violent conflict between human action and Gaia's reaction to it. In contrast to these approaches, Choquehuanca Céspedes and Rist 2023, Chapter 3 in this volume) begins with the notion of harmony.

> We will promote opposing coincidences to seek solutions between the right and the left, between the rebelliousness of the youth and the wisdom of the grandparents, . . . Our truth is very simple, the condor takes flight only when its right wing is in perfect balance with its left wing. . . . [T]he task of forming us as balanced individuals was brutally interrupted centuries ago, . . . the time of the era of the ayllu, community, is already with us. It demands that we be free and balanced individuals *to build harmonious relationships with others and with our environment.*
>
> (Chapter 3, Choquehuanca Céspedes and Rist 2024:43, in this volume, emphasis added)

The metaphor of the condor who can only fly if its right and left wings are in perfect harmony seems persuasive. However, it does not consider that these wings belong to one body steered by one brain. This is not the case regarding the societal conflicts which manifest themselves in different bodies guided by different, sometimes contradicting goals. To remain in the metaphor, the representatives of these conflicting goals want to fly in opposite directions. However, neither of those opposed pairs mentioned by Choquehuanca Céspedes can realize their goals by getting rid of the other. Thus, the need for an equilibrium, at least a compromise, is clear.

The question arises, how such an equilibrium could be achieved. Choquehuanca Céspedes' presentation of the indigenous vision of Vivir Bien and Madre Tierra presupposes a harmony between all living beings:

> The path of life is the path of integrality between human beings and nature, living in the totality of the cosmos and Mother Earth as part of a great cosmic–telluric family. Nature is the primordial substance, the principle of all being. We are an intrinsic part of nature and nature is a constitutive part of us. In the "cosmobiocentric" vision of millenary and ancestral cultures, the human being coexists in harmony and balance with the taqpacha (all that exists), where human beings constitute with nature a single living being.
>
> (Chapter 3, Choquehuanca Céspedes and Rist 2024:44–45, in this volume)

The harmony of all living beings is a vision about the past as well as a state that needs to be recovered for the future, since it has been destroyed through colonialism and capitalism.[6] A struggle is needed to re-create the lost harmony. This struggle is defined by Choquehuanca Céspedes not only as a struggle against capitalism, colonialism, and the way these have colonized the minds of the people, but predominantly a struggle *for* something. This "for" is outlined in terms of broadening the aims of the struggles that have taken and take place in the Global North:

> Living Well fights against capitalism and goes beyond socialism, since for Living Well the most important thing is to build a culture of life. . . . However, when we talk about the search for complementarity we are talking about the life and freedom of all living beings raised by Mother Earth and not only about the freedom of human beings. In the Living Well, each one is the size he/she is; no more and no less, and thus, he/she has to respect and complement the other.
>
> (Chapter 3, Choquehuanca Céspedes and Rist 2024:53, in this volume)

6 The vision that indigenous knowledge has been in harmony with nature is debated by some anthropologists (Krech 2005). But research by the FAO shows that indigenous populations are the "best guardians" of the forests (Forest governance by indigenous and tribal peoples: An opportunity for climate action in Latin America and the Caribbean (FAO and Filac 2021).

While the harmony between all living beings as a point of departure and its recovery as the objective of the struggles for life becomes clear, what is missing is a differentiation between human and non-human species in terms of the different ways in which they transform natures to meet their needs. Haraway and Stengers both have noted the differences between the activities of human and non-human natures. However, neither they nor Choquehuanca Céspedes has discussed the specific character of the human activity: work. In the image (figure 1) that represents the lost and aimed for harmony (see Chapter 3 in this volume), labor does not feature.

In the next sections, I will discuss different perspectives on work. I will begin with an ecofeminist perspective on "reproductive labour" exemplified by Ariel Salleh who connects Marxist, eco-socialist, feminist, and decolonial perspectives.

Nature and reproductive labor: an ecofeminist approach

Ariel Salleh (2017) has developed the notion of meta-industrial workers to denote different kinds of non-waged work that is not integrated into the capitalist production system but is necessary for its functioning. They include subsistence workers (men and women) and the unpaid work of predominantly women in the domestic sphere.

Starting from Marx' analyses of the contradictions of the capitalist production process, Salleh contends that "an ecofeminist perspective is readily compatible with this materialist analysis but seeks to re-frame these contradictions using a different lens" (Salleh 2020:42).[7] This lens is the shift from "production toward reproduction in its several senses" (ibid.). Reproduction is defined as the work that is done toward humans to either reproduce their capacity to work or to care for those who cannot work. Salleh argues that Marx has not been sufficiently explicit on the role of reproductive labor. While feminists have criticized this in the 1970s, their analyses did not include an understanding of the relationship between labor and nature. This is where ecofeminists come in to critically analyze the dualism between "humanity and nature" and the "split between productive versus reproductive labors, as reflecting a profound alienation embodied in the social construction of masculine gender identity and the social construction of its thought products" (ibid.). What masculine gendered conceptualizations of production do not undertake is "giving equal weight to the interrelated entities of man, woman, nature" (ibid.:43). Reproductive labor is central, Salleh argues, because "it is necessary, integral, and attuned to general causal processes within the ecosystem" (ibid.:44). An example she

7 Salleh has further developed this argumentation in a range of works, her most influential being *Ecofeminism as Politics*, which first appeared in 1997. An updated version was published in 2017 (Salleh 2017).

gives for such reproductive labor is the work of the Australian Aborigines. Their work, and the hunter-gathering mode in general,

> is really "reproductive" in that it does not take more than it needs. . . . [T]he seasonal walk is made in knowledge that each habitat will replenish and provide again on the return.
>
> (ibid.:46)

Salleh's ecofeminist perspective of embodied reproductive labor has the advantage of focusing on ways in which human labor can be performed in alliance with nature. However, meta-industrial labor is described as if it takes place outside of the dominant relations of production, untouched by the dominant logic of capitalist profit-making. Its marginalization seems to protect it from alienation and a destructive approach to nature. This becomes obvious where Salleh describes how women's reproductive labor and its relation to nature is different from "capital" and thus, in her view, from men's labor:

> A first difference involves experiences mediated by female body organs in the hard but sensuous labours of birthing and suckling. A second difference follows from women's historically assigned caring and maintenance chores that serve to "bridge" men and nature, a third difference involves women's manual work in making goods as farmers, cooks, herbalists, potters, and so on. A fourth difference involves creating symbolic representations of "feminine" relations to "nature" – in poetry, in painting, in philosophy, and everyday talk. Through this constellation of lay labours the great majority of women around the world are organically and discursively implicated in life-affirming activities, and they develop gender-specific knowledges grounded in this material base.
>
> (ibid.:44)

Salleh is clear in distinguishing her description of women's work and their capabilities and worldviews as developing through practices from essentialist ideas that women are per definition closer to nature. Her argument is one of embodied materialism claiming that it is through practices that we form our views of the world and develop specific competences and knowledges. Yet her description does not consider the mode of production within which women's work takes place; their work seems to be transhistorical. It can be valuable to first analyze the character of human work independent of the specific mode of production and societal context in which it takes place, but then it is necessary to analyze how the general character of human work changes under different conditions. The split (between production and reproduction) which Salleh analyses is, as she says, already the result of an alienation "embodied in the social construction of masculine gender identity". In other words, the split that assigns reproductive work predominantly to women is historically specific. Thus, it is not adequate to analyze women's work as if it would take place in a societal vacuum, since the fact that it is women's work is already the result of historically specific societal developments.

When Salleh writes that "while industrialized productive labor is historically contingent", "reproductive labor is necessary, integral, and attuned to general causal processes within the ecosystem", she argues on two different levels of historical specificity. She writes about reproductive labor "as such" independent of place and time and of productive labor at a specific historical moment: industrialization. If she would investigate reproductive labor in industrialized societies, she would see that it takes place under different gendered, racialized, relations of class power across the world. In most countries today, it takes place within the logic of (varieties of) capital. There is no reason why reproductive work within capitalist societies should not be subjected to similar forms of alienation as labor in the sphere of capitalist production – even if it takes place in the domestic sphere and not under the direct control of capitalist ownership.

To clarify this point, I want to confront two descriptions, one of reproductive, one of productive work. The first is a description of domestic work, quoted by Salleh, the second a description of productive work by workers in an industrial plant constructing buses. I begin with Salleh quoting Ulla Terlinden, an ecology activist:

> "Housework requires of women [or men] a broad range of knowledge and ability. The nature of the work itself determines its organisation. The work at hand must be dealt with **in its entirety** . . . The worker must possess a high degree of personal synthesis, initiative, intuition, and flexibility". Salleh comments: "Contrast this total engagement with the fragmented industrial division of labour and the numb inconsequential mindset to which it gives rise:.
>
> (ibid.:45, emphasis in the original)

Workers in Volvo plants with whom we talked would not agree with Salleh's description of their work:[8] a woman in a Mexican Volvo plant describes her work in the factory as being similar to her domestic work:

> The satisfaction of doing a job that was done by two people before, and now I am doing this all alone, . . . This gives me the satisfaction that yes, I can do this. As women, we really finish what we start with. We are more agile. Like in the house. The women come home, get going with the household stuff and think of tomorrow. We have a rhythm. And the man hasn't.
>
> (Räthzel et al. 2014:158)

8 See Räthzel, Mulinari, and Tollefsen 2014 for a discussion of the workers' "producer's pride".

A man in the same factory explained:

> When you are a metal cutter you must do things well because there are specifications. And you cannot ignore those specifications because if you do, this will result in a bad product. Responsibility is very important; it is very serious. Because you make the holes where the bus is going to be connected. And if these holes are not precise, you cannot connect the bus. I like this responsibility very much.
>
> (Räthzel et al. 2014:201)

All three descriptions portray the concrete character of people's work from their point of view. While Salleh sees differences between women's reproductive work in the domestic sphere and industrialized work, the woman working at Volvo experiences both kinds of work as similar. She can do both successfully due to her qualifications as a woman, and she enjoys both.

Both Volvo workers do their work with care and comprehension. Both know what they are doing, are proud of it, and cannot be described as having a "numb inconsequential mindset". How can it be that Salleh agrees that reproductive domestic labor requires high qualifications while she sees industrialized labor as lacking not only qualifications but even understanding or sensitivity? Salleh thinks of reproductive labor as *detached from* the societal conditions within which it takes place, while she thinks of productive labor *only* in terms of the mode of production within which it is performed. To get a clearer picture of both "reproductive" and "productive" labor, it is necessary to analyze in *both cases* how work can be a source of satisfaction and requires competence. But it is equally necessary to analyze both kinds of practices within the societal contexts in which they take place. We will then be able to see that both productive and reproductive work are characterized by the dialectics of the producer's pride and the producer's alienation.

In the case of wage workers in industrialized production processes, Marx has described the workers' subordination as an alienation from the product, the production process, and ultimately their humanity since they cannot control the work processes in which they are engaged. They are forced to sell their work power to those who own the means of production. What Marx did not analyze is that despite this alienation, workers can and need to develop their skills and they can still be proud to "do a job well for its own sake" (Sennett 2008). Similarly, what Salleh does not analyze clearly is that even if reproductive work does not take place under the control of privatized production processes but in the domestic sphere or in public services, the societal conditions, today the logic of capitalist profit-making, individualization, and competition, do shape the way in which this work is undertaken similar to the way they shape industrial labor.

The reason why the experiences of "reproductive labor" appear to be different from the experiences of "productive labor" is due to the different theoretical levels on which Salleh discusses them. Describing the various forms in which workers are alienated from the production process she talks about waged labor within capitalism. Their work consists of what Marx calls: "concrete labor" and "abstract labor". Concrete labor is the specific work that produces the use value of a commodity (e.g., the concrete work of a carpenter producing a table). Abstract labor is the societally necessary time needed to create a commodity, producing the exchange value of products. It is called abstract labor, because it does not take the concrete production process into consideration, nor the qualifications of the worker but only the socially necessary time needed for producing the product.

What is important from the point of view of capital is that a product is produced in the shortest amount of time. The less the time needed to produce a commodity, the less it costs capital to produce it (in terms of wages), which means the more profit can be made by selling it to a lower price than its competitors or maybe at the same price as the competitors but having invested less money. Thus, the two dimensions of work, concrete labor needed to produce a useful product and abstract labor (the time needed to do this) are in constant tension. Minimizing the time within which a commodity can be produced means to sacrifice to a certain degree the use value of the product. This hierarchical relation between concrete labor and abstract labor impacts on the labor–nature relationship. Working, transforming nature in a way that allows it to regenerate costs more time, therefore reduces profit.

When Salleh speaks about reproductive labor as non-alienated labor, she talks only about concrete, useful labor. In turn, when she talks about alienated labor directly subsumed under capital, she talks only about abstract labor, the time pressure in the production process. However, care work, even when it takes place at home or in public services, consists of concrete labor and abstract labor, the time needed to create useful care. Because of the way in which "efficiency" dominates daily life, a tension between the concrete work of caring and the time in which it can or has to be done, exists here as well.

The concept of reproductive labor and its relation to nature is an important contribution of ecofeminists to the question of the society–nature relationship because they introduce the notion of human work as a central category to understand this relationship. However, by describing this work without considering how reproductive work is also shaped by the mode of production, they run the risk of idealizing reproductive labor as an escape from the destructive character of work within capitalist relations of production. I will discuss this further in the next section, where I develop my understanding of work as a mediator between human and non-human natures.

Work as a conflictual mediator between human and non-human natures

The approaches discussed earlier, when describing human and non-human natures as co- constructing each other, have alerted to but not specified the differences between the activities of human and non-human natures. Salleh has described the regenerative character of reproductive labor while dismissing productive labor as neither necessary nor regenerative. In what follows, I argue that it is by developing the concept of work that we can better grasp the society–nature relationship.

I begin with Marx's definition of work:

> [I]n the first place, a process in which both, human beings and nature participate, and in which human beings of their own accord start, regulate, and control the metabolism (Stoffwechsel) between themselves and non-human nature. Humans face natural matter as a force of nature. They set in motion the natural forces of their own bodies, arms and legs, head, and hands, in order to appropriate natural matter in a form suitable to their own lives. By thus acting on the external world through their movements and changing it, humans simultaneously change their own nature. They develop their own slumbering capacities and subject their forces to their sway.
>
> (Marx 1887/1998: Capital, chapter 5)

This process is what Marx calls an eternal necessity for humans to survive. At this level of analysis, the relations of production within which work takes place are not considered. No matter the mode of production, work is and will always be a process in which humans transform nature to meet what they define as their needs. Marx' formulations have motivated authors to interpret this passage as a description of humanity's domination of nature. One can see it that way if one emphasizes the description of humans *beginning* the process of work. Marx is here congruent with Moore's and Haraway's representations of humans as the originating actors. But since human nature and non-human natures both change their nature in the process, both are active parts of the process. One can think of processes in which non-human natures are the originators, for instance, when bacteria attack the human body or when humans begin to plant seeds because they have observed that nature produces edible plants from them. However, these activities are different from the work process itself in which humans transform non-human and human natures.

If we consider work on this general level irrespective of the societal relations within which it is practiced (irrespective of the power relations between humans) and irrespective of the concrete ways in which non-human natures and human natures are transformed, there is no difference between a worker creating a tool out of stones or metal, a worker farming the land, a worker suckling a baby, cooking a meal, or caring for another person. Each of these labors does not only transform natures (human and non-human) in a productive way, since they create something new, they must also be reproductive to start the work process again the next day, week, month. All workers, those who produce tools, housing, clothes, food, that is, "the means of life", and those who produce human beings through procreation, caring for their emotional, social, and physical well-being, are simultaneously productive and reproductive. All of them are necessary workers dependent on each other. Salleh's meta-industrial workers need tools, clothes, housing, culture, art, science, etc. There is no difference in terms of necessity between the (re-)production of the means of life (commonly called production) and the (re-)production of life (commonly called reproduction). However, even on this general level, there is already a conflicting relationship between human work and non-human natures due to the specificity of human work.

In *Capital*, Marx explains the specificity of human work distinguishing it from the work of animals. In difference to the best bees, he claims, the worst architect has an image of the product they are about to create in their minds before they start working. Given research on the activities of animals and plants, this description does not ring true anymore (e.g., Prieur and Pika 2020; Mancuso et al. 2015). However, even knowing that the roots of plants need to decide whether they grow toward a place with more water or one with more minerals, depending on what the plants need; even knowing that, for instance, gorillas create and keep tools, which means they need to have a mental image of the relationship between the natures they need and the activities necessary to get them, there remains a fundamental difference between human work and the activities of animals and plants. In the Economic-Philosophical manuscripts, Marx has formulated this difference this way:

> The life of the species, both in humans and in animals, consists physically in the fact that humans (like animals) live on organic nature; and the more universal humans (or the animals) are, *the more universal is the sphere of inorganic nature on which they live*. Just as plants, animals, stones, air, light, etc., constitute theoretically a part of human consciousness . . . so also in the realm of practice they constitute a part of human life and human activity. Physically, humans live only on these products of nature, whether they appear in the form of food, heating, clothes, a dwelling, etc. The universality of humans appears in practice precisely in the universality which makes *all nature their inorganic body* – both in as much as nature is (1) their direct means of life, and (2) the material, the object, and the instrument of their life activity. Nature is human's inorganic body – nature, that is, insofar as it is not itself human body. That humans live on nature – means that nature is their body, with which they must remain in continuous interchange if they are not to die. That human's physical and spiritual life is linked to nature means simply that nature is linked to itself, for humans are part of nature.
>
> (Marx 2011 [1844], emphasis added)

This resonates with the oneness of all living beings as outlined by Choquehuanca Céspedes and with Haraway and Moore when they emphasize the "work" of animals and plants. The difference between non-human natures and the work of humans that Marx describes here is not one of consciousness versus instinct, but one of universality versus locality. Humans make potentially (and as history has shown in reality) all nature the object of their needs – independent of the mode of production within which they labor. In difference to humans, animals and plants remain mostly within their respective habitats since their capabilities to survive are dependent on the ecological system within which they thrive. Humans, in contrast, can adapt virtually all ecological systems to their own needs. This universality of human labor constitutes an advantage for their survival but also the potential of destroying the "inorganic body" on which their survival depends. Humans cannot destroy nature, but they can destroy nature as the life support system for themselves and other species. Capitalism alienates workers from nature as their inorganic body, hampering the continuous interchange between human and non-human natures, thereby threatening the survival of humans and other species.

However, work as a process in which both human and non-human natures interact and change their natures also bears the potential for a labor–nature relationship in which labor transforms nature in a way that it can regenerate – as Salleh demands. Production of the means of life and of life is a learning process in which non-human natures can teach humans who they are and what they need. Through the work process in which human and non-human natures constitute each other, humans develop new knowledge (as well as animals and plants in their specific ways) that enables them to organize their potentially destructive labor into a constructive and regenerative form of labor. This is where the response-ability of humans lies: in using their knowledge and their universal capacities to adapt natures to fit their needs while simultaneously respecting the needs of nature. This would be a process in which traditional forms of knowledge developed through different ways of working within nature can be connected to modern scientific knowledge as Choquehuanca Céspedes suggests. Whether that happens is an empirical question and depends on the mode of production within which work takes place. We need to leave the level of "work as such" independent of its societal contexts and come back to the world of capitalism within which we live.

Thinking about human work at a general level, independent of its societal context is necessary to understand that when we discuss the question whether we live in the Anthropocene or in the Capitalocene (Moore 2016), we need to take the (self-)destructive potential of human labor into account. Simultaneously, we need to understand that workers are globally forced by capitalism to sell their labor power without being able to decide whether the production process in which they engage is one that regenerates human and non-human natures or destroys them.

Workers can and do fight against extractivism, they can and do fight to change production processes in a way that protects their own and the lives of other living beings (Räthzel et al. 2021; Juan Martinez-Alier 2002). However, these struggles do not change the dominant exploitative capitalist, patriarchal, colonialist, and racialized power relations, they can only limit their destructiveness. Thus, the term Capitalocene is necessary to avoid blaming those who are the victims, not the perpetrators of the crisis of life we are experiencing.

Yet we also need to avoid the illusion that the end of capitalism would automatically create harmony between human and non-human natures. Choquehaunca's (Chapter 3 in this volume) insistence that struggles need to go beyond confronting capitalism and beyond fighting for socialism but need to be struggles for life is central here. To recognize that human labor because of its universal and transformative power is already potentially destructive helps us to understand why the environmentalism of the poor and labor-environmentalism remain limited. The relationship between human and non-human natures is conflictual irrespective of the mode of production within which humans transform non-human natures. Martínez-Alier (2012) makes this point when he argues that all economic systems, including the former state socialist ones, were destructive of the environment because they relied on economic growth. The labor–nature relationship can only be harmonized through the conscious, societal creation of a nature– labor alliance. Such an alliance would need to start from the insight that human needs depend on respecting the needs of non-human natures.

Bringing schools of thought together

Exploitation and appropriation

I want to suggest that connecting the thoughts of Jason Moore and the Marxist-Feminist scholar Frigga Haug could create a framework that serves to better understand the differences but also the similarities of the different areas of work, which are usually investigated independent of each other: the production of life (aka reproduction) and the production of the means of life (aka production).[9]

Moore synthesizes "the core insights of Marxism and environmental historiography" (2015: chapter 1). The result is a theory of the accumulation of capital that interweaves Marx's theory of the exploitation of waged labor with the theory of the appropriation of unpaid labor and nature outside of the commodification process. For this, he draws on feminist and ecofeminist work.

Moore argues:

> [V]alue as abstract labor cannot be produced except through unpaid work/energy. . . . The "commodification of everything" can only be sustained through incessant revolutionizing – yes, of the forces of production, but also of the relations of reproduction. The relations of reproduction cut across the paid/unpaid work and human/extra-human

9 Space impedes to include the broad literature on Marx' concept of value and capital accumulation, nor the equally extensive literature on gender relations and work.

> boundaries. In this, the historical condition for socially necessary labor-time is socially necessary unpaid work.
>
> (ibid.:46)

In other words, the value form, the representation of the societally necessary time needed to produce a commodity (=abstract labor, irrespective of any qualification and complexity of the concrete work needed to produce that commodity) does not include *all* the work needed to produce the commodity, namely, the work that does not form part of the process of commodification and is therefore not included in the exchange value – like the unpaid work of humans (mostly women and racialized workers) and non-human natures. Capitalist relations of production include not only the unpaid activity of non-human natures but also unpaid work of humans.

Feminist scholars have argued that the cost of this work is "externalized" by capital. Moore maintains that unpaid work is part and parcel of the *internal* process of capital accumulation: exploitation of waged labor and appropriation of non-waged labor are two dimensions of the same process. Within this understanding, the construction of the category of women, indigenous peoples, and people of color as "nature" (critically analyzed by feminist scholars) can be understood as part of the process of *appropriation*. From the point of view of capital, since nature can be appropriated for free, everything that is appropriated for free is constructed as nature. Following this argumentation would mean that appropriation of unpaid work (of humans and the activities of non-human natures) is an essential part of the creation of profit. Dependent on the degree in which this work is commodified, new forms of appropriation would need to be developed to secure the continued creation of profit through the appropriation of unpaid work.

Throughout his work Moore talks about "most women" being produced as nature together with colonized, racialized peoples and non-human nature. However, women as unpaid workers in the household are not only producers of life, but they are also expected to be "symbolic border guards", "embodiments of the collectivity", and "cultural reproducers" of the nation (Yuval-Davis 1997).

The notion that most women are produced as nature is only a half-truth. It ignores the essential role that women as a category play in reproducing cultures, national identity, and dominant ideologies and it also ignores the way in which women are included in different sectors of the production of the means of life (Räthzel et al. 2014). We are dealing with a dialectical process that assigns women contradictory tasks for the reproduction of societies. Feminist scholarship concentrates predominantly on care work, ignoring the ideological work of women.

Gender relations as relations of production

To understand the role of gender relations, Frigga Haug (Haug 2015; Haug 2005) conceptualizes gender relations as relations of production. Haug finds some "flashes of inspiration" (Haug 2005:284) in Marx work, which are close to Moore's idea of the double process of exploitation and appropriation:

> However, it still remains true that to replace them [the commodities functioning as elements of capitalist industrial production] they must be reproduced, and to this extent the capitalist mode of production is conditional on modes of production lying outside of its own stage of development. But it is the tendency of the capitalist mode of production to transform all production as much as possible into commodity production.
>
> (MECW 36, 108, quoted in Haug 2005:284)

The examples Marx gives for work existing outside of the capitalist mode of production is the work of slaves, serfs, or communities. The unpaid work of women does not feature. In another "flash of inspiration", he differentiated as follows:

> On the one hand, the production of the means of life, of the objects of food, clothing and shelter and the tools necessary for that production; on the other hand, the production of human beings themselves, the propagation of the species.
>
> (MEW 21, 27, quoted in Haug 2005:287)

These points are taken up by Haug to develop her theory of gender relations as relations of production. The production of life and the production of the means of life are seen as both being productive and reproductive, which is why "reproduction" is replaced with "the production of life". While recognizing the differences between the production of life and the production of the means of life, understanding gender relations as relations of production allows understanding both areas as internal to the capitalist relations of production. To speak in Moore's terms, the work of producing life is appropriated for free by capital, while the work of producing the means of life is exploited.

Women have been historically assigned to and are still dominant in the area of the production of life, but they are also active in the production of the means of life, mostly in subordinated positions and in low-paid areas that are similar to unpaid domestic work. If we understand the concept of gender relations as relations of production in a broader sense, we can analyze the ways in which the production of life and the production of the means of life depend on each other and how men and women are positioned differently in both areas. Domestic work (predominantly conducted by women)[10] is appropri-

10 "Estimates based on time-use data . . . for 64 countries . . . show that time spent in unpaid care work (own-use provision of services) accounted for 16.4 billion hours per day, with women contributing more than three fourths 76.2 per cent of the total. . . . This is equivalent to 2.0

ated as unpaid work in its entirety, waged labor is exploited by paying workers only for the time they need to reproduce their work power. Given an 8-hour working day, let's assume workers create the equivalent of what they need to reproduce their work power within 4 hours. The remaining 4 hours are unpaid work, the exchange value produced during that time is appropriated by capital as surplus value.[11] Moore's differentiation between exploitation and appropriation needs qualification: unpaid work that is appropriated in its entirety is not totally subsumed under the logic of capitalist production. Wage workers are exploited through their subsumption under the logic of capitalist production and their unpaid work is appropriated to create surplus value. Once we understand gender relations as relations of production, it becomes clear that without the production of life, there would be no workers to produce the means of life; while producing life is impossible without producing the means of life. Whatever happens in one area has repercussions in the other. They constitute two communicating vessels.

Connecting the conceptual frames of Moore and Haug allows us to resolve lacunae in each of them. Moore omits the concrete differences between appropriated unpaid work of producing life and the exploited work of producing the means of life. Producing the means of life and producing life (whether paid or unpaid) are different processes if we analyze them as concrete labor. As Haug (and others, e.g., Adam and Groves 2011) explains, the production of life has a different rhythm and logic of time, it is slower, long-term, and when subordinated to the rhythms of profit-oriented efficiency, cannot achieve its goals. Understanding gender relations as relations of production within the process of capital accumulation requires an analysis of the production of life in the same dialectical way as Marx analyzed the production of the means of life, as a conflict between the production of use value, using the qualifications, materials, and tools needed to create a useful commodity (=concrete labor) and the production of exchange value, where only the socially necessary time needed to produce that commodity counts (= abstract labor). In a capitalist production process, the "efficiency logic" rules, that is, producing a commodity within the minimum of time. This has repercussions for concrete labor. In our research, we have seen how workers wanting to do their job well run up against the efficiency logic. Depending on their position in the hierarchy of the workplace, they have possibilities to resist those pressures:

> And I have explained to my supervisor that I take one cab at a time. So, I don't care if there are 40 cabins waiting behind me. . . . It is better to make sure the cabs come out right. . . . You can just ask me whether I

billion people working on a full-time basis, that is, 40 hours per week without pay" (International Labour Office 2018:43).

11 For a detailed analysis of the production of surplus value, see Marx' Capital, Volume 1, Chapter VII.

> should do the cabs so they are ready to be sold or shall we produce crap. Those are the alternatives.
>
> (Räthzel et al. 2014:210)

The production of the means of life and the production of life are connected. Time to care for each other is squeezed to enable success on the labor market – a success that is needed for material survival. The logic of efficiency and success invades the process of the production of life even where it is not directly subsumed under the control of capital: when parents desperately wanting their children to succeed economically aim for the best schools, in the best urban areas with no "immigrants" in them, everyday racism is reproduced. The dominance of efficiency and competitiveness not only shapes the way in which the production of life is performed but also creates practices of exclusion through the accompanying, legitimizing ideologies.

To sum up, the production of life and the production of the means of life constitute each other. One is directly and the other indirectly subsumed, under the logic of capitalist efficiency. They both are shaped by the conflict between concrete labor, the desire to "do a job well for its own sake", and abstract labor, the pressure to achieve this within the minimum time demanded by the rules of competitiveness. Both production processes incorporate non-human natures as necessary "raw materials" and transformed natures as tools and machines. However, in terms of their concrete dimension, both forms of production are also different: the production of life is characterized by a slower time rhythm, its goal of recovery, regeneration, care, and a direct reciprocity between the worker and those who are the "objects" of their work.

The similarities and differences between the production of life and the production of the means of life need to be analyzed within a conceptual framework that recognizes their interdependency and the ways in which gender relations, and the relations of racism, colonialism, and patriarchy govern their position within the labor–nature relationship. If we understand gender relations as relations of production and the production of life and the production of means of life as being both productive and reproductive, we need not theoretically replicate the split enforced between these areas through capitalist, patriarchal, and racist power relations. We need to recognize that the reproduction of societal power relations occurs through all dimensions of re-production.

Conclusions

To think the society–nature relationship, we need to introduce the concept of work as a distinctive human activity. Its distinctiveness lies in the fact that it is potentially universal, it can adapt non-human natures to its needs on a global level. Work has thereby the capacity to destroy nature as a life support system for humans and other species. No other kind of work has that capacity, which is why I would suggest distinguishing between human work and the activities of non-human natures.

There is no preordained harmony between humans and non-human natures. If there was a time and if there is a space today where humans transformed/transform natures in a way that allows for their regeneration as Salleh describes the work of meta-industrial workers, this may be due to the fact that they did not/do not have the means to expand the transformation of natures in a way that made it/makes it more destructive of natures than constructive.

As the critical psychologists Holzkamp and Holzkamp-Osterkamp (Holzkamp 2013) have argued, humans create as a "second nature" the built environment that transforms nature on a scale that is not possible to do for animals and plants. This "second nature" constitutes the inventory of human knowledges and capabilities which makes the development of humans a dynamic process in contrast to the development of non-human natures whose accumulated knowledges can only pass to the next generation through unmediated relationships. Subordinated to the capitalist mode of production whose goal is profit, not use-values and a good life for all, the destructive potential of work is dominating and destroying its constructive possibilities. Whether producing the means of life or life itself, workers are forced in the capitalocene to become the agents of their own destruction.

Thinking in terms of the production of life and the production of the means of life as subsumed in different ways under the logic of capital allows connecting Marxist, feminist, and ecofeminist perspectives, the perspectives of environmental labor studies, environmental justice, and the environmentalism of the poor. We need to analyze the interdependencies between the environmentalism of the poor, resisting appropriation of their labor and lands by capital, often led by women (Martínez-Alier et al. 2016), and labor environmentalism of workers and unions in industries, resisting the exploitation of their labor and threats to their health. We cannot contend ourselves with the idea, that "the technologies which mediate daily needs in industrial societies numb people to their organic embodiment as nature" (Salleh 2020:47).

Most workers labor in industrialized mass production (including industrialized agriculture) and all humans are dependent on the products produced in these areas. Reducing CO_2 is also dependent on new technologies, even if technological fixes are not the solution to the ecological emergency. Industrialized work needs transformation the most because it is the main source of ecological destruction. The alienation of workers from nature in these processes is a result of their position as waged workers without control of the means of production and therefore without any possibility to decide how and for what they are produced (Marx 2011 [1844]). But neither are workers producing life mistresses/masters of their fate.

Within Capitalism "sympoiesis", humans and non-human natures co-producing the means of life and life is a reality imposed on both and thus destroying both. We need to unpack this co-production by defining the different kinds of activities in which both parts of the process engage. Human work, the production of the means of life and the production of life, is the mediator between human and non-human natures.

Looking at the production of the means of life and at the production of life as independent domains makes it difficult to see how the capitalist mode of production creates and subordinates gendered, racialized, colonized workers, whom it sets in competition to each other in both domains. We diminish our capability of understanding the interdependencies between the different domains when we compartmentalize them into separate sections, whereby "social reproduction" is necessary and useful, while "industrialized production" is contingent and damaging.

Seeing the interdependencies between all domains does not imply becoming blind to their different contexts and situated knowledges, as Salleh rightly emphasizes. These knowledges lead to different forms of resistance against environmental degradation – or the lack of them. Where workers in agriculture and fishing can more easily develop a holistic concept of the labor–nature relationship, especially if they are part of cultures that defend the rights of "Madre Tierra", workers in industrial processes need additional knowledge that goes beyond their everyday working practices to be able to connect experiences of draught, floods, weather changes, and occupational hazards to the ecological crisis (Räthzel and Uzzell 2019; Räthzel 2021). However, all workers, wherever and whatever they produce, are connected to each other along global value chains. Therefore, to confront the destructive capitalist system and develop a labor–nature alliance, they need to connect and create a common ground of resistance that can overcome their differing and contradicting interests.

I want to give one example that illustrates the connections and contradictions between different categories of workers and non-human natures locally and globally: Earthsight (2020) published a report showing that major European car producers receive the leather for the seats of their luxury cars from legal and illegal forest clearing in the Gran Chaco of Paraguay by cattle owners (not that legal clearing is any less destructive). Cattle rearing threatens the lands of indigenous communities and the existence of protected animal species like the jaguar and the giant anteater. The forests of the Paraguayan Gran Chaco are disappearing faster than any others on earth. Every 2 minutes an area of the size of a football field is cleared, mostly for cattle to satisfy the international demand for beef and leather.

We can imagine the contradictions between the interests of workers in the cattle business and neighboring production processes, indigenous communities, and industrial workers in the car and beef industry. There is no safe place in which indigenous communities can go about their livelihoods as meta-industrial workers. These conflicting interests are based on existential needs of all the workers and natures involved. They seem unsurmountable. But it is also possible to pose a different set of questions. How could workers, geographically separated and yet existentially connected, transform their connection into a common strength? The trade union movement has experience in actions of solidarity across nations and across the Global North and Global South. Indigenous communities have experiences in connecting across borders to fight against the dispossession of their lands and livelihoods and women are often at the forefront of such struggles (Tran et al. 2020; Tran 2021). What if these experiences could be combined, as opposed to creating fractures between these workers? It is ultimately the same power that exploits, dispossesses, and appropriates their lands and labor. These separations and conflicts are existential and constantly reproduced by political, ideological, and cultural apparatus that serve the interests of the powerful to divide and rule. But ultimately the existence of workers and the non-human natures they live on and within depend on the ability of all workers to cooperate, to create an alliance among themselves and with nature. Such a cooperation needs to include the workers producing the means of life and the workers producing life. A comprehensive theoretical framework helps to understand the real contradictions but also the real connections between workers in all areas of production across the globe.

Acknowledgments

I thank Patrick Bottazzi, David Uzzell, Diana Mulinari, and Jan Rehmann for their encouraging critical comments on a previous version. Errors and misunderstandings remain my responsibility.

References

Adam, B., and Groves, C., 2011. Futures tended: Care and future-oriented responsibility. *Bulletin of Science, Technology & Society*, 31(1), 17–27.

Barad, K. M., 2007. *Meeting the Universe Halfway: Quantum Physics and the Entanglement of Matter and Meaning*. Durham: Duke University Press.

Bennett, J., 2010. *Vibrant Matter: A Political Ecology of Things*. Durham: Duke University Press.

Bloch, E., 1973. *Das Prinzip Hoffnung 3 3*. Frankfurt am Main: Suhrkamp.

Castree, N., and Braun, B., 2001. *Social Nature. Theory, Practice and Politics*. London: Wiley-Blackwell.

Choquehuanca Céspedes, D., and Rist, St., 2023. The great disruption – A culture that understands that everything is interrelated, that nothing is divided, and nothing is outside. In: Rist, S., Jacobi, J., and Bottazzi, P., eds. *Critical Sustainability Sciences: Intercultural and Emancipatory Perspectives*. London: Routledge.

FAO and Filac, 2021. *Forest Governance by Indigenous and Tribal Peoples. An Opportunity for Climate Action in Latin America and the Caribbean*. Santiago. FAO. https://doi.org/10.4060/cb2953en

Haraway, D. J., 2008. *When Species Meet*. Minneapolis: University of Minnesota Press.

Haraway, D. J., 2016. *Staying with the Trouble: Making Kin in the Chthulucene*. Durham: Duke University Press.

Haug, F., 2005. Gender relations. *Historical Materialism*, 13(2), 279–302.

Haug, F., 2015. Marxistische Refundierung des Feminismus, feministische des Marxismus. In: *Wege des Marxismus-Feminismus*. Hamburg: Argument Verlag, 517–526.

Holzkamp, K., 2013. *Psychology from the Standpoint of the Subject: Selected Writings of Klaus Holzkamp*. Basingstoke: Palgrave Macmillan.

International Labour Office, 2018. *Care Work and Care Jobs for the Future of Decent Work*. https://www.ilo.org/global/publications/books/WCMS_633135/lang--en/index.htm

Krech, S., 2005. Reflections on conservation, sustainability, and environmentalism in Indigenous North America. *American Anthropologist*, 107(1), 78–86.

Latour, B., 2000. When things strike back: A possible contribution of 'science studies' to the social sciences. *The British Journal of Sociology*, 51(1), 107–123.

Mancuso, S., Viola, A., Benham, J., Pollan, M., and Mancuso, S., 2015. *Brilliant Green: The Surprising History and Science of Plant Intelligence*. Washington, DC: Island Press.

Martinez-Alier, J., 2002. *The Environmentalism of the Poor: A Study of Ecological Conflicts and Valuation*. Cheltenham: Elgar.

Martínez-Alier, J., 2012. Environmental justice and economic degrowth: An alliance between two movements. *Capitalism Nature Socialism*, 23(1), 51–73.

Martínez-Alier, J., Temper, L., Del Bene, D., and Scheidel, A., 2016. Is there a global environmental justice movement? *The Journal of Peasant Studies*, 43(3), 731–755.

Marx, K., 1998. *Capital. A Critique of Political Economy* (1887th ed.). London: ElecBook.

Marx, K., 2011. *Economic and Philosophic Manuscripts of 1844*. (1844th ed.) Blacksburg, VA: Wilder Publications.

Moore, J. W., 2015. *Capitalism in the Web of Life: Ecology and the Accumulation of Capital* (1st ed.). New York: Verso.

Moore, J. W., 2016. *Anthropocene or Capitalocene? Nature, History, and the Crisis of Capitalism*. Oakland, CA: PM Press.

Prieur, J., and Pika, S., 2020. Gorillas' (Gorilla g. gorilla) knowledge of conspecifics' affordances: Intraspecific social tool use for food acquisition. *Primates*, 61(4), 583–591.

Räthzel, N., 2021a. Society-labour-nature: How to think the relationship? In: *The Palgrave Handbook of Environmental Labour Studies*. Cham, Switzerland: Palgrave Macmillan, 793–814.

Räthzel, N., 2021b. Trade union perceptions of the labour – Nature relationship. *Environmental Sociology*, 7(4), 267–278.

Räthzel, N., Mulinari, D., and Tollefsen, A., 2014. *Transnational Corporations from the Standpoint of Workers*. Basingstoke and New York: Palgrave Macmillan.

Räthzel, N., Stevis, D., and Uzzell, D. L., eds., 2021. *The Palgrave Handbook of Environmental Labour Studies*. Cham, Switzerland: Palgrave Macmillan.

Räthzel, N., and Uzzell, D., 2019. The Future of work defines the future of humanity and all living species. *International Journal of Labour Research. ILO*, 9(1–2), 145–171.

Salleh, A., 2017. *Ecofeminism as Politics: Nature, Marx and the Post Modern* (2nd ed.). London: Zed Books.

Salleh, A., 2020. Ecofeminism as (Marxist) sociology. In: *Marxist-Feminist Theories and Struggles Today*. London: Zed Books, 367.

Sennett, R., 2008. *The Craftsman*. New Haven: Yale University Press.

Stengers, I., 2010. *Cosmopolitics*. Minneapolis: University of Minnesota Press.

Stengers, I., 2015. *In Catastrophic Times: Resisting the Coming Barbarism*. Open Humanities Press. http://openhumanitiespress.org/books/download/Stengers_2015_In-Catastrophic-Times.pdf.

Swyngedouw, E., 1999. Modernity and hybridity: Nature, *Regeneracionismo*, and the production of the Spanish waterscape, 1890–1930. *Annals of the Association of American Geographers*, 89(3), 443–465.

Tran, D., 2021. A comparative study of women environmental defenders' antiviolent success strategies. *Geoforum*, 126, 126–138.

Tran, D., Navas, G., Martinez-Alier, J., and Mingorria, S., 2020. Gendered geographies of violence: A multiple case study analysis of murdered women environmental defenders. *Journal of Political Ecology*, 27(1).

Yuval-Davis, N., 1997. *Gender & Nation*. London and Thousand Oaks, CA: Sage Publications.

9 Regenerative work

From commodity to collective action

Patrick Bottazzi

Introduction

The 21st century is characterized by growing evidence that the capitalist economic system is contributing to its own destruction (Rockström et al. 2009). Instead of being a generator of well-being and collective utility, economic activity – marked by the constant pressures of globalization and productivity gains – leads ineluctably to the deterioration of the conditions of existence of humans and non-humans alike. What has been called the ecological paradox of work can be summarized in three points. First, the search for unlimited growth and productivity gains and the globalization of productive functions leads to constant pressure on workers – who must work ever harder and faster – but also on the ecosystems from which humans derive resources and well-being. Second, this doubly extractive economy (the extraction of natural resources and of human vitality) leads to the degradation of both working conditions and the capacity of ecosystems to provide resources over time. Instead of producing well-being and public utility, the increased pressure on human labor leads to a general degradation of the living conditions of both humans and non-humans. Third, the degradation of socio-ecosystems and their capacity to provide well-being (e.g., food, care for relatives, contemplation) requires an increase in collective work effort to compensate for this loss, which places the work/welfare/utility metabolism in a vicious cycle of degradation.

In other words, the more we work for the capitalist economy, the more we contribute to the destruction of our fundamental ecological assets and the more we have to work to compensate for this loss. This contradiction is very well-represented by the image of Chinese manual pollinators on almond tree plantations, who must compensate for the consequences of years of bee exterminations (Partap and Ya 2012). Despite these contradictions, most economic policies continue to magnify their ambition to "create new jobs", a demagogic argument used indistinctly by political parties, both left and right, across the world. Such policies are usually supported by arranged statistical counts of the "working poor", who have never ceased to grow in recent decades (Andress and Lohmann 2008). How can we explain this continuing societal obsession with more work despite its negative consequences for human and non-human

DOI: 10.4324/9781003043577-9

well-being? What type of ontological adjustment would be required to change our conception of work and align it more with a regenerative rather than an extractive economy?

To respond, we build on the idea – shared by an increasing number of heterodox economists and other social scientists – that the contemporaneous double social and ecological crisis could better be addressed by a reconceptualization and reassigning of the place and role given to work (Gollain 2004; Gorz 2015; Frayne 2016; Ryle and Soper 2016). Work can be considered as a set of culturally defined perceptions, institutions, and practices and is a core interface between social and ecological systems' transformations (Gollain 2004); it should therefore be given more attention from the perspective of the critical sustainability sciences. After deconstructing the current modern conception of productive work in the so-called "extractive economy" based on the ontology of the dominant neoliberal capitalism model, we will progressively introduce the core components of what we consider some fundamental axioms of regenerative work.

The extractivist foundation of our societies

Contrary to what contemporary neo-liberal ideology would have us think, the concept of work is not a sui-generis anthropological category but the product of a complex and sometime contradictory historical construction. An object of contempt by the Greeks and associated with torture by the Roman, work has also been glorified by Christianism and Marxism and become progressively a utopia in the industrialized and rationalist European society. During this latter stage, work has been reified as the only way for humans to face uncertainty and fulfil infinitely growing needs. According to French sociologist of work Dominique Méda, the reification of work corresponds to the necessity in industrialized societies to find an ideological justification to wage workers' sacrifice in front of growing competition and search for productivity gains (Méda 1996).

Anthropologists have noted that some preindustrial societies, such as hunter-gatherers, do not always have a corresponding word for "work" or "labor". It is also wrong to think that primitive societies were crushed by subsistence tasks (Descola 2014). They only devoted a tiny part of their time to tedious tasks, which led Marshall Sahlins to write in his famous book *Stone Age Economics* (1973) that forager societies dedicated the greatest part of their time to cultural and socializing activities and had a better quality of life than agrarian societies (Sahlins 1972). Although the satisfaction of their basic needs was principally covered by "extractive" rather than productive activities, such as hunting and gathering, they had relatively low impact on the ecosystems. Foraging societies relied on hunting, fishing, and gathering wild plants for their subsistence. In this sense, they can be qualified as more extractive than productive. Nevertheless, past and contemporaneous foraging societies tend to have a deep understanding of and connection to the natural and spiritual world as well as complex systems of resource management that ensure the sustainable use of resources (Berkes 2017). Such human groups are usually small and do not accumulate useless material goods, which allows them to be sufficiently mobile and change environment when the pressure on natural regeneration is becoming unsustainable (Graeber and Wengrow 2021). Under these conditions, their foraging economies had lesser impact on their nearby environment, although the role of hunting in the extinction of mammal species has been recognized and is still an ongoing debate (Ceballos and Ehrlich 2002).

Humanity has spent more than 90% of its existence relying principally on foraging activities, with little necessity to change the pattern to the active regeneration of their natural capital. Despite the historical technical transformations leading to the Neolithic, industrial, and, more recently, digital revolutions, modern human societies have retained some characteristics of foraging societies and become essentially extractive. The mid-20th-century Green Revolution was the culminating point of the dominant economic model, completely dependent on fossil energy, synthetic inputs, and the commodification of human and non-human labor controlled by large corporate power. In such model, the production of goods and services is artificially separated from the periodicity, the conditions and the complexity of the regeneration of life. Agricultural systems in extractive societies are comparable to a stock from which we would draw without ever contributing to its replenishment. The overdependence on fossil energy has exacerbated the principles of extractivism in all aspects of social, economic, and political life. Over decades, all our habits and institutions have been crafted on those principles. The seminal book of anthropologist and historian Timothy Mitchell, *Carbon Democracy: Political Power in the Age of Oil* (2011), has shown how Western "democracies" are products of the ways fossil energy have been extracted and exploited. With the military control of oil extraction, political life became anemic, the social peace and prosperity rested on Middle Eastern authoritarianism, and unlimited growth was transformed into a religion through the conceptual enforcement of the neoclassical economic paradigm.

"Extractivism" has had a significant impact on how we understand the world as well as on the institutions responsible for producing knowledge. The social sciences, particularly the classical economic paradigm, have been heavily influenced by extractivism. Their foundations were built on a Darwinian vision of social evolution, which is based on the principle of competition between individuals, an unrealistic idealization of unlimited growth, and a complete disregard for the limits of social–ecological systems to generate new resources.

Practically, extractivism results in short-term economic gain and the overexploitation of resources. It is often linked to the displacement of local communities, destruction of natural habitats, and pollution of the environment. It generally leads to the domination of a small group of people, the concentration of wealth and power in the hands of a small elite. In an extractivist system, workers are considered as mere inputs in the production process, favoring the concentration of wealth in the hands of the owners of the means of production. Following this logic, the more the economic activity generates precarious working conditions, the more the economic operators can underpay workers and profit from their competition. This is the Marxist notion of "the reserve army of labour", referring to the concept that in capitalist societies, there is a surplus of unemployed or underemployed workers who are available to take jobs as they become available. According to Karl Marx, this reserve army of labor creates a pool of surplus labor that serves to keep wages low as employers are able to hire workers at lower wages than would otherwise be necessary. The reserve army of labor also serves as a source of labor for capitalist employers to draw upon during periods of expansion. Private property is the corresponding institution that sustains this system, protecting the interests of the owners of capital. The commodification of land and labor also plays a role at a global level due to the "new international divisions of labor, with the concomitant globalization of ecological degradation" (Escobar 1996:329).

From an extractive to a regenerative economy

In recent years, alternative models have started to emerge in the minds of so-called heterodox economists to try to define the epistemological basis of a more ecologically sustainable and socially just economy. Several fields of research have been exploring the links between socioeconomic inequalities, working relations, and the environment. First, the field of ecological economics began by analyzing material and energy flows during the economic process of the transformation of nature from the perspective of system ecology and thermodynamic approaches. It has been extended to other aspects, combining social justice, sustainable incomes, and working conditions. Second, political ecology followed ecological economic by focusing on ecological distribution conflicts and the unequal exercise of power to gain access to natural resources by considering social characteristics such as social class, caste, gender, and ethnicity (Robbins 2012). Third, environmental justice has been increasing in popularity since the 1980s and is based on the principle that socioeconomic inequalities determine the distribution of the costs and benefits of environmental degradation (Martinez-Alier 2012). Forth, a social and solidarity-based economy radically shifted the main aims of an economy and changes some of its core indicators and explanatory models (Laville 2010). It acknowledged the plurality of economic forms while relativizing the importance of financial profit versus the other non-monetary benefits of the economy. Reciprocity and redistribution were presented as fundamental functions of the solidarity economy in parallel with the market.

First inspired by philosophy, and then combining the above mentioned approaches, the concept of a "contributive economy" has recently gained attention as an alternative economic model that focuses on the value created by individuals and organizations who freely contribute their skills, knowledge, and resources without relying solely on financial compensation (Béraud et al.

2012). This type of economy operates on the margins of the traditional market economy and emphasizes the non-monetary value of human contributions to society (Bottazzi 2019). In a contributive economy, human work is viewed not just as a commodity but as an opportunity to improve individuals' capabilities and as a value in itself. This perspective was first described by Paul Gomberg, who argued that people naturally seek to experience the social and psychological gratification of self-accomplished tasks. Gomberg introduced the concept of "contributive justice", which advocates for a socioeconomic system that promotes the right to meaningful work for all (Gomberg 2007). Under the principle of contributive justice, capabilities are not just a means to perform work but also an objective to improve people's awareness and social inclusiveness. A contributive economy is therefore an economy based on contributors' (or workers') sense of social and environmental justice, which seeks to create common value for all members of society based on intrinsic motivations such as self-esteem, legitimacy, and peer-to-peer recognition.

All of the aforementioned approaches form the foundation of what can be referred to as a "regenerative economy". Unlike an extractive economy, which transforms life and nature into waste, a regenerative economy is based on the principle that economic activity must prioritize the protection of its factors of production, particularly natural capital and labor. A regenerative economy is focused on creating value through the restoration and regeneration of natural resources rather than exploiting them. It promotes sustainable practices that ensure that natural resources are not depleted but rather replenished over time. This approach also emphasizes the importance of creating fair and equitable working conditions for all members of society and promoting social and environmental justice (Morseletto 2020).

The regenerative economy promotes the self-feeding, self-renewal, and adaptive learning processes that natural systems use to sustain their capacity to thrive over long periods of time. This approach requires significant investments in human, social, natural, and physical capital (Fath et al. 2019). Permaculture, for example, aligns perfectly with this philosophy as it enriches the soil without exhausting it, using minimal resources from outside the farm, while also requiring a significant amount of knowledge and labor. This holistic view of the economy can be applied to all sectors, such as renewable energy and waste management, and can lead to long-term improvements in yield. A regenerative economy is one that can produce positive externalities through a permanent awareness of the impacts on other human and non-human entities. It is an economy that can produce social–ecological resilience by seeking synergies among human and non-human entities. Human work is at the heart of this equation as it is both a means of production and a human living experience that can contribute to individual and collective well-being. The regenerative economy must not only regenerate natural resources but also participate in the regeneration of human experience and resilience through a regenerative relationship to work.

Decommodifying work as a regenerative condition

An objective condition for the liberation of work and to allow it to become fully regenerative is its decommodification (Gerber and Gerber 2017). This has been a contentious issue, particularly since the publication of Karl Polanyi's *The Great Transformation*, a work that critiques the commodification of land and money (Polanyi 1957). Similarly, the commodification of labor reflects a strong fetishism since it directly involves human beings. This is exemplified by the lexical abuse of "labor market", which has been incorporated into popular language to reflect the level of acceptance or indoctrination to which we have all been subjected. The generally accepted idea of "wage labor" contributes to the same aberration of being able to buy and sell human activities.

Philosopher Andre Gorz in the 1980s proposed an emancipation from wage labor to free work from its commodity condition, allowing individuals to generate value(s) according to their own autonomous principles (Gorz and Turner 1999). In his view, if work were freed from the constraints of wage labor, the autonomous individual could choose on which aspects they concentrate their efforts and contribute to individual or collective utility. Gorz believed that by liberating people from the dictates of wage labor, for example, by becoming owners of their means of production, or by working less, people would become more aware of the ecological and societal challenges they face, and therefore they would be more involved in trying to solve them through voluntary collective action.

Contributive economy, which operates on the margins of commodified work, is therefore an interesting place from which to generate social innovation and resistance and could become a driving force for regeneration. The question remains as to what type of institutional arrangements can best lead to a reevaluation of work in society.

Human work as "commons": between individual autonomy and collective values

Research on social–ecological system sustainability has long since captured the importance of "commons" in proposing alternative institutional models to private versus public management rights. Nobel Prize winner Elinor Ostrom highlighted the complexity and efficiency of the community-based management of certain resource systems, such as grazing lands, traditional fishing areas, and forests (Ostrom 1990). She showed that many resources, such as common-pool resources, cannot be appropriated in an individual and private way as they are embedded in complex and irreducible social–ecological mechanisms. Building on her legacy, many scholars have applied her framework to different kinds of physical commons, such as air, water, and forests, as well as immaterial commons, such as knowledge and cultural heritage (De Angelis and Harvie 2014). However, few have taken into account the existence of the intangible social structures consisting essentially of social ties, skills, intelligence, and human effort that are often encapsulated in the notions of work or labor. Whether it is considered individually or collectively, human work contains a multitude of direct and indirect connections with resources, knowledge, and action schemes that are part of what can be considered a set of "common goods and services" provided by the larger social–ecological system. Such a complex background should preclude work being considered through the single prism of a per-hour or per-task commodity and could or should be shown in a more holistic manner and potentially be considered as "commons".

First, human work is almost always the result of a collective effort over several generations based on the experience and knowledge of known and unknown people's efforts and errors. Ancestral knowledge in agriculture is the result of peasant experimentation and intergenerational transmissions dating back several millennia. Farming exists as a professional activity in the continuity of plural agronomic knowledge co-constructed by diverse actors and adjusted empirically through the ages and the evolution of agroecosystems in the form of practical knowledge. This embodied knowledge often exists through its application in agricultural work activities and is transmitted through working experience across generations. These historical and collective properties of work should preclude considering work as a single individual input or as a commodity – rather, it should be considered a human and cultural patrimony that requires protection from fragmentation or depletion, most often due to a neoliberal conception.

Second, the reproduction of the labor force of one worker is most often done at the expense of other people in their surroundings, such as the members of their family (especially women), who are barely recognized and rewarded for their collateral efforts. The famous research of Claude Meillassoux on the African domestic modes of production has highlighted how the capitalist system takes profit from both the productive and reproductive efforts made by women without recognizing and rewarding their role in the remuneration scheme (Meillassoux 1982 [1975]). This collective effort to support individual work also appears more broadly in societal institutions, such as social services, hospitals, and healthcare institutions, which are not accounted for in production costs. The most direct example is the public contribution to people's education and professional training. Employers who provide a salary to an individual benefit from the individual effort of their employees but also from the collective effort of those who have contributed to supporting them outside the professional sphere. Following this logic, individual work cannot be seen as a private contribution to the capitalist economy but rather as emerging from a collective or public effort.

Third, human work is never isolated but is part of a structure of social links involving mechanisms of reciprocity (solidarity and informal exchanges), collective norms, and shared resources (communication channels, meeting interfaces). These networks of workers provide aggregated value through their interconnectedness, which is more than the sum of the individual values. Food value chains are good examples of combined material and immaterial commons resulting from long-term human effort (Vázquez and Gallo 2021). Taken separately, the elements of food systems do not provide any value but must be considered dynamically as a whole to provide their additional value to human societies. Upon closer inspection, these value chains are constituted by both formal and informal norms that maintain the links between groups of producers and groups of consumers. These value chains fulfil functions such as providing food for people and leading to food security. In most cases, these value chains have taken decades to be created and are based on social ties involving trust and reciprocity. With the extension of globalized forms of capitalist dispossession, these "immaterial commons" crafted by the repetitive work activities of traditional societies are subject to depletion. Recent initiatives to build local alternative food systems depend on the importance of joint value production through work and social ties. These systems generate incomes, ecological practices, and social ties while fulfilling the fundamental function of producing healthy food.

Changing the conception of work from an extractive to a regenerative economy

To summarize, this explorative chapter seeks to propose a parallel comparison between two opposite conceptions of work that we present as archetypes of a more complex reality. The first represents the current dominant capitalist model of work based on the principle of the extractive economy (see Figure 9.1). It is centered on the idea that economic activities are based on the main objective of generating financial profits for the owners of the production means and capital investors. In this model, work is considered as a simple means of production disconnected from its social and ecological basis. It is substitutable by any other technical options as soon as these options are more productive and efficient in terms of financial income and should ideally lead to an unlimited accumulation of money and capital. This idealized scenario is based on the correlated idea of an unlimited consumption of goods and services, which in turn boosts productive sectors and leads to increased resource extraction. To be productive, workers should preferably be hyperspecialized in their field of activity and put in competition against each other to increase their productivity and performance. Unlimited technical progress and economic growth are the leitmotifs of the extractive economy and are presented as societal aims to motivate policies and other collective action scenarios. In this competition, both human and non-human entities are overexploited until their productive capacity ends, leading to a dramatic situation of poverty and work-related diseases and illnesses. Individuals or social categories that cannot be productive and competitive enough are left aside and marginalized. To avoid social uprising and rebellion and to try to contain criminality and other societal pathologies, extractive economic models are based on simple

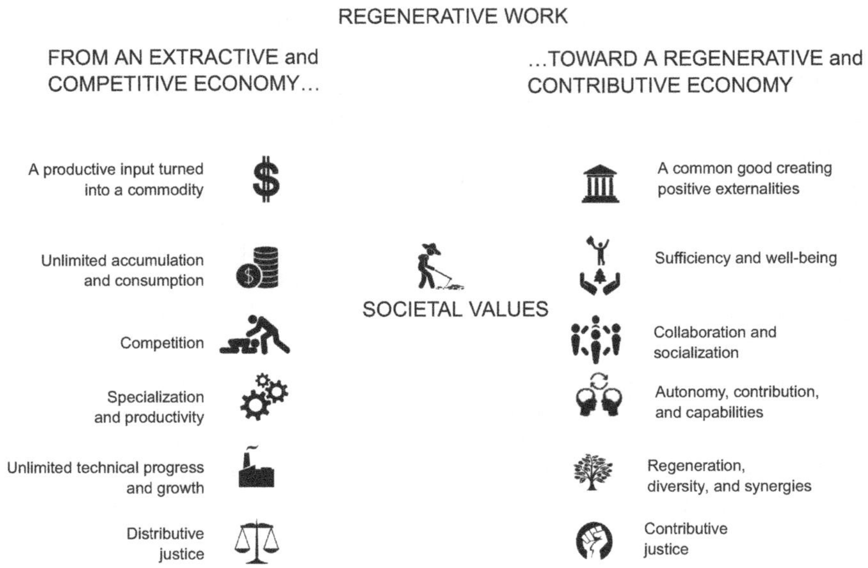

Figure 9.1 Key transforming aspects toward regenerative work.

Source: Adapted from Bottazzi, P. (2019).

principles of financial redistributions (Ferguson 2015). These neoliberal responses to social inequalities inspired by John Rawls's idea of justice do not intend to transform the model but only to reduce its impact by allowing part of the surplus economy to flow to the most affected individuals.

In the opposite conceptual model of a regenerative and contributive economy, work is considered as deep human experience that aims to contribute to the creation of individual and collective value. Starting from this general principle, a number of other collateral principles can be defined. Work is considered as a common whose main societal aim is not unlimited financial accumulation dependent on the excessive consumption of goods and services but well-being that can emerge through simple aspects of life, such as living together, eating healthy food, and having a harmonious life, including freedom and time for care and self-healing. Consumption is not a leitmotif but a socially constructed idea based on sufficiency, which invites us to redefine what makes most sense in personal and collective life. Work organization is oriented toward the capacity to collaborate and the resulting emulation between individuals and socialization. People get to know each other during work activities and share their experiences. This allows the creation of collective values while improving the autonomy of each individual, considered as integral persons and not scattered components of a production process. Contribution to the generation of common or collective value and capabilities is at the center of the equation. The more the individuals and collectives gain in capability and autonomy, the more the socio-ecological system gains in resilience. The protection and regeneration of life and ecosystems, including human and non-human entities, is proposed to be the main leitmotif of collective action processes by supporting the diversity of living and non-living entities and by finding synergies between them (e.g., the image of permaculture or agroecology). In contrast to distributive justice, contributive justice distributes not only the *surplus* value of the richest to the poorest but also the *means to produce* value and to generate capabilities. Meaningful and tedious tasks are therefore the central resources to be redistributed, not the surplus value of work. The rights of people to contribute to their society and to feel they are recognized and valued are at the center of a regenerative economy.

It goes without saying that both models are "ideal types" in the Weberian sense of the term, meaning they are extreme positions that help in thinking about the possible directions of an ontological change in the way work is conceptualized.

Innovative practical experiences of regenerative work are emerging all over the world at the margin in response to the commodified capitalist economy. The agroecology movement is an interesting example of such initiatives. In a pioneering paper, C. Timmerman defends the idea that agroecology is a "vehicle to contributive justice" as it requires more complex knowledge and a stronger solidarity among farmers than conventional agriculture (Timmermann and Felix 2015). In this view, farmers adopting agroecological practices have a greater capacity to acquire and develop capabilities due to the complex tasks they have to achieve, are better rooted in social networks of collective action, experience less tedious tasks during their work, and gain in bargaining power and self-determination compared to conventional farmers. Numerous examples from the literature show how agroecology and other transition initiatives are a growing expression of this paradigm shift to a regenerative economy. The French agroecological association "colibri" (*flybird*), headed by charismatic leader Pierre Rabhi, uses the image of a small bird throwing a water drop on a forest fire to signify that a big change can happen with a small contribution from everyone (Rabhi 2010). "Transition Cities", a now-famous translocality movement launched by Rob Hopkins (Hopkins 2008), has demonstrated that social and ecological resilience could be obtained in critical economic contexts through the mobilization of local networks by inventing new ways of working and creating new connections among peoples in the context of a post-wage economy (Méda 2017). Contemporary consumer-oriented movements such as the *Fédération des Artisans du Monde* and the *Associations pour le Maintien d'une Agriculture Paysanne* are other examples of the engagement of voluntary workers in mobilizing consumers through responsible consumption but also by taking part in the production itself (i.e., participative gardening). These social movement are engaged in support of environmental and social values of production and consumption but also more generally in proposing new ways of life based on "sufficiency" and "happiness frugality" (Dubuisson-Quellier et al. 2011).

Regenerative work and agroecological transition are closely linked as both concepts aim to promote collective action to protect biodiversity and other natural elements. However, agroecological approaches are often limited to small, isolated successes or small projects supported by international cooperation and lack a broader political perspective to address the root causes of vulnerability (Gonzalez de Molina 2013; Bottazzi and Boillat 2021). A critical sustainability perspective such as the one developed in this collective book is therefore necessary to identify and address the root cause of the degradation of both working conditions and agroecological systems. Globalization and the power of large transnational corporations limit the potential of agroecology to a simple substitution of inputs (or industrial organic inputs) rather than a more holistic change that incorporates biophysical dynamics such as agroforestry, livestock integration, water and soil conservation, and social and economic criteria to protect the most vulnerable food system workers. To be fully regenerative, initiatives for agroecological transition therefore need to be supported by proactive social movements engaged in collective action and where work and worker may play a central role.

Conclusions

As the French sociologist of work Dominique Méda states, our current conception of work is the product of a long-lasting historical and ideological process dominated by neoliberal and capitalist views. This has led human societies from the idealization of work to its commodification (Méda 2010). The global environmental crisis and the perspectives of post-growth societies require an urgent change in our conception of work. What has been called the "end of work" (Rifkin 1995) is the consequence of having reached the limit of a system that constantly seeks to increase productivity while replacing human contributions with automated and digitalized devices and algorithms. To be regenerative, work requires a completely opposite scenario based on the evaluation of human contributions to social–ecological systems' resilience. It requires a complete shift away from a productivist vision of work – which considers economic activity as the generation of profit based on technical progress and the continuous extraction of resources – to a regenerative economy based on other mechanisms of the evaluation of human and non-human contributions to well-being. Changing our conception of work and institutionalizing is not an easy task due to considerable barriers, such as corporate capitalist power and many other lock-ins preventing change in the system. This ambitious task will require strong mobilization from people and civil society organizations, where the workers themselves might have to play the most significant role.

References

Andress, H.-J., and Lohmann, H., 2008. *The Working Poor in Europe: Employment, Poverty and Globalisation*. Cheltenham; Northampton, MA: Edward Elgar Publishing.

Béraud, P., Du Castel, V., and Cormerais, F., 2012. Open innovation, economy of contribution and the territorial dynamics of creative industries. *Journal of Innovation Economics & Management*, 10(2), 81–105.

Berkes, F., 2017. *Sacred Ecology*. New York and London: Routledge.

Bottazzi, P., 2019. Work and social-ecological transitions: A critical review of five contrasting approaches. *Sustainability*, 11(14), 3852.

Bottazzi, P., and Boillat, S., 2021. Political agroecology in Senegal: Historicity and repertoires of collective actions of an emerging social movement. *Sustainability*, 13(11), 6352.

Ceballos, G., and Ehrlich, P. R., 2002. Mammal population losses and the extinction crisis. *Science*, 296(5569), 904–907.

De Angelis, M., and Harvie, D., 2014. The commons. In: *The Routledge Companion to Alternative Organization*. London: Routledge, 304–318.

Descola, P., 2014. *Les lances du crépuscule*. Paris: Plon.

Dubuisson-Quellier, S., Lamine, C., and Le Velly, R., 2011. Citizenship and consumption: Mobilisation in alternative food systems in France. *Sociologia Ruralis*, 51(3), 304–323.

Escobar, A., 1996. Construction nature: Elements for a post-structuralist political ecology. *Futures*, 28(4), 325–343.

Fath, B. D., et al., 2019. Measuring regenerative economics: 10 principles and measures undergirding systemic economic health. *Global Transitions*, 1, 15–27.

Ferguson, J., 2015. *Give a Man a Fish: Reflections on the New Politics of Distribution*. Durham, NC: Duke University Press.

Frayne, D., 2016. Stepping outside the circle: The ecological promise of shorter working hours. *Green Letters*, 20(2), 197–212.

Gerber, J.-D., and Gerber, J.-F., 2017. Decommodification as a foundation for ecological economics. *Ecological Economics*, 131, 551–556.

Gollain, F., 2004. *A Critique of Work: Between Ecology and Socialism*. London, UK: IIED.

Gomberg, P., 2007. *How to Make Opportunity Equal: Race and Contributive Justice*. Malden, MA: John Wiley & Sons.

Gonzalez de Molina, M., 2013. Agroecology and politics. How to get sustainability? About the necessity for a political agroecology. *Agroecology and Sustainable Food Systems*, 37(1), 45–59.

Gorz, A., 2015. *Le fil rouge de l'écologie*. Paris: Éditions EHESS, 95–96.

Gorz, A., and Turner, C., 1999. *Reclaiming Work: Beyond the Wage-Based Society*. Cambridge: Polity Press.

Graeber, D., and Wengrow, D., 2021. *The Dawn of Everything: A New History of Humanity* London: Penguin.

Hopkins, R., 2008. *The Transition Handbook*. Devon: Green Books.

Laville, J.-L., 2010. The solidarity economy: A plural theoretical framework. *Economic Sociology: The European Electronic Newsletter*, 11(3), 35–32.

Martinez-Alier, J., 2012. Environmental justice and economic degrowth: An alliance between two movements. *Capitalism Nature Socialism*, 23(1), 51–73.

Méda, D., 1996. New perspectives on work as value. *International Labour Review*, 135, 633.

Méda, D., 2010. *Le Travail. une valeur en voie de disparition?* Paris: Flammarion.

Méda, D., 2017. L'emploi et le travail dans une ère post-croissance. In: Cassiers, I., Maréchal, K., and Méda, D., eds. *Vers une société post-croissance. Intégrer les défis écologiques, économiques et sociaux*. La Tour d'Aigues: Edition de l'Aube, 35–75.

Meillassoux, C., 1982 [1975]. *Femmes, greniers et capitaux*. Paris: Maspero.
Mitchell, T., 2011. *Carbon Democracy. Political Power in the Age of Oil*. London: Verso.
Morseletto, P., 2020. Restorative and regenerative: Exploring the concepts in the circular economy. *Journal of Industrial Ecology*, 24(4), 763–773.
Ostrom, E., 1990. *Governing the Commons: The Evolution of Institutions for Collective Action*. Cambridge: Cambridge University Press.
Partap, U., and Ya, T., 2012. The human pollinators of fruit crops in Maoxian County, Sichuan, China: A case study of the failure of pollination services and farmers' adaptation strategies. *Mountain Research and Development*, 32(2), 176–186.
Polanyi, K., 1957. *The Great Transformation*. Boston: Beacon Press.
Rabhi, P., 2010. *Vers la sobriété heureuse*. Paris: Acte Sud.
Rifkin, J., 1995. *The End of Work*. New York, NY: G.P. Putnam's Sons.
Robbins, P., 2012. *Political Ecology: A Critical Introduction* (2nd ed.). Oxford: John Wily & Sons.
Rockström, J., et al., 2009. A safe operating space for humanity. *Nature*, 461(7263), 472.
Ryle, M., and Soper, K., 2016. Introduction: The ecology of labour. *Green Letters*, 20(2), 119–126.
Sahlins, M., 1972. *Stone Age Economics*. Chicago, IL: Aldine-Atherton.
Timmermann, C., and Felix, G. F., 2015. Agroecology as a vehicle for contributive justice. *Agriculture and Human Values*, 32(3), 523–538.
Vázquez, A. M., and Gallo, G. S., 2021. Sustainability and immaterial commons: Rentier appropriation and intermediation in the artisanal fishing space of southern Chile. *Journal of Cultural Economy*, 14(2), 209–224.

10 Food, food systems, and sustainability

Elements of the "real food" debate in Brazil

Renato S. Maluf

Introduction

Food systems occupy a central place in contemporary debates on sustainability, inequalities, and malnutrition in a context marked by the confrontation with climate change. These debates have mobilized and generated positions on the part of social organizations and movements, academics, the private commercial sector, and governments in many countries, as well as international organizations. The discursive repertoires express different conceptions and proposals for action, shaping an environment in which disputes over narratives on sustainability, justice, and adequate and healthy food, as well as on related, necessary transformations in current food systems and the ways to carry them out, gain relevance. In this sense, the perspective of promoting just transitions toward sustainable and healthy food systems that contribute to climate change mitigation is widely adopted at the international level, a formulation that is also subject to controversies and contradictions on key issues.

In this context, two international events are relevant, for their importance and scope. One of them was the United Nations Food Systems Summit (UNFSS) convened and held in 2021. It came under strong criticism from international civil society networks, due to the capture of its agenda by agribusiness corporations and private foundations made operational through an "backroom agreement" between the UN and the World Economic Forum (WEF). Critics denounced the resistance to the need for radical transformations of the dominant food systems to end the related serious social problems, damages to the environment and health, and the erosion of food cultures. The second event was the 26th Conference of the Parties to the United Nations Framework Convention on Climate Change (COP 26), held in 2021. It was also criticized for the predominance of perspectives that do not go beyond the "greening" of harmful food systems. This stance ignores the need to change the fundamentals of the production and consumption trends that caused the impending climate catastrophe, in addition to the social and health implications. In both cases, we observe the predominance of responses based on private economic mechanisms and a mercantile logic, in a direction clearly contrary to the understanding of broad social sectors – promoting the right

DOI: 10.4324/9781003043577-10

of food and food sovereignty – according to which food (eating) and nature (climate) are part of the list of common goods to be decommodified.

In the framework of the above, the argument developed in this book, in which this chapter is inserted, is very timely, concerning the need to develop critical sustainability sciences, based on critical, intercultural, and emancipatory thinking that takes into account the perceptions, ontologies, and the corresponding political approaches of the numerous social, political, and cultural movements for social, environmental, and cognitive–cultural justice. It suggests a collective and transdisciplinary construction of conceptual self-determination, whose ontological foundations are relational and value people's lived experience and counter-hegemonic practices that deconstruct dominant impositions and visions. In this chapter, I explore the question, how conceptions of food can contribute to the development of critical visions of sustainability, based on conceptual constructions of food and eating[1] that originate in social movements that challenge the dominant trends of production and consumption in food systems.

In these terms, I will address the conceptual and political dimensions of the notion of "real food" formulated by a broad spectrum of Brazilian social actors (Consea 2015). This concept is part of a dispute of narratives that has at its core the condition of food and eating as central elements of social formations, as well as the determining role of food systems in achieving the joint promotion of environmental sustainability, social justice, and healthy living, while respecting cultural traditions of peoples. This brings us to the field of *food politics* to consider the political dimension of food and eating. The latter is reflected in the relationships between social actors whose convictions, interests, actions, and political practices generate tensions and conflicts, not only between the private and public spheres, but also in the dispute for control of the State and public policies (Paarlberg 2010).

Using Mouffe's (2005) formula, antagonisms are constitutive of the *political* dimension of food, while *politics* of food refer to the practices and institutions that organize coexistence in a context of conflict. Societies with high social inequality, such as Brazil, are characterized by asymmetries of economic and political power that affect the prevailing choices in terms of actors' strategies and public policies. Note that narratives are part of corporate political action and social organizations, being also present in the sometimes-conflicting orientations of public policies (Mialon et al. 2015; Burlandy et al. 2021).

The ideas and proposals presented in the following chapter indicate possible ways out of the unsustainable and antidemocratic situation in which the dominant food systems find themselves today, taking account of the alternatives proposed by actors from different strands of food-related social and political movements. With their demands, they contribute to renewing science

1 Note the joint use of "food" to refer to goods, and "eating" corresponding to the appropriation of these goods, here considered analytically inseparable dimensions as explained at the end of the chapter.

and influencing policy. This allows us to verify how the politics of food, including the dispute of narratives present in it, promotes or counterposes the above understanding of food and nature as "commons". That is, "the common" as a counterpoint to the one-dimensional valuation of food as commodities (Vivero-Pol 2017, 2018), and as a political principle (Dardot and Laval 2019).

To meet these objectives, the chapter is divided into three parts. The first presents the context in which the debates on food systems, environment, and health are framed globally and in Brazil, and the referential framework used to approach them. The next part provides an overview of the social mobilizations in Brazil around the reference to "real food" in the countryside and in the city based on food rights and sovereignty. The third part highlights elements of the Brazilian debate on food, eating, and sustainability in view of some challenges to achieve food sustainability, considering the articulation between food, politics, and democracy. The concluding remarks indicate the possible contributions of the proposed analytical perspective on food and eating for the critique of conventional conceptions of sustainability, as well as drawing more general conceptual and political implications from the reference to "real food".

Context and conceptual frameworks

The global food system can be described as the interconnection of standardized and specialized processes of production, marketing, and consumption, organized in the form of far-reaching agri-food chains under the control of multinational corporations; in this case, it is important to highlight the soybean and meat value chains, given Brazil's prominent role. However, demands for just transitions toward sustainable and healthy food systems in the context of climate change have gained relevance at the international level while incorporating justice parameters in the evaluation of transformation processes (Kaljonen et al. 2021). For the purposes of this chapter, the concept of food justice is explored to identify systemic determinants of injustice, inequalities, and inequities in food systems (Gottlieb and Josh 2010; Cadieux and Slocum 2015), with a view to developing the perspective of promoting social justice through food (Maluf et al. 2022), as will be argued later.

The format and dynamics of global value chains and food systems threaten all three dimensions of justice: distributive, procedural, and cognitive justice (Fraser and Honneth 2003). International alliances of civil society actors, such as *La Via Campesina* and the Civil Society and Indigenous People Mechanism of the United Nations Committee on World Food Security, as well as global scientific panels, have further stated that the dominant global food system poses serious threats to the planet, leading to the term of a "global syndemic" expressing the synergy of three pandemics: obesity, malnutrition, and climate change (IPES-FOOD 2017; Swinburn et al. 2019; Willett et al. 2019; IPCC 2019).

We assume for analytical purposes that food systems should be defined and characterized based on food and eating as the center of observation that sheds light on the actors, processes, and tensions that shape food politics, adopting a multidimensional and multiscale approach. Thus, food systems are complex sets of actors and activities, characterized by interdependent flows that evolve with complementarities, conflicts, and contradictions and, consequently, require public and private mechanisms to coordinate components of systems that do not function harmoniously, which can take a variety of directions. The multiscale approach presupposes the plurality of food systems at different scales (international, national, and subnational levels). Their coexistence also implies interdependence, complementarity, and conflicts under multiple territorial dynamics with important repercussions on food provisioning (Maluf 2021a).

It is well-known that Brazil stands out among the largest producers and exporters of soybeans and meat in the world, together with the United States and Argentina, with China and the European Union as the main final destinations for these products. The exponential increase in the area for both soybean growing and cattle raising and the consolidation of the respective value chains have always had strong governmental support, with repercussions on food systems at different levels, from the global to the local. The deforestation of the "Amazon" and "Cerrado" biomes is accompanied by the continuity of the historical concentration of land ownership and economic and political power. On the one hand, political coalitions are formed at the local–territorial and national levels, bringing together the interests of large landowners and agroindustrial capital, supported by macroeconomic and agricultural policies, oriented toward internationalization. On the other hand, the country has an enormous and diversified contingent of around four million farms classified as family farming, which face a little more than 500,000 farms taken as patronal agriculture. The former are part of a social construction of strategies driven by a wide range of actors in a different direction from the predominant trends in food production, circulation, and consumption. From here on, we will approach this construction as a product of the political camp for food and nutrition sovereignty and security (FNSS) and the human right to adequate and healthy food (RtF) – in short, "FNSS/RtF political camp" – that emerged on the country's political scene as a counter-hegemonic movement, in the 1990s (Maluf 2021b).[2]

Against this backdrop, we will insert the "real food" movement, which represents one of the recent initiatives of the FNSS/RtF political camp that had important repercussions on food policy in Brazil, since it goes beyond the one-dimensional valuing of food as commodities and, consequently, nuances the debate with the actors representing the private commercial sector and governmental sectors. Before, however, it is necessary to add some more references. We mentioned in the introduction the political dimension of food

2 Following Dubois (2021), this is a limited use of the concept of "camp" for the purposes of a sociology of public actions without fully applying Pierre Bourdieu's theoretical system and analytical method.

and eating, to which we should add the factors that, according to Herring (2015), give food a distinct political nature: although being a commodity, food is confronted with the urgency of its provision, cultural norms, and ethical standards that – when ignored – give rise to growing perceptions of crisis and uncertainty.

Vivero-Pol (2018) presents an analogous argument by highlighting the six dimensions of food, which prevent its one-dimensional valuing as a commodity, that is, to its non-economic attributes, such as being an essential promoter of life and a human right, reflecting cultural determinants, being a tradable good that should be subject to fair trade and sustainable production, and being able to be constituted as a public good. Incidentally, this author emphasizes the revival of the notion of commons with multiple narratives, to which he adds the alternative concept of food as common, in the face of the processes of enclosures and commodification. The same author concludes that food is both a relevant agent of change, and an important determinant of planetary destruction, that is, it is the cause and solution of multiple crises affecting humanity (Vivero-Pol 2018). This conclusion confirms the importance of the approach proposed here, which takes food and eating, more specifically food politics, as one of the gateways to the critique of conventional approaches to sustainability.

To what has already been mentioned on broader scales, I add the scope suggested by Carolan (2011) with respect to people's lived experiences and the relational basis of knowledge about food. What he calls embodied food politics acts on material relationships or connectivity, while affecting the worldview shaped by such connections. This entails verifying skills, knowledge, and feelings produced by shared social practices in food processes and their transformative potential. Here also fits Dardot and Laval's (2019) perspective of the "commons" as a political principle and basis of the 21st-century revolution. In their words, "Common is the political principle from which we must build the commons and to which we must refer in order to preserve, expand and bring them to life" (own translation).[3]

Social mobilization for "real food" in Brazil

The V National Conference on Food and Nutrition Security, held in November 2015 in Brasilia (DF), adopted the slogan "real food in the countryside and in the city: for rights and sovereignty" as the organizing reference for the debates and proposals approved by the nearly 2,000 participants from all regions of the country, representing a broad and diverse spectrum of sectors

3 Regarding the idea of revolution, Dardot and Laval (2019) propose to understand it in its purest sense as the explicit restitution of society by the collective and autonomous activity of society itself, or at least of a large part of it, with a view to developing a project of radical transformation of society.

and social groups (Consea 2015). This resulted in a political manifesto toward Brazilian society affirming that "real food" is a safeguard of life. It is healthy, both for the human being and for the planet, contributing to the reduction of the effects of climate change. The slogan for "real food" seeks to guarantee human rights, the right to land and territory, food quality, and adequate quantity throughout life. It respects the rights of women, the diversity of indigenous peoples, Quilombola communities (rural black population descended from slaves), traditional peoples of African origin, gypsy peoples, forest and water peoples, and other traditional and peasant peoples and communities.

As for the intercultural perspective – a central component in critical sustainability sciences – "real food" means protecting and promoting food cultures, socio-biodiversity, ancestral practices, the management of herbs and traditional medicine, and the sacred dimension of food, beginning with breastfeeding. These are foods produced by family farming, agroecologically based, using creole and native seeds and through the proper management of natural resources, while considering the principles of sustainability, traditional knowledge, and regional specificities. This agriculture is free of pesticides, transgenics, chemical fertilizers, and other types of contaminants.

"Real food" is an expression of food sovereignty and protects cultural and genetic heritage, recognizes memory, aesthetics, knowledge, flavors, actions and discourses, identity, rites involved, and autochthonous technologies and innovations. It considers water as food. It is produced under decent working conditions, is socially just, and is not subject to the one-sided economic interests of the private commercial sector. "Real food" is characterized by fresh and minimally processed food; therefore, it is opposed to ultra-processed products. Food must be accessible, physically and financially, bringing production closer to consumption. It must contemplate people with special dietary needs. The "real food" is shared with emotions and in harmony. It promotes healthy eating habits in the countryside, in the forest, and in the city. Eating is a political act. The "real food" is that which recognizes the role of women and respects the principles of integrality, universality, and equity. It kills neither by poison nor by conflict. It is that which eradicates hunger and promotes healthy eating, preserves nature, and promotes health and peace among peoples (CONSEA 2015).

In summary, the slogan "real food" was intended to highlight the sociocultural dimensions of food and nutrition security to

- combining food production and consumption;
- building bridges between urban and rural;
- valuing agrobiodiversity, fresh and regional foods;
- respecting the black and indigenous ancestry, Africanity and traditions of all traditional peoples and communities;
- rescuing identities, memories, and food cultures of the Brazilian population.

So "real food" means protection of life and the planet, health, environmental justice, and human right (Consea 2015). Nonetheless, this recent formulation has a history of the aforementioned social construction of the FNSS/HRtF political camp (Maluf 2021b). There is an ancient tradition of politicizing hunger inaugurated by Josué de Castro in the 1940s (Castro 1946) who attributes the causes of hunger to human action and development models. The emergence of the FNSS/HRtF camp, since the 1990s, made explicit the antagonisms with the dominant views on food production, commercialization, and consumption.

For the purposes of this chapter, it is important to highlight three characteristics of the social construction that led to the formulation "real food': it is a reference that agglutinates a broad and heterogeneous set of actors and social segments, provides elements of rupture with the dominant conceptions of food in the context of food politics in Brazil, and contributes to the current debate on sustainable and healthy food systems. Of course, the slogan expresses a clear denunciation of "fake food" in its various manifestations, such as ultra-processed foods.

The periodic holding of National Conferences constituted a remarkable exercise of social participation in Brazilian public policies, until it was interrupted by the authoritarian rupture of policymaking, starting with the rise of Bolsonaro, affecting the country since 2019.[4] The process that consecrated the formulation "real food" in the countryside and in the city – for rights and sovereignty", from the municipal level, passing through the state, until reaching the V National Conference – mobilized representatives of a diverse set of social segments, such as family farmers, peasants, fishermen, indigenous peoples, Quilombola communities, categories classified as traditional peoples and communities, LGBTI+ community, women, youth, black populations, professionals from various sectors, academics, environmentalists, and non-governmental organizations. Drawing attention to this characteristic is intended to provoke reflection on the ability of this formulation to welcome and express such diverse visions and aspirations, therefore, on the depth of its cultural and symbolic meanings, no doubt, but also on the requirements for a food to be "real" and pleasurable (Consea 2015).

However, it is worth noting the proximity of the notion of "real food" with that of adequate and healthy food with wide international circulation, also subject of conceptualization approved at the III National Conference in 2007, according to which

> Adequate and healthy food is the realization of a basic human right, with the guarantee of permanent and regular access, in a socially just manner, to an eating practice appropriate to the biological and social aspects of individuals, according to the life cycle and special food needs, based on the local

4 It is important to mention that the rupture has not paralyzed the organizations and social movements that continue to meet with a view to holding a National Democratic, Popular and Autonomous Conference as soon as sanitary conditions permit.

> traditional reference. It must take into account the principles of variety, balance, moderation, pleasure (taste), the dimensions of gender and ethnicity, and environmentally sustainable forms of production, free of physical, chemical, biological pollutants and genetically modified organisms.
>
> (Consea 2007) (own translation)

As a conceptualization coming from a wide range of social actors, it is possible to note the breadth of the aspects involved and the implicit criticism of technical visions such as "nutritionism" or "productivism" and, even worse, of increasingly artificialized food, disseminated by the industry. As important as the origin of the concept are the disputes over the appropriation of contested referents by food corporations. To eat "real food", one needs to know the truth about food (Consea 2015). As underlined by Azevedo (2014), in the face of food risks and doubts surrounding the plate and the mind of the contemporary eater, the right to information and the truth about "the story behind the food" becomes increasingly relevant (Thompson et al. 2008; Azevedo 2014).[5] The same author points out that the concept of healthy eating is pluralized (polysemic character) and interacts with assumptions of tradition and other elements, such as the perspective of sustainability and activism, the rescue of traditional medical rationalities, risk, uncertainty, and scientific controversies. Defining this concept became an arduous task, in the light of late modernity, so that it reflects the complexity and reaffirms the plurality of lifestyles as a contemporary condition (Azevedo 2014). Still regarding knowledge about food, the relational and subject-decentered approach proposed by Carolan (2011) assumes the relational basis of knowledge about food and the material connectivity involved in food choices that are produced and maintained through practice. His argument for an embodied food politics decenters individual decisions to position collective relationships among those who participate in food processes.

An important moment in the arduous, but also conflictive, task of conceptualizing healthy eating occurred when the Ministry of Health launched the Food Guide for the Brazilian Population in 2006, and reissued it in 2014, in the face of strong opposition from the food industry and agribusiness (Ministério da Saúde 2014). The Guide makes recommendations in favor of in natura or less processed foods, and an explicit condemnation of ultra-processed foods, while valuing the contribution of different knowledges, respect for cultures, and people's autonomy of choice. It is possible to correlate this conception of healthy eating developed in the Food Guide with that of "real food" that emerged later, as noted in the set of options offered in the Guide to compose the three usual daily meals in Brazil. It is difficult to summarize them here, given the

5 Criticizing the premise that choices are personal responsibility and that they are free and informed, Nestle (2013) draws on the focus on the "politics of food choices" to propose an ethics of choice.

great socio-spatial and cultural diversity, which is characteristic of the country; as an illustration, I refer to some components, common to the compositions suggested for breakfast (coffee with milk, corn cake or wheat bread, fresh cheeses, and typical fruits of the region), lunch (rice, beans, meats, vegetables, cassava, and salad), and dinner (similar composition of lunch or soup).

It is important to underline that in the same Guide, as well as in Martinelli and Cavalli (2018), it is argued that an unsustainable food system cannot produce healthy food for consumption. Food can only be considered healthy if it is also sustainable and if it goes beyond the purely nutritional perspective. This leads us to important questions regarding food systems, which are currently strongly challenged through the following critical questions: Who produces/distributes "real food"? How to ensure availability and access to "real food" in contexts of high inequality and inequity, as is the case in Latin America? What are the implications when connections between food, sustainability, health, and justice are established?

It was mentioned that the dispute between narratives plays an important role in food politics. We therefore try to give some critical contributions to established concepts, starting from the notion of "real food". The questions refer to limited conceptualizations of healthy eating, proposals to promote sustainable and healthy food systems, and the connection between food justice, social justice, and human health. This requires a look at the social actors, their ideas, and their interests involved, hence at the politics underlying the correlation between food, food systems and the issue of sustainability, which is the subject of this chapter.

Food justice is a key concept in the analysis of the systemic determinants of injustices, inequalities, and inequities, involving questions related to the following: (a) How, where, what, and by whom is food produced, processed, sold, and consumed? (b) How are benefits and related values shared, and how and by whom are decisions about food systems made? (Gottlieb and Joshi 2010; Gilson and Kenehan 2018). Cadieux and Slocum (2015) argue that doing food justice means finding ways to intervene against structural inequalities and promote transformative change, with the specific goal of institutionalizing equity in the food system, and expanding social control over food production and consumption, given the power of food corporations, industry, and supermarket chains, with the decisive support of the media. Maluf et al. (2022) understand food justice as the promotion of social justice through food. Environmental justice, in turn, not only refers to inequalities in the distribution of environmental risks, but also expands the concept of social justice by understanding the environment and nature as conditions for social justice (Schlosberg 2013).

Food, politics, and democracy: challenges for a critique of sustainability

This part attempts to insert the movement in Brazil as part of a broad global movement, which questions the current format of dominant food systems and advocates for their radical transformation, driven by the criticism of corporate grabbing of the agenda of the UN Food Systems Summit finalized in September 2021. In other words, the aim is to establish the threads of critical propositions, emerging in the international debate on the necessary transformations of the predominant food systems, in view of sustainability and health; more specifically, we are considering the political paths of the discussions in Brazil and whose interests, visions, narratives of different actors on food, food systems, and sustainability are opposed.

As highlighted by Maluf et al. (2022), the narratives of national and international civil society organizations, as well as national policies and plans built with social participation, emphasize social, environmental, and power issues. It recurs the argument of the need for paradigm shifts in food systems toward the consolidation of agroecological and peasant modes of food production in which social relations, values, and practices should reorient the relationship with the territories and promote climate and environmental justice. Based on a wide range of experiences around the world, the document called "*Voices from the Ground*" launched by the Civil Society and Indigenous Peoples Mechanism linked to the United Nations Committee on World Food Security (CSM 2020) strongly associates sustainability with expressions of diversity, such as "agrobiodiversity", "sociobiodiversity", and "diversity of food cultures". They emphasize the power inequalities that strengthen large corporations and the dominant, monocultural, agroindustrial food models, and hence weakening the possibilities to give visibility to rural social segments and highlighting their diversity: family farmers, peasants, indigenous peoples, traditional peoples and communities, landless peasants, rural workers, etc.

In synthesis, the critiques coming from food-related social movements value the knowledge and practices involved in the ways of life, in the production, extraction, processing, distribution, and consumption of food, with agroecology as a key notion. Advocating for the autonomy of peasants to choose what and how to produce and consume food, managing their own seeds is a crucial aspect, in the face of seed trade and genetic modification by large corporations. This perspective is reflected in the concept of food sovereignty, launched by segments of the rural world as an alternative for society as a whole. Thus, the new paradigms will be based on rights and justice, food sovereignty, socio-biodiversity, ethnic–racial and gender equality, reduction of power asymmetries, also as true and just solutions to the climate crisis. The technological innovation component, emphasized by corporations, becomes the use of "socio-technologies" (LVC 2020).

It can be said that almost all the concepts, values, and practices mentioned in the international debates are found in the principles and proposals defended by the movements around the notion of "real food" in Brazil. Notwithstanding, attention was drawn to their counter-hegemonic character and the environment of disputes and power asymmetries, which shape food politics in the country. It is worth highlighting the dimension of violence surrounding the struggle for rights and for the defense of territories (land and natural resources), against the expansion of monocultures. As Paarlberg (2010) emphasizes, the food politics of each country is very similar to other types of politics of their respective political regimes. Therefore, it is not superfluous, to stress the importance of the democratic conditions needed in the construction of sustainable and healthy alternatives. How far have these movements come in the co-creation of sustainable alternatives?

Behind the disputes of narratives, social practices, and conflicts, there is the one-dimensional valuing of food as commodities, which opposes the approach of Vivero-Pol (2017, 2018) presented earlier, on the differentiated political nature of food; in particular, this refers to his proposition of applying the notion of "commons" to food. It seems evident that the movement analyzed here shares the perspective of building options of collective meaning of food, or of valuing the communalities of food, for example, when it establishes common principles, or with the very idea of knowing how to feed oneself. The actions promoted by this, and other social movements, involve initiatives aimed at "recommunalizing" food, for example, within the framework of negotiation processes between decision-makers, at different levels and in different links of value chains and food systems. In this regard, it would be interesting to assess the extent to which they are configured as food communities, and what dynamics they propose for managing the commons.

Related to the above, one could think of joining the search for autonomy, one of the main subjects of this book, with the perspective of retaking or recovering power over food, based on "the common" as a political principle (Dardot and Laval 2019). Some of the political propositions made by Dardot and Laval (2019) from this conception seem quite appropriate, namely, creating institutions of self-government that allow the development of common action; property, subordinated to the right of use; organizing the social to enable deliberations in the public sphere with participation and co-obligation; partnership in the economy; establishing a federation of the commons.

As a conclusion of this part in dialogue with the more general theme of the book, it is worth mentioning the notion of food sustainability as defined by Jacobi et al. (2020) which synthesizes almost everything that has been discussed in the chapter; it involves, in addition to environmental performance, the reduction of poverty and inequality, the right to food, food security, and socio-ecological resilience. This understanding of food sustainability implies that the assessment of the sustainability of food systems includes their links with changes in the environment and social context. Hence, food sustainability refers to food and eating based on human rights, in particular the right to food, the reduction of poverty and inequality contributing to greater equity and better livelihoods for the actors involved, and finally the protection of the environment and increasing resilience within the systems (Tribaldos et al. 2018).

Concluding remarks

Instead of drawing conclusions from the complex picture presented in the chapter, analytical perspectives are rescued for the necessary critique of conventional views on sustainability. Through the analysis of food systems and the implications of food politics, in terms of stimulus or obstacle, we propose to engage in radical transformations that articulate sustainability with the confrontation of inequalities, promotion of social justice, adequate and healthy nutrition, and mitigation of climate change. Of course, the recourse to interdisciplinary dialogue, in the face of disciplinary transpositions and transgressions, observed in the real world of practices, stands out (Goodman and Sage 2014). The multiplicity of contemporary border crossings or transgressions of knowledge, practices, and politics around food give meaning to contemporary, broadly transgressive, food politics. It shapes and is shaped by the economic, social, and ecological systems from which multiscale production and consumption networks originate.

As is well-known, we live in a civilization centered on cities, with food systems capable of providing food in large quantities, over great distances, and at a allegedly decreasing monetary cost, but with evident collateral effects that are widely questioned. Thus, food and nutrition have gone from "episodic problems" to permanent issues on the public agenda. Going further, it can be said that there is a food issue in the development of countries that involves a complex and conflicting relationship of business, power, basic needs, and rights (Maluf 2021a). It is in this context that we look toward obtaining environmentally sustainable and socially equitable means of production, with consumption based on adequate and healthy eating habits, respecting the cultures of peoples.

The aim of this chapter was to extract some meanings and implications of the emergence of the notion of "real food in the countryside and in the city: for rights and sovereignty" as a mobilizing referent of a political camp in a context of conflicts and intense dispute over narratives in Brazil, not dissociated from debates at the international level. The capacity of this notion to harbor different aspirations of a very diverse set of social sectors is striking, since it directly confronts the many meanings that make food real. Among its merits, we find that it is not separating products from their use, as was intended to be demonstrated in a previous work with the use of the double denomination food and eating (Maluf 2021a). It is proposed to differentiate, and at the same time consider together goods or products (food) and the ways in which individuals, families, or social groups appropriate them. The intention is to supplant the productive mercantile logic that subsumes what is eaten, as well as the focus on nutrient intake, without taking into account what food it is. Integrating both aspects, which in fact should complement each other, is an important achievement of the notion of "real food".

We have shown in a previous paper (Maluf et al. 2022) how underlying tensions and conflicts arise from the inequalities and injustices inherent in food systems at territorial, national, and international levels, which are not adequately addressed by the predominant narratives, proposals, and technical-economic trends. In fact, the narratives and documents from the private commercial sector, governments, and international bodies argue that inequalities would be overcome with the generation of employment and the consequent improvement of living conditions, especially in the countryside. They are in line with narratives that prioritize private mechanisms and economic responses, based on innovations and technical progress, reducing the notion of sustainability to the reduction of greenhouse gases emissions.

In turn, civil society documents and many academic papers highlight the determinants of inequalities located in food systems and in iniquitous social formations, such as the Brazilian one. They call for radical changes in food systems and demand ethnic–racial democracy, gender equality, and the guarantee of human rights. The technological innovation component translates into the use of social technologies. The instruments proposed by transnational corporations and several governments are denounced as false solutions, such as the carbon market and the so-called green economy.

It is in this context that the voices coming from social organizations and movements, such as those expressed through the notion of "real food in the countryside and in the city: for rights and sovereignty", acquire importance. In open contestation to the production and consumption trends coming from private food business, it constitutes a path that allows to unite the criticism of the prevailing responses to the contemporary challenge of sustainability with the perspective of retaking or recovering the power over our food, based on emancipation and peoples' cultures.

References

Azevedo, E., 2014. Alimentação saudável: uma construção histórica. *Revista Simbiótica*, 7(1), 83–111.

Burlandy, L., et al., 2021. Reflections on ideas and disputes in the context of promoting healthy eating. *Reports on Public Health*. Rep. Public Health 37 (1:e00195520), 1–20.

Cadieux, K. V., and Slocum, R., 2015. What does it mean to do food justice? *College of Liberal Arts All Faculty Scholarship*. Paper 3. http://digitalcommons.hamline.edu/cla_faculty/3 (Accessed 9.2.2022).

Carolan, M. S., 2011. *Embodied Food Politics*. Abingdon: Routledge.

Castro, J., 1946. *Geografia da fome. O dilema brasileiro: pão ou aço*. R. Janeiro: O Cruzeiro.

CONSEA, 2007. *Relatório Final da 3ª Conferência Nacional de Segurança Alimentar e Nutricional*. Brasília, DF: CONSEA.

CONSEA, 2015. *Relatório Final da 5ª Conferência Nacional de Segurança Alimentar e Nutricional*. Brasília, DF: CONSEA.

CSM (Civil Society and Indigenous People Mechanism), 2020. *Voices from the Ground – From COVID-19 to Radical Transformation of Our Food Systems*. Rome: CSM/WG.

Dardot, P., and Laval, C., 2019. *Common – On Revolution in the 21st Century*. London: Bloomsbury Academics.
Dubois, V., 2021. Os espaços sociais para a produção de políticas públicas – a noção de campo como ferramenta da sociologia da ação pública. In: Olivera, O. P., and Hassenteufel, P., eds. *Sociologia política da ação pública: teorias, abordagens e conceitos*. Brasília, DF: Enap, 135–152.
Fraser, N., and Honneth, A., 2003. *Redistribution or Recognition? A Political-Philosophical Exchange*. London and New York: Verso.
Gilson, E., and Kenehan, S., eds., 2018. *Food, Environment, and Climate Change: Justice at the Intersections*. London: Rowman & Littlefield International.
Goodman, M. K., and Sage, C., 2014. Food transgressions: Ethics governance and geographies. In: Goodman, M. K., and Sage, C., eds. *Food Transgressions: Making Sense of Contemporary Food Politics*. Surrey: Ashgate Publishing, 1–14.
Gottlieb, R., and Joshi, A., 2010. *Food Justice*. Cambridge, MA: MIT Press.
Herring, R. J., 2015. How is food political? Market, state, and knowledge. In: Herring, R. J., ed. *The Oxford Handbook of Food, Politics, and Society*. Oxford: Oxford University Press.
Intergovernmental Panel on Climate Change-IPCC, 2019. *Climate Change and Land. Summary for Policymakers*. www.ipcc.ch/report/srccl/ (Accessed 9.2.2022).
International Panel of Experts on Sustainable Food Systems-IPES-Food, 2017. *Unravelling the Food – Health Nexus: Addressing Practices, Political Economy, and Power Relations to Build Healthier Food Systems*. Geneva, Switzerland: The Global Alliance for the Future of Food and IPES-Food.
Jacobi, J., et al., 2020. A new understanding and evaluation of food sustainability in six different food systems in Kenya and Bolivia. *Nature Scientific Reports*, 10(1), 1–14.
Kaljonen, M., et al., 2021. Justice in transitions: Widening considerations of justice in dietary transition. *Environmental Innovation and Societal Transition*, 40, 474–485.
La Via Campesina (LVC), 2020. *Annual Report 2019*. Zimbabwe.
Maluf, R. S., 2021a. Decentralized food systems and eating in localities: A multi-scale approach. *Revista de Economia e Sociologia Rural*, 59(4), e238782.
Maluf, R. S., 2021b. Participação social e política dos alimentos no Brasil: elementos para uma reflexão a partir do CONSEA. In: Schubert, M. N., Schneider, S., and Tonin, J., eds. *Desafios e tendências da alimentação contemporânea: consumo, mercados e ação pública*. P. Alegre: Ed. UFRGS, 61–88.
Maluf, R. S., et al., 2022. Sustainability, justice and equity in food systems: Ideas and proposals in dispute in Brazil. *Environmental Innovation and Societal Transitions*, 45, 183–199.
Martinelli, S. S., and Cavalli, S. B., 2018. Alimentação saudável e sustentável: uma revisão narrativa sobre desafios e perspectivas. *Ciência & Saúde Coletiva*, 24(11), 4251–4261.
Mialon, M., Swinburn, B., and Sacks, G., 2015. A proposed approach to systematically identify and monitor the corporate political activity of the food industry with respect to public health using publicly available information. *Obesity Reviews*, 16(1), 519–530.
Ministério da Saúde (Brasil), 2014. *Guia alimentar para a população brasileira*. Brasília, DF: Ministério da Saúde (Brasil).
Mouffe, C., 2005. *On the Political*. Abingdon: Routledge.
Nestle, M., 2013. *Food Politics – How the Food Industry Influences Nutrition and Health* (3rd ed.). Berkeley, CA: University of California Press.

Paarlberg, R., 2010. *Food Politics – What Everyone Needs do Know*. Oxford: Oxford University Press.

Schlosberg, D., 2013. Theorising Environmental Justice: The Expanding Sphere of a Discourse. *Environmental Politics* 22 (1): 37–55.

Swinburn, B. A., et al., 2019. The global syndemic of obesity, undernutrition, and climate change: The Lancet Commission report. *The Lancet* – published on line, 27 January.

Thompson, E., Harper, A. M., KRAUS, S., 2008. *Think globally-Eat Locally: San Francisco Foodshed Assessment*. Available at: http://www.farmland.org/programs/states/ca/Feature%20Stories/document/Think GloballyEatLocally-FinalReport8-23-08.pdf

Tribaldos, T., Jacobi, J., and Rist, S., 2018. Linking sustainable diets to the concept of food system sustainability. *Future of Food: Journal on Food, Agriculture and Society*, 6(1), 71–84.

Vivero-Pol, J. L., 2017. The idea of food as commons or commodity in academia. A systematic review of English scholarly texts. *Journal of Rural Studies*, 53, 182–201.

Vivero-Pol, J. L., 2018. Alimentos como bens comuns: uma nova perspectiva sobre a narrativa do sistema alimentar. In: Correa, L., ed. *Diálogos sobre direito humano à alimentação adequada*. Juiz de Fora, MG: Fac. Direito/Universidade Federal de Juiz de Fora, 1–40.

Willett, W., Rockström, J., Loken, B., and Springmann, M., 2019. Food in the anthropocene: The EAT – lancet commission on healthy diets from sustainable food systems. *The Lancet*, 393(2), 447–492.

11 Agroecology as a transformative approach to sustainable food systems

Florencia Spirito, Viviana Blanco, and René Montalba Navarro

Introduction

The old colonial structures installed from the 16th century in Latin America, and their logic of dehumanization of indigenous peoples and enslaved African populations (Dubois and Turits 2019), were key to the conformation of the current agro-extractivist model[1] (Svampa 2012, 2019). The consolidation of sugar and grain plantations established a regime of exploitation where the resources of the Global South were traded as raw materials for predominantly Western profit. The plantation model proposed by Burnard and Garrigus (2016) gave way to the first Western institutions that implemented industrial-type production, deploying hundreds of enslaved workers, as in a factory. However, it was not until the European Industrial Revolution of the late 18th century to early 19th century that the foundations of Industrial Agriculture as we know it today were laid (Shiva 2000). This style of agriculture was born out of the need to increase land productivity in order to free up labor for industry. After World War II, widespread use of mechanization and the use of petrochemical inputs began, including the use of highly toxic synthetic pesticides, such as Agent Orange and DDT, both previously used as weapons of war (Ascuasiati 2012). The Green Revolution of the 1960s, together with the encouragement of international organizations and nation states, both central and peripheral, allowed the globalization of industrial production models, with the consequent establishment of transnational corporations (Jain 2010). The use of technologies promoted by the Green Revolution has generated cumulative negative effects, such as environmental degradation of common resources, concentration of land ownership, control of food systems in the hands of a few actors (Jacobi and Llanque 2018), exclusion of rural communities, and rural–urban migration leading to emptying of the countryside (Pengue et al. 2018). Likewise, these technologies have contributed to major environmental and human health problems in rural and peri-urban territories, such as air, soil, and water pollution, and intoxication by fumigation and pesticides

1 Based on Svampa (2019), we define extractivism as the special mode of accumulation, based on the plunder of the commons, product of the hierarchization between colonies and empires, and the expansion of capitalism.

DOI: 10.4324/9781003043577-11

resulting in chronic ailments in most cases (Andreotti et al. 2018; Steffen et al. 2011, 2018).

In this context, agroecology is positioned as a science and as a practice that questions the capitalist logics of appropriation of life as unsustainable and lacking in resilience (Bennett 2016; Ceballos et al. 2017). Thus, it has become an opportunity, both in social movements and in academia, to transform conventional food systems, proposing other methods of food production, distribution, and consumption as urgent emancipatory models. Agroecology is positioned as a critical and pluriepistemological paradigm, given that it recognizes both scientific and non-scientific knowledge in the hands of farmers and communities (i.e., biocultural diversity) (Maffi 2007). This is the meeting point with the epistemologies of the South proposed by De Sousa Santos (2011:8), both of which are characterized by the pursuit of sustainable production processes, valorization of both scientific and non-scientific knowledge, and new relationships between different types of knowledge, especially knowledge held by those who have suffered systematic oppression. The pluriepistemological perspective is inseparable from transdisciplinarity, which implies processes of co-creation of knowledge where different actors have participation and influence (Max-Neef 2005; Hadorn et al. 2006); but our epistemology extends beyond academia, since it gathers the local, situated, and empirical knowledge of those who work on the land.

Actions of adaptation and transformation from agroecology (Servigne and Stevens 2015) imply an extraordinary effort of holistic interpretation, based on Western scientific evidence, but also from local communities and indigenous peoples. It also requires imagination to find emancipatory alternatives to dependence on the dominant conventional food system. Agroecology in its pluriepistemological construction is nourished by ecology, economics, sociology, anthropology, political science, and feminism, but its strength, as discussed earlier, lies in recognizing, valuing, and incorporating a corpus of knowledge, wisdom, and locally adapted practices, which are in the hands of peasant, riverine, quilombola communities, and indigenous people. Agroecological-based knowledge promotes the rights of all living beings to a healthy environment and democratic access to healthy, rich, varied, culturally appropriate, and sovereign food. We open from here a debate on the elements that in our opinion explain how conventional models of food production are unsustainable, and then propose agroecology as a transformative and surpassing alternative. Throughout this chapter, we will make use of the analysis of principles of food sovereignty and socio-ecological resilience proposed by Giraldo and Rosset (2021), whose framework integrates conditions of social, ecological, and economic justice into discussions of sustainability.

The failure of industrialized food systems and the crisis of sustainability

Our planet is immersed in a global, systemic, and civilizational crisis driven by the capitalist development of the last 200 years (Giraldo 2014; Herrero

López and Rodríguez Muñoz 2017). This crisis has a close relationship with food production. The industrial agricultural model has generated indisputable negative consequences which contribute to the seriousness of the crisis, but will also, if this model persists, imperil the ongoing quantity and quality of food production, and, consequently, the very survival of the human species. It should be clarified that, although the sustainability of the agri-food system as a whole is in question, for the purposes of this chapter, we will focus on specific aspects of primary production and its effects on rural spaces. In order to fully understand the origin of the crisis, we must understand that agriculture,[2] as a human activity has been with us for more than 10,000 years, and today it is the main source through which humanity obtains its food. By definition, agriculture implies the modification and management of natural environments, transforming them into complex systems, founded on the superposition of the natural and social subsystems. Moreover, precisely because of the purpose it serves, agriculture is an essential activity for human survival, and as such is the main transforming force of habitable terrestrial ecosystems (50% of the habitable land on the planet is devoted to agriculture) (Porter et al. 2017). Therefore, when analyzing food production, we must understand that human action on natural ecosystems will always generate impacts, what we must discuss and address is the nature and the magnitude of these impacts.

The environmental impacts of conventional food production

The physiognomy of the agricultural production model in Latin America, as we know it today, began to take shape as a result of the project of modernization driven by the United States, which emerged in the aftermath of World War II (Chrispeels et al. 1994). The underlying theory of modernization assumed the inevitable transformation of traditional societies into modern ones based on a process of differentiation that would allow "backward" nations to reach "maturity", taking developed countries as a model. The mechanisms for this transformation were based on diffusionism, which consisted of the transmission of cultural, economic, and social characteristics from the dominant (in this case Western) society. Policies based on this theory promoted reforms in three key areas: the family, to solve what was considered a structural problem of excessive population growth; education, to solve illiteracy; and, finally, to transform and modernize the rural world (Huergo 2012). In the field of agricultural extension, diffusionism consisted of placing the focus on external advisors, their knowledge, the information they have to transmit, their purposes, and their ways of acting, thus taking it away from the target people and their cultural practices (Rogers 1958). Thus, the diffusion of innovations in rural areas promoted the incorporation of knowledge and technologies generated in

2 For the purposes of this chapter, we understand the term agriculture in a broad sense, encompassing the production of agricultural raw materials, including livestock and poultry.

developed countries by peasant communities and traditional farmers, known as the Green Revolution (Rogers 2003).

Thus, the hegemonic model of agricultural production, known as the modern or conventional model or industrial agriculture, brought about a revolutionary change in the way of producing food and fiber, given that up to that time farmers had cultivated species that were adapted to the ecological conditions of their local production systems. From the Green Revolution technologies, new genotypes with high yield potential were disseminated, which required appropriate conditions for this potential to be expressed, that is, the environment had to be adapted to the new genotypes, and this was achieved based on energy inputs, mainly derived from fossil fuels, in the form of fuels, fertilizers, pesticides, and plastics, among others (Jain 2010). In brief, the Green Revolution involved a technological model for food and fiber production based on the simplification of production systems (reductionism), rejecting the complexity that defines them. It was grounded in productivism, in that its main measure of success is based on yield per hectare, which favors short-term goals and momentary profit, and was based on the contribution of inputs and knowledge external to the production system and did not take into account social and gender diversity or environmental justice. The result was the simplification and homogenization of agricultural ecosystems and a myriad of negative consequences, both environmental and social (e.g., Pingali 2012; Sarandón and Flores 2014).

Multiple factors have favored the validation of the conventional model as the (only) possible production model; in this chapter, we will point out only three. First, the analytical framework of modernity entails the idea of separation between humans and nature, conceiving nature as inexhaustible, and as a resource to be exploited for the satisfaction of individual needs (Carrasco Aquino 2006; Morales Jasso 2016; Castillo Sarmiento et al. 2017). This view remains practically unchanged to the present day, defining modes of appropriation (and destruction) of common resources, aggravated especially since the industrial revolution and the development of capitalism (Price et al. 2011; Lewis and Maslin 2015; Zalasiewicz et al. 2016; González Márquez 2020; González Reyes 2020). Second, and linked to the positivist paradigm, an excessive confidence in modern science and technology and their ability to solve any problem has been generated in societies, ignoring that they can also be the cause of such problems; this sanctification of modern science and technology denies at the same time other types of knowledge and wisdom, different from those obtained through the scientific method (Sarandón and Flores 2014; Sarandón 2019). Finally, the contributions of critical feminist studies on modern science show how this paradigm excluded half of the population, women, confining them to the domestic-reproductive sphere, ignoring and making invisible both their contributions to productive work and their knowledge and needs in terms of equipment, machinery, implements, practices, and chores (González Márquez 2020; González Reyes 2020; Gupta 2020). According to Soler et al. (2014), the technological model derived from modern science that sustains this productive paradigm is tinged with the same biases as Western thought: anthropocentrism, ethnocentrism, and androcentrism. These authors question and describe how anthropocentrism focuses on the human being above all species, and strictly speaking, above nature, so that humanity can conquer, dominate, and, in our opinion, destroy it. Additionally, ethnocentrism rejects, delegitimizes, and erodes cultures different from the Western one, for being "backward" or "savage". Clearly, this rejection implies the discrediting of other forms of knowledge, which are qualified as irrational, and the elevation of knowledge obtained from the Western rational method to a unique category. Finally, the androcentric bias leads us to consider how Western society was structured socioculturally, economically, and politically from the masculine perspective, placing males (white, adult, resource-owner, and heterosexual) as the standard and at the center of existence, and makes women invisible and feminized identities, arguing, among other things, their lack of rationality (Navarro and Gutiérrez 2018). These three biases place humanity, strictly speaking, man (male, adult, resource-owner, white person), above the non-human and the rest of humanity (non-white, male, proprietary and adult); and Western culture and all that it produces above other cultures, which it considers non-rational. And to dominate the untamed, for example, nature, it makes use of modern science. From it, everything that is not rational is discarded, thus justifying why certain cultures and certain identities are inferior. In this sense, it is interesting to note the criticism of Boaventura de Sousa Santos (cited by Soler et al. 2014), who defines scientific rationality as a totalitarian model through which only the knowledge derived from science is valid and rejects others as "irrational" (Soler et al. 2014).

Currently, there is consensus, even in broad sectors of academia, on the failure of agro-industrial production systems developed under the conventional model, due to the unsustainable environmental, socioeconomic, cultural, and political externalities produced (Pengue 2009; Sandhu et al. 2019). To cite some examples, environmental pollution and degradation (water, soil, air); loss of biodiversity; contribution to the generation of greenhouse gases; gene erosion and culture; rural–urban migration and emptying of rural spaces; exclusion, marginalization of those who cannot access technologies as well as women, youth, and other identities (indigenous, Afro, gender dissident, etc.); concentration of power in the hands of transnational corporations supplying inputs and their local allies; and problems in the health of the populations involved, due to both collateral effects of this process (contamination, intoxication, chronic, or cumulative effects) and the consumption of food produced under these models. All these negative effects are part of the current epistemic and civilizational crisis that we face as humanity, derived from the ideas that emerged in modernity and the positivist paradigm in science, but also partly due to the crisis that capitalism is going through. The creation of new participatory epistemologies is urgent, in which the political character of all thought, of all ideas is recognized, where "all social actors and actresses have something important to say about the object and product, thus generating an extended community of evaluators" (Martínez Alier cited in Sevilla Guzmán and Soler Montiel 2010:199). Perhaps most conclusively, the Green Revolution, with its founding tenet of ending world hunger, has failed even in its own terms, as today about 12% of the world's population still suffers from malnutrition, and almost 30% suffer from obesity induced by the type of diet promoted by the industrial agri-food system (Latham 2021).

Agroecology as an emerging emancipatory response to the crisis of civilization

Agroecology was born as a critical perspective on the ecological and social impacts of industrialized agriculture. The first use of the term dates back to 1930; however, it was not until the 1970s that the idea began to be popularized by academics such as Stephen Gliessman and Miguel Altieri, among others, who began to outline a theoretical and methodological framework to address the socio-environmental problems generated by the advance of modernization and industrial agriculture in rural areas (e.g., Altieri 1989; Gliessman 1990). Based on the contributions of Sevilla Guzmán, we can recognize the bases of agroecology in agrarian social thought and in the social movements that emerged as a response to the first processes of agricultural industrialization, becoming a counterposition against capitalist modernization (Sevilla Guzmán and Woodgate 1997). Over the years, this idea was transformed into what we define today as agroecology:

> The science that studies the processes and functions of agroecosystems and food systems, as they relate to biological, ecological, biophysical, social, cultural, economic and political characteristics (scientific dimension); as a set of technical-productive designs and practices that allow obtaining fiber and food in a more sustainable way (methodological dimension); and as a political and social movement that fights for the existential and conceptual self-determination of human and non-human beings and for generating fairer, sovereign food systems with local governance (political dimension).
>
> (Rosset and Altieri 2017)

In brief, agroecology promotes the ecological management of agroecosystems through forms of collective social action and participatory proposals to address the crisis generated by modernity. In the historical development of its theoretical construction, agroecology has changed its focus, moving from the scale of the farm to address the problems of the entire food system, including the stages of production, processing, transportation, financial intermediation, marketing, distribution, and consumption. The potential of agroecology is based on its pluriepistemic origin, where the local dimensions – cultural and biological diversity, and the local, empirical, and situated knowledge carried by peasant and native communities – are not excluded, but rather complement academic perspectives. It therefore provides a direct contrast to the hegemonic model, which is antagonistic to local and unique epistemologies, ecology, agrodiversity, and communities insofar as it proposes a single, hegemonic, and homogeneous global society disconnected from local community logics (Giraldo 2014). In addition, agroecology contains a key element, the questioning of capitalist relations of production and of capitalism in general. Therefore, those who seek to pigeonhole agroecology as an applied science without a sociopolitical context limit its transformative potential, its ability to contribute to create more just, egalitarian, and inclusive systems of production, distribution, and consumption of healthy and wholesome food. In Section "The Political Dimension of Agroecology", we will return to this topic in order to deepen our understanding of the transformative scope of agroecology.

The pluriepistemological power of the agroecological approach

The agroecological paradigm is built on the premise that academic knowledge obtained through the scientific method is not the only form of knowledge, since the original communities, peasants, and farmers of the world possess a wealth of knowledge, wisdom, and practices of local, situated, and empirical character, the result of years (sometimes thousands) of coevolution with their environments. In recent years, a significant number of proposals have arisen to build spaces of thought and practice based on the wisdom of the indigenous peoples of the continent, inviting us to think about the pluriverse, where multiple forms of knowledge are possible (De Sousa Santos 2011; Leff 2012; Escobar 2014, 2020). These efforts are directed toward the construction of a relational ontology that allows us to overcome the abyss that separates us from the world, to reconnect with nature and other human beings. Indigenous peoples' expressions such as *Sumak Kawsay* translated as *Good Living* ("*Buen Vivir*") in the Andes or the concept of *Living richly* ("*Vivir Sabroso*") in El Cauca are part of relational ontologies. These concepts are promoted by social movements to vindicate ancestral cultures based on the principles of reciprocity, complementarity, and cooperation, relativizing the capitalist economy through recognizing, valuing, and promoting non-capitalist forms of production and consumption (Escobar 2015). In the Latin American proposal, the heart constitutes the epistemic and ontological core of a rationality based on the worldviews of the native peoples of the region, of feelings (Borda and Moncayo 2009), as a different paradigm of thought and construction of knowledge, closely linked to the defense of land and nature (Rosset and Barbosa 2021).

Within this plural conformation, agroecology proposes the implementation of methodologies that value and consider other knowledge, encouraging political proposals that seek to change the terms and conditions of conversation between modern knowledge and alternative and indigenous knowledge (Restrepo and Rojas 2010). This dialogue, understood as a dialogue among different types of knowledge ("*Diálogo de saberes*")[3] (Torres and Rosset 2016), is also a dialogue of lived experiences, which includes practices and techniques, but also senses, meanings, histories, and affections. This dialogue also transforms academia, because it challenges it to address complexity, and relies on transdisciplinarity. Therefore, instead of transferring decontextualized technologies, transformative social methodologies begin by revaluing agricultural knowledge, traditional diets, traditional healthcare, and vernacular forms of construction based on one's own culture and spirituality, and then includes them in dialogue, so that collective processes of problem-solving and transformation of reality can be activated. One of the worst impacts of industrial agriculture was to subtract the ability to act with the body, to trust the senses, to commune with nature and open up to its mysteries, to find, through bonds of brotherhood, concrete solutions to common problems, making use of oral tradition and direct experience as the most effective means of learning and living together. It is then through this pluralistic approach, this diverse dialogue, that agroecology is positioned as a critical and propositional paradigm that builds ways of linking people from horizontal and egalitarian spaces, establishing links between the past and the future through the rescue of agricultural management practices and strategies that have been shaped over thousands of years, in a coevolution between environment and humanity, marked by adaptation to changing climatic conditions, providing resilience to these productive systems (Altieri and Nicholls 2012; Montalba et al. 2015).

Agroecology as a pluriepistemological science is a life project with an enormous capacity to transform hearts, to regenerate community ties, to reinsert culture into the ecological order of the inhabited place, which far exceeds the economistic understanding that dominates decision-making institutions (Rosset and Altieri 2017). Agroecology and its territorial transformations have profound ontological implications, since landscapes of diversity are the fertile soil where the spiritual transformations we need to reunderstand ourselves as interdependent, hyper-related beings belonging to the webs of life can sprout (Capra 1999; Linsalata 2020). Ontological changes – from an isolated Being to an interconnected one – cannot take place independent of the conditions of the place where we live, since they occur within the habitats that we forge with our actions. The cry of nature calls for the transformation of the territories without breaking the biotic equilibrium, nor transgressing the natural niches of plants and animals, and their symbiotic relationships; but it is also a cry to understand ourselves as an emergence of interrelationships, interdependencies, and complementarities that we inhabit and that inhabit us.

3 At this point, it is important to emphasize that it is not possible to practice the dialogue of knowledge without recognizing and acting on the different types of discrimination or biases that affect all people directly or indirectly: sexism, racism, classism, xenophobia, homophobia, transphobia, age discrimination, physical and linguistic ability, etc. Action on the different forms of discrimination is a great and permanent challenge.

Contributions from ecology, ecological economics, and feminisms

Another distinctive element of the epistemological construction of agroecology is the convergence in its theoretical corpus and praxis of a set of disciplines that contribute to the understanding of a complex whole. Here, we describe some of these disciplines, which by their influence, contribute to the singularity of agroecology.

Agroecology is inspired by nature to promote processes of diversity, energy efficiency, and nutrient cycling (Sarandón and Flores 2014; Sarandón 2021). Drawing from ecology, it proposes complex relationships between agroecosystem components, which translates into agroecological principles proposed by Nicholls et al. (2017): (1) plant and animal diversification within the agroecosystem; (2) recycling of nutrients and organic matter; (3) management of organic matter and the stimulation of soil biology to provide optimal provision for crop growth; (4) minimizing water and nutrient loss by maintaining soil cover, controlling erosion, and managing the microclimate; (5) adopting preventive measures to control insects, pathogens, and weeds; and (6) taking advantage of synergies and symbioses that emerge from plant–animal interactions. The design and management of agroecosystems under these criteria makes it possible, among other things, to reduce the enormous impact of conventional agriculture on natural environments; reduce greenhouse gas emissions; reverse the loss of biodiversity, organic matter, soil erosion, and desertification processes; optimize the use of water, soil, and nutrients; guarantee the supply of healthy, nutritious, and accessible food for the entire population; and generate systems that are adapted to local conditions and are also resilient. In this way, by strengthening ecological principles and freeing the productive sector from the corset imposed by transnational input suppliers, more inclusive and redistributive systems are also generated.

Another discipline that contributes significantly to the epistemological construction of agroecology is Ecological Economics (Martínez Alier 2001). The emergence of this field is accompanied by the questioning of the concept of unlimited growth, which is the basis of capitalism, and the explicit recognition of the finite limits of our planet, in an effort to interpret the relationship between ecological–productive systems, thermodynamics, and economics (Carpintero 2006). This represents an interesting change, breaking with the idea of the economic system as a closed system (Naredo 1994). On the contrary, Ecological Economics interprets the economic system as a subsystem of the ecological system, and, therefore, open, complex, and finite, with a unidirectional flow of energy, whose source is the sun (Martínez Alier and Roca 2000; Martínez Alier 2005). From this discipline also arises the questioning of the idea of weak sustainability and strong sustainability (Martínez Alier 2005). In simple terms, weak sustainability means that the depletion of resources is not an impediment to continued production, since these can be replaced by human capital and technology. Strong sustainability implies an advance over the previous definition, since it considers the need to preserve the non-replaceable elements of the environment, questions the idea of not altering the objective of economic growth and of generating partial solutions to a global problem (Medellín-Milán et al. 2011). Faced with this panorama, Ecological Economics focuses on the performance of communities, mainly in Latin America, which are recognized as the

> main actors in maintaining natural resources . . . it means that in conflicts caused by the use and distribution of resources and environmental services, the interests and values of communities – organically linked to their territory – lead them to favor the conservation of ecosystems and generate, in some cases, real niches of sustainability.
>
> (Medellín-Milán et al. 2011:11)

Finally, another novel contribution of Ecological Economics is that it incorporates concepts such as social metabolism, endo- and exosomatic consumption, and virtual water, and proposes methodologies such as the ecological footprint and water footprint to highlight aspects made invisible by neoclassical economics, such as the idea of ecological debt that central countries maintain with peripheral countries, based on the extractivist appropriation of natural goods and the accumulation of environmental liabilities in the peripheral countries (Guzmán et al. 2018).

On the other hand, the contributions of feminist studies show that, although agroecology is proposed as a plural epistemological construction, the result of the androcentric bias is that the dialogue that is promoted is often limited to that among men, and excludes the accumulated knowledge held by women. Although much remains to be explored, there is enough academic evidence to affirm that rural women have different and complementary knowledge to that of men, as well as a different referential framework, where the care tasks socially assigned to them generate, for example, a different perception of risk in the use of pesticides (Kunin 2019; Larrañaga 2021), as well as empathy and respect for nature (Noseda 2018). On the other hand, returning to the effects of the biases of the Western gaze, a relationship between masculinity and technology is observed. Thus, as masculinity is constituted around mastery over technologies, these are constructed from an ideal of masculinity (Estebanez et al. 2013), reinforcing and legitimizing the exclusion of women from the world of agricultural production. Finally, we highlight that gender oppression is often intertwined with other oppressions, termed by feminists as intersectionality, where issues of race, ethnicity, class, and gender converge, to realize that it is not the same to be a woman, a poor woman, a migrant, or a rural woman. All these prejudices make women and gender-diverse farmers, peasants, and indigenous women even more invisible, even so, they are often the ones who lead emancipatory movements in defense of land rights, even sacrificing their lives in the face of the extractivist advance (e.g., Osorio Pérez 2008; Le Tran et al. 2020).

Political dimensions of agroecology

In Latin America and the Caribbean, unlike in other continents, agroecology is defined from a political perspective, with a markedly sociocultural character and driven by and from social movements (Caporal and Petersen 2011; Altieri and Nicholls 2012; Rosset et al. 2021; Petersen 2022). In this sense, agroecology is perceived as a way of being and of living on the land, which translates into resistance and survival of food and agricultural cultures, in which agroecosystems are biocultural repositories of traditional knowledge systems, but without a romanticized vision of an idealized ancestral past, rather in dialogue with modernity and current forms of development (Royero-Benavides et al. 2019). The agroecological transition of a farm cannot be fully realized without the development of a sociocultural and political context from which collective proposals that transform the structures of social dependence, such as the concentration of land and power, are based, in order to promote community governance.

Critical agroecological thinking, which questions the consequences of the conventional production model, has guided the last four decades of capitalist restructuring and counter-hegemonic initiatives in Latin America and the Caribbean. In Cuba, for example, agroecology went from being a minority trend in the 1980s to the solution after the exit of the Soviet Union (1991) and has proven to be able to feed the Cuban people safely, with a peasantry with access to land, education, health, and markets (Altieri and Funes-Monzote 2012). Latin American agroecologists emphasize the need and right of communities and nations to self-determination, self-sufficiency, autonomy, and change from below (Giraldo and Rosset 2018). The processes are of long duration, but the objective of revolutionary agroecological struggles is always the same: that the communities, social movements, and organizations themselves collectively direct their own production, their distribution and consumption mechanisms, and their conditions of existence, in order to recover the collective power to control their own life and territory (Rosset and Barbosa 2021). This autonomy implies articulation and assemblage: weaving networks between community autonomy and external allies such as governments or universities, the recovery of local knowledge, and the collective creation of new knowledge, thus freeing the capacity of the peasantry and the people who practice agriculture to take charge of their most immediate problems without depending on government subsidies, or the imposition of technologies and management by agronomists or external consultants who teach the peasantry how to be peasants and who imply forms of knowledge and sociocultural organization alien to the territory and to the farmers.

Agroecology also becomes political through promoting novel ways of inhabiting the land, in conjunction with popular knowledge and social movements, in order to build sovereignty, identity, and collective action for transformation, toward the concept of *Good Living*. Rural social movements composed of peasants, artisanal fisherfolk, pastoralists, indigenous peoples, and urban farmers have insisted that agroecology must be revolutionary in the sense that it is capable of radically transforming both the food system and its structures and the adverse local realities faced by rural communities (La Via Campesina 2011, 2015):

> Popular pressure has urged many multilateral institutions, governments, universities and research centers, some NGOs, corporations and others to finally recognize "agroecology". However, they have tried to reduce the concept to a mere proposal of technologies to offer some tools to soften the sustainability crisis of industrial food production without challenging existing power structures. Agroecology is political, it requires us to challenge and transform the power structures in society.

The Declaration of the Nyéléni International Forum on Agroecology (La Via Campesina 2015) was the first collective expression of global social movements on the importance of agroecology. The declaration was explicit in assuring that the struggle of social organizations around the world is not for a superficial reform of the food system, but for a peasant- and indigenous-based agroecological revolution that challenges and profoundly transforms power structures, and puts control of the means of production, primarily land, in the hands of the people. Defending, recovering, and redistributing land is the first step to initiate or resume any agroecological process. There is no agroecological revolution without an agrarian revolution. However, recovering land is not enough (Rosset et al. 2021). As peoples are increasingly exposed to the siege of land grabbing, the threat of mega infrastructure projects, extractivist investments, concessions to national and foreign corporations, and all other forms of territorial dispossession by capital, the struggle for land must also include the defense of territory. Quoting Giraldo and Rosset (2018), "agroecology, as a growing social movement, comes to propose a political project of life in clear opposition to political projects of death". This involves a radical political critique of monopolies, but also of heteronomous structures that are administered under centralized power. Agroecology is political because it accompanies and strengthens the initiatives of social movements that seek the transformation of the food system in all its stages: production, distribution, and food consumption. The proposal of food sovereignty of La Via Campesina is a clear example of such a movement, developed under principles consistent with agroecology and from a popular, peasant, indigenous, and feminist perspective (Val et al. 2019).

The adoption of the agroecological paradigm has become so widespread and profound in Latin America that Altieri and Toledo (2011) have called it an "agroecological revolution". Scaling or territorialization of agroecology refers to those socio-environmental processes that make possible its geographical and organizational scaling, resulting in agroecology being adopted and practiced by a growing number of rural and urban families and communities, resulting in wholly agroecological food systems (Mier y Terán Giménez Cacho et al. 2018). Scaling is a transformative strategy to address the multiple current crises, recognizing that agroecology at the plot level, although important, has limitations to transcend and generate processes of transformation of national and global productive and food systems if it does not scale and massify in the territories through social pedagogies. The meanings, codes, and values that circulate among peers are intangible aspects shared in processes such as Peasant to Peasant (Rosset et al. 2011). Of course, there will be an agroecological practice that enters into dialogue, but also a sensitive communication that motivates and mobilizes, that incorporates the body as a whole, that revives the senses, that creates changes.

In the face of the relentless pursuit of profit by global capital, through land grabbing, the displacement of people from small-scale agriculture and the patenting of seeds, knowledge, and technologies developed over generations of agricultural practice, the second continental meeting of the Americas in 2011 issued a declaration: "Agroecology is ours and is not for sale. In this context we reaffirm that indigenous, peasant and family agroecological agriculture can feed the world and cool the planet" (La Via Campesina 2011). Currently, agroecology is facing a bid for the conventionalization (or co-optation) of the term by large corporate companies, which empty it of content to the extent that it does not internalize the different forms of sexist, environmental, social, etc. injustice, such that it will not be transformed as an alternative to the conventional model. Quoting Grosfoguel (2016):

> [I]f you organize against capital in a sexist, racist, occidentalocentrist, Cartesian, heterosexist, ecologicist, etc. way, you reproduce all the civilizational logics of the capitalist model. You reproduce all the civilizing logics of domination of modernity and the struggle ends up being corrupted.

In this sense, Giraldo and Rosset (2021) analyze the political, economic, organizational, methodological, pedagogical, and philosophical difference between "false agroecologies" and "emancipatory agroecologies" and propose six principles for building transformative and revolutionary agroecological processes: (1) question and transform structures, not reproduce them; (2) shape economies based on use value, not exchange value; (3) strengthen organicity and think in collective processes, not individualized projects; (4) build horizontal processes, not hierarchies; (5) train to struggle and transform, not to conform; and (6) act from culture and spirituality (Toledo 2022), not from productivism. The key to any agroecological transformational process is that the protagonism and social control be in the hands of the peasantry, the community, and other internal agents, and not external entities such as NGOs, foundations, international organizations, universities, religious entities, political parties, and state institutions, which usually try to organize the lives of their beneficiaries through hierarchies and values of change. What we are talking about is self-government, self-governance: that the people, through their organizations, conduct their lives and make their own political decisions based on collective deliberation and the implementation of common agreements. The self-management of popular organizations of the countryside and the city is the tool that the peoples have to successfully carry out their own agendas, among which is increasingly the deepening understanding and expansion of agroecology, to strengthen the Living Well of its members and Mother Earth.

Final reflections

By considering emancipation as a way of becoming aware of the power of the mechanisms of oppression and recognizing the forms of domination to which we are exposed (Boltanski 2011), agroecology is an emancipatory practice, born from the encounter with the *other*, which enables reflexivity to fortify and coordinate individual and community action as a fundamental element in the construction of counter-hegemonic forms of self-determination, resulting in the transfer of the ideals of solidarity, justice, democratic participation in decision-making, and self-determination of the political and economic sphere of life (Gros 2018). To speak of emancipation does not imply denying or devaluing all the knowledge accumulated under modern science, but that by recognizing that other forms of knowledge are possible, we are recognizing other ways of linking ourselves to knowledge, exposing the power structures that are woven behind every link and the productivist vision of life (Giraldo 2014). In order to feed ourselves, reproduce, develop our existences, and fulfill ourselves as human beings, we depend on other beings, human and non-human, as well as on the abiotic elements that constitute our environments. All life is made of interdependencies, which are built as a result of interrelationships and interactions between different species (Moore 2020). Agroecology emerges as a new paradigm, a new episteme, of a critical, global, holistic, systemic, and pluriepistemological nature, an overcoming proposal, a profound change that implies a transformation in the ways of linking ourselves with nature, of inhabiting territories, of designing and managing agroecosystems, and a questioning of the capitalist system as the only possible economic model. Agroecology claims the ancestral culture based on the principles of solidarity, cooperation, and organization of work and production, relativizing the capitalist economy through recognizing, valuing, and promoting non-capitalist forms of production and consumption, based on collective action to weave sovereignty and guarantee rights to food, inextricably linked to the right to life, as an exercise in transforming the world (Holt-Giménez and Altieri 2013). Agroecology is a tool within the framework of elements that must be generated for this civilizational transformation; it is one of the instruments, of the approaches, that must be intertwined with other struggles, in particular the union with other groups of the oppressed and exploited of the world, along with other environmentalist struggles, and of all kinds, to install new narratives that break with the logic of the world as given and narrate from the care, the sensitive, the solidarity, the emancipation.

References

Altieri, M. A., 1989. Agroecology: A new research and development paradigm for world agriculture. *Agriculture, Ecosystems & Environment*, 27(1–4), 37–46.
Altieri, M. A., and Funes-Monzote, F. R., 2012. The paradox of Cuban agriculture. *Monthly Review*, 63(8), 23–33.
Altieri, M. A., and Nicholls, C. I., 2012. Agroecología: única esperanza para la soberanía alimentaria y la resiliencia socioecológica. *Agroecología*, 7(2), 65–83.
Altieri, M. A., and Toledo, V. M., 2011. The agroecological revolution in Latin America: rescuing nature, ensuring food sovereignty and empowering peasants. *Journal of Peasant Studies*, 38(3), 587–612.
Andreotti, G., et al., 2018. Glyphosate use and cancer incidence in the agricultural health study. *JNCI: Journal of the National Cancer Institute*, 110(5), 509–516.
Ascuasiati, A. A., 2012. *Plagas Domésticas: Historia Patologías Plaguicidas Control*. España: Palibrio.
Bennett, J., 2016. Global ecological crisis: Structural violence and the tyranny of small decisions. In: Brauch, H. G., Spring, U. O., Bennett, J., and Serrano Oswald, S. E., eds. *Addressing Global Environmental Challenges from a Peace Ecology Perspective*. Cham: Swiss, Springer, 55–75.
Boltanski, L., 2011. *Sociology of Emancipation*. Cambridge: Polity.
Borda, O. F., and Moncayo, V. M., 2009. *Una sociología sentipensante para América Latina*. Bogotá: Siglo del hombre.
Burnard, T., and Garrigus, J., 2016. *The Plantation Machine: Atlantic Capitalism in French Saint-Domingue and British Jamaica*. Philadelphia: University of Pennsylvania Press.
Caporal, F. R., and Petersen, P., 2011. Agroecologia e políticas públicas na América Latina: O caso do Brasil. *Agroecología*, 6, 63–74.
Capra, F., 1999. *La trama de la vida. Una perspectiva de los sistemas vivos*. Barcelona: Editorial Anagrama.
Carpintero, 2006. *La bioeconomía de Georgescu Roegen*. Barcelona: Montesinos.
Carrasco Aquino, R., 2006. La naturaleza y sus formas de apropiación en contradicción. *Revista Mundo Siglo XXI*, 2(6), 55–65.
Castillo Sarmiento, A., Surarez Gelvez, J., and Mosquera Tellez, J., 2017. Naturaleza y sociedad: relaciones y tendencias desde un enfoque eurocéntrico. *Revista Luna Azul*, 44, 348–371.
Ceballos, G., Ehrlich, P. R., and Dirzo, R., 2017. Biological annihilation via the ongoing sixth mass extinction signaled by vertebrate population losses and declines. *Proceedings of the National Academy of Sciences*, 114(30), E6089–E6096.
Chrispeels, M. J., Sadava, D. E., and Mulligan, B., 1994. *Plants, Genes, and Agriculture* (No. 04; SB123. 57, C4). Boston: Jones and Bartlett Publishers.
De Sousa Santos, B., 2011. Introducción: las epistemologías del sur. *IV Training Seminar en Dinámicas Interculturales*, 26–28 January 2011 Barcelona.
Dubois, L., and Turits, R. L., 2019. *Freedom Roots: Histories from the Caribbean*. Chapel Hill, NC: UNC Press Books.
Escobar, A., 2014. *Sentipensar con la tierra. Nuevas lecturas sobre desarrollo, territorio y diferencia*. Medellín: Ediciones UNAULA.
Escobar, A., 2015. Territorios de diferencia: la ontología política de los "derechos al territorio". *Cuadernos de Antropología Social*, 41, 25–38.

Escobar, A., 2020. Sentipensar with the Earth: Territorial struggles and the ontological dimension of the epistemologies of the South. In: Escobar, A., ed. *Pluriversal Politics*. Durham, NC: Duke University Press, 67–83.

Estebanez, M. E., Turkenich, M., and Sued, G., 2013. Tecnología y género. Aportes de la sociología de la innovación al análisis de la agricultura familiar. *X Jornadas de Sociología*, 1–6 July 2013, Facultad de Ciencias Facultad de Ciencias Sociales, Universidad de Buenos Aires, Argentina.

Giraldo, O. F., 2014. *Utopías en la era de la supervivencia: una interpretación del buen vivir*. Ciudad de México: Editorial Itaca.

Giraldo, O. F., and Rosset, P. M., 2018. Agroecology as a territory in dispute: Between institutionality and social movements. *The Journal of Peasant Studies*, 45(3), 545–564.

Giraldo, O. F., and Rosset, P. M., 2021. Principios sociales de las agroecologías emancipadoras. *Desenvolvimento e Meio Ambiente*, 58, 708–732.

Gliessman, S. R., 1990. Agroecology: Researching the ecological basis for sustainable agriculture. In: Gliessman, S. R., ed. *Agroecology*. New York: Springer, 3–10.

González Márquez, E., 2020. Crisis civilizatoria: hacia una transformación profunda. *Diálogos ambientales*, 1(1), 17–22.

González Reyes, L., 2020. *Colapso del capitalismo global y transiciones hacia sociedades ecocomunitarias, mirando más allá del empleo*. Bilbao: Manu Robles Arangiz Fundazioa.

Gros, F., 2018. *Desobedecer*. Brasilia: Taurus.

Grosfoguel, R., 2016. Caos sistémico, crisis civilizatoria y proyectos descoloniales: pensar más allá del proceso civilizatorio de la modernidad/colonialidad. *Tabula Rasa*, 25, 153–174.

Gupta, N., 2020. *Women in Science and Technology: Confronting Inequalities*. Thousand Oaks, CA: Sage Publications Pvt. Limited.

Guzmán, G. I., et al., 2018. The agrarian metabolism as a tool for assessing agrarian sustainability, and its application to Spanish agriculture (1960–2008). *Ecology and Society*, 23(1).

Hadorn, G. H., et al., 2006. Implications of transdisciplinarity for sustainability research. *Ecological Economics*, 60(1), 119–128.

Herrero López, Y., and Rodríguez Muñoz, V. M., 2017. Un nuevo currículo para afrontar la crisis civilizatoria. *Cuadernos de Pedagogía*, 477, 62–65.

Holt-Giménez, E., and Altieri, M. A., 2013. Agroecology, food sovereignty, and the new green revolution. *Agroecology and Sustainable Food Systems*, 37(1), 90–102

Huergo, J., 2012. Desafíos de la Extensión desde la perspectiva cultural. *Revista Dialoguemos*, 8(14), 11–15.

Jacobi, J., and Llanque, A., 2018. "When we stand up, they have to negotiate with us": Power relations in and between an agroindustrial and an Indigenous food system in Bolivia. *Sustainability*, 10(11), 4001.

Jain, H. K., 2010. *Green Revolution: History, Impact and Future*. Houston: Studium Press LLC.

Kunin, J. R., 2019. *El poder del cuidado: mujeres y agencia en la pampa sojera argentina*. Thesis (PhD). Universidad Nacional de San Martín.

La Via Campesina, 2011. *Obtenido de*. http://viacampesina.org/es/index. php/temas-principales-mainmenu-27/soberanalimentarycomercio-mainmenu-38/314-que-es-la-soberania-alimentaria

La Via Campesina, 2015. *Agroecología campesina, por la soberanía alimentaria y la madre tierra: experiencias de La Vía Campesina*. Zimbabwe: La Vía Campesina. Recuperado de. https://viacampesina. org/es/wp-content/uploads/sites/3/2015/11/ CUADERNO, 207

Larrañaga, A., 2021. *Aplicación del enfoque de género en el análisis de la percepción de la peligrosidad del uso de pesticidas en la región hortícola platense*. Thesis (Undergraduate). Universidad Nacional de La Plata.

Latham, J., 2021. The myth of a food crisis. In: Kassam, A., and Kassam, L., eds. *Rethinking Food and Agriculture*. Sawston: Woodhead Publishing, 93–111.

Leff, E., 2012. Pensamiento ambiental latinoamericano: patrimonio de un saber para la sustentabilidad. *Environmental Ethics*, 34, 97–112.

Le Tran, D., Martinez-Alier, J., Navas, G., and Mingorria, S., 2020. Gendered geographies of violence: A multiple case study analysis of murdered women environmental defenders. *Journal of Political Ecology*, 27(1), 1189–1212.

Lewis, S. L., and Maslin, M. A., 2015. Defining the anthropocene. *Nature*, 519, 171–180.

Linsalata, L., 2020. ¡Nuestra lucha es por la vida! Apuntes críticos sobre la reorganización capitalista de la condición de interdependencia. *Trabalho necessário*, 18(36), 44–68.

Maffi, L., 2007. Biocultural diversity and sustainability. In: Pretty, J., Ball, A. S., Benton, T., Guivant, J. S., Lee, D. R., Orr, D., Pfeffer, M. J., and Ward, H., eds. *The Sage Handbook of Environment and Society*. Thousand Oaks, CA: SAGE Publications Ltd, 267–277.

Martínez Alier, J., 2001. Ecological economics. *International Encyclopedia of the Social and Behavioural Sciences*, 4(9).

Martínez Alier, J., 2005. *El ecologismo de los pobres: conflictos ambientales y lenguajes de valoración*. Barcelona: Editorial Icaria.

Martínez Alier, J., and Roca, J. J., 2000. *Economía ecológica y Política Ambiental*. Ciudad de México: Fondo de Cultura Económica.

Max-Neef, M. A., 2005. Foundations of transdisciplinarity. *Ecological Economics*, 53(1), 5–16.

Medellín-Milán, P., Avalos-Lozano, J. A., and Nieto-Caraveo, L. M., 2011. Más allá de la Economía Ecológica, la construcción de nichos de sostenibilidad. *Polis*, 10(29), 227–259.

Mier y Terán Giménez Cacho, M., et al., 2018. Bringing agroecology to scale: Key drivers and emblematic cases. *Agroecology and Sustainable Food Systems*, 42(6), 637–665.

Montalba, R., et al., 2015. Determinación de los niveles de riesgo socioecológico ante sequías en sistemas agrícolas campesinos de La Araucanía chilena. Influencia de la diversidad cultural y la agrobiodiversidad. *Papers. Revista de Sociologia*, 100(4), 607–624.

Moore, J. W., 2020. *El Capitalismo en la Trama de la Vida. Ecología y Acumulación de Capital*. Madrid: Traficantes de Sueños.

Morales Jasso, G., 2016. La apropiación de la naturaleza como recurso. Una mirada reflexiva. *Revista Gestión y Ambiente*, 19(1), 141–154.

Naredo, J. M., 1994. Fundamentos de la economía ecológica. In: Aguilera Klink, F., and Alcántara, V., eds. *De la Economía Ambiental a la Economía Ecológica*. Barcelona: Fuhem e Icaria, 231–252.

Navarro, M. L., and Gutiérrez Aguilar, R., 2018. Claves para pensar la interdependencia desde la ecología y los feminismos. *Revista Bajo el Volcán*, 28, 45–57.

Nicholls, C. I., Altieri, M. A., and Vazquez, L., 2017. Agroecological principles for the conversion of farming systems. In: Wezel, A., ed. *Agroecological Practices for Sustainable Agriculture: Principles, Applications, and Making the Transition.* Singapur, NJ: World Scientific, 1–18.

Noseda, C., 2018. *Lógica y saberes campesinos en la zona norte del Alto Paraná, Misiones y su aporte a una propuesta de desarrollo rural sostenible.* Thesis (Master). Universidad Nacional de La Plata.

Osorio Pérez, F. E., 2008. Forced displacement among rural women in Colombia. *Latin American Perspectives*, 35(6), 29–40.

Pengue, W. A., 2009. Agrofuels and agrifoods: Counting the externalities at the major crossroads of the 21st century. *Bulletin of Science, Technology & Society*, 29(3), 167–179.

Pengue, W., et al., 2018. Eco-agri-food systems: Today's realities and tomorrow's challenges. In: *TEEB for Agriculture & Food: Scientific and Economic Foundations.* Geneva: UN Environment, 57–109.

Petersen, P. F., 2022. Agroecología política: crítica de la ecología política al capitalismo agroalimentario. *Agrocienc Uruguay*, 26(NE3), 972.

Pingali, P. L., 2012. Green revolution: Impacts, limits, and the path ahead. *Proceedings of the National Academy of Sciences*, 109(31), 12302–12308.

Porter, J. R., Howden, M., and Smith, P., 2017. Considering agriculture in IPCC assessments. *Nature Climate Change*, 7(10), 680–683.

Price, S. J., Ford, J. R., Cooper, A. H., and Neal, C., 2011. Humans as major geological and geomorphological agents in the Anthropocene: The significance of artificial ground in Great Britain. *Philosophical Transactions of the Royal Society of London*, A369, 1056–1084.

Restrepo, E., and Rojas, A., 2010. *Inflexión decolonial: Fuentes, conceptos y cuestionamientos.* Popayán: Editorial Universidad del Cauca.

Rogers, E. M., 1958. Categorizing the adopters of agricultural practices. *Rural Sociology*, 23(4), 346–354.

Rogers, E. M., 2003. *Diffusion of Innovations.* New York: Free Press.

Rosset, P. M., and Altieri, M. A., 2017. *Agroecology: Science and Politics.* London: Practical Action Publishing.

Rosset, P. M., and Barbosa, L. P., 2021. Autonomía y los movimientos sociales del campo en América Latina: un debate urgente. *Aposta: Revista de Ciencias Sociales*, 89, 8–31.

Rosset, P. M., Barbosa, L. P., Val, V., and McCune, N., 2021. Pensamiento Latinoamericano Agroecológico: The emergence of a critical Latin American agroecology? *Agroecology and Sustainable Food Systems*, 45(1), 42–64.

Rosset, P. M., Machin Sosa, B., Jaime, A. M., and Lozano, D. R. A., 2011. The campesino-to-campesino agroecology movement of ANAP in Cuba: Social process methodology in the construction of sustainable peasant agriculture and food sovereignty. *Journal of Peasant Studies*, 38(1), 161–191.

Royero-Benavides, B., et al., 2019. Desarrollo y buena vida en la mixteca alta: El caso de una organización campesina oaxaqueña. *Agricultura, sociedad y desarrollo*, 16(1), 19–41.

Sandhu, H., et al., 2019. The future of agriculture and food: Evaluating the holistic costs and benefits. *The Anthropocene Review*, 6(3), 270–278.

Sarandón, S. J., 2019. Potencialidades, desafíos y limitaciones de la investigación agroecológica como un nuevo paradigma en las ciencias agrarias. *Revista de la Facultad de Ciencias Agrarias de la Universidad Nacional de Cuyo*, 51(1), 383–394.

Sarandón, S. J., 2021. Agroecología: una revolución del pensamiento en las ciencias agrarias. *Ciencia, Tecnología y Política*, 4(6), 55.

Sarandón, S. J., and Flores, C. C., 2014. *Agroecología*. La Plata: Universidad Nacional de La Plata (EDULP).

Servigne, P., and Stevens, R., 2015. *Comment tout peut s'effondrer. Petit manuel de collapsologie à l'usage des générations présentes*. Paris: Éditions du Seuil.

Sevilla Guzmán, E., and Soler Montiel, M., 2010. Agroecología y Soberanía alimentaria: alternativas a la globalización alimentaria. In: Instituto Andaluz del Patrimonio Histórico, ed. *Patrimonio Cultural en la nueva ruralidad andaluza*. Sevilla: Junta de Andalucía, 191–217.

Sevilla Guzmán, E., and Woodgate, G., 1997. Sustainable rural development: From industrial agriculture to agroecology. In: Redclift, M., and Woodgate, G., eds. *The International Handbook of Environmental Sociology*. Chaltenham: Edward Elgar, 83–100.

Shiva, V., 2000. *Stolen Harvest. The Hijacking of the Global Food Supply*. Cambridge: South End Press.

Soler, M., Pérez, D., Siliprandi, E., and Zuluaga, G., 2014. Alimentación, agroecología y feminismo: superando los tres sesgos de la mirada occidental. In: Siliprandi, E., and Zuluaga, G., eds. *Género, agroecología y soberanía alimentaria*. Barcelona: Icaria, 17–40.

Steffen, W., et al., 2018. Trajectories of the Earth system in the Anthropocene. *Proceedings of the National Academy of Sciences*, 115(33), 8252–8259.

Steffen, W., Grinevald, J., Crutzen, P., and McNeill, J., 2011. The anthropocene: Conceptual and historical perspectives. *Philosophical Transactions of the Royal Society A: Mathematical, Physical and Engineering Sciences*, 369(1938), 842–867.

Svampa, M., 2012. Resource extractivism and alternatives: Latin American perspectives on development. *Journal fur Entwicklungspolitik*, 28(3), 43–73.

Svampa, M., 2019. *Las fronteras del neoextractivismo en América Latina: conflictos socioambientales, giro ecoterritorial y nuevas dependencias*. Bielefeld: Bielefeld University Press.

Toledo, V. M., 2022. Agroecology and spirituality: Reflections about an unrecognized link. *Agroecology and Sustainable Food Systems*, 46(4), 626–641.

Torres, M. E. M., and Rosset, P. M., 2016. Diálogo de saberes en la vía campesina: soberanía alimentaria y agroecología. *Revista Espacio Regional*, 1(13), 23–36.

Val, V., et al., 2019. Agroecology and La Via Campesina I. The symbolic and material construction of agroecology through the dispositive of "peasant-to-peasant" processes. *Agroecology and Sustainable Food Systems*, 43(7–8), 872–894.

Zalasiewicz, J., et al., 2016. The geological cycle of plastics and their use as a stratigraphic indicator of the Anthropocene. *Anthropocene*, 13, 4–17.

12 Through the veil

A relational and participatory perspective to knowledge production and sustainability

Hugo Marcelo Zunino

The need to revisit (once again) our research practices

The aim of this chapter is to reflect about the epistemological standpoint taken by mainstream academia, and about how it permeates subtly the production of knowledge. Further, we suggest a research perspective that thinks the world from another side, giving due attention to the power of inner experience and social encounter in acquiring knowledge about the world and peoples. For critical sustainability sciences aiming at integrating diverse cultural perspective in a common project, it appears necessary to question our epistemological foundations to open space to imagine ways in which societies and people can emancipate from and reshape the social and economic institutions that regulate our lives.

The system of knowledge built on the grounds of the intellectual enlightenment (16th and 17th centuries) shapes the way in which current social problems are framed and what we consider as solutions. Indeed, our contemporary society can be understood as a crystallization, and condensation, of a system of knowledge. For instance, when we scrutinize the institutional responses to the events surrounding the expansion of COVID-19 and the increasing lever of vulnerability people are experiencing all over the globe, it is not difficult to distinguish a global bureaucracy at work, which is colonizing institutions and other social arrangement. It is founded in a mechanical, abstract, unartistic way of thinking that permeates our daily life to its finest details.

Current academic discussions informing state-led responses to the current multidimensional crisis appear devoid of life, distant from human experience and creativity. We can witness how, in the first few decades of the 21st century, technocracy and Auguste Comte's idea of knowledge as a rational prediction holds authoritative grip over decision-making. In fact, vast numbers of studies have been conducted and incommensurable resources have been destined, for example, to measure and model the dissemination of the COVID virus around the world and to design the regulatory social protocols to battle against it. To validate the truthfulness of this intervention, we are led to an impressive array of external evidence, which is based exclusively on what can be seen with our senses and counted with our analytical mind. It is not difficult for people to

DOI: 10.4324/9781003043577-12

realize intuitively that the current world-situation is not the result of a single event, but the expression of a world-situation, marked by unhealthy relations of humans with their surrounding social and natural world.

As we argue in this chapter, the external world provides us with only one-half of the reality. The other half must be penetrated through other means: by acknowledging our human capacity to engage with the world with the fulness of our inner life, not only with abstract, intellectual thinking. To examine the epistemological gridlock that confines us to the visible and to shed some light on how we can advance in investigating the other half of reality is the purpose of this chapter.

What happened to the academic revisionist impulse elicited by the work of Jean-François Lyotard and many others, more than 40 years ago? (see Lyotard 1979). There is a veil that has not been crossed. It remains firmly in place confining us to one side of reality. Despite that many academic discourses informed by poststructuralist and postmodernist thinkers have reshaped important aspects of our external social and cultural life, they are still marginal in shifting the critical global research agenda toward the examination and crossing of this veil. In Latin America, in recent years, a significant number of proposals have been raised to incorporate the holistic and integrative perspective, contained in the wisdom of the indigenous peoples of the continent; they convert toward the construction of a world in which different worldviews can coexist in a pluriverse (De La Cadena 2010; De Sousa 2012; Escobar 2005, 2011, 2015; Gudynas 2011; Leff 2004; Walsh 2007, 2009, 2010; among others).

However, as white researchers harden by the modern outlook of science, it is very difficult for us to *experience* what indigenous worldviews convey us about the connection of human to nature, the cosmos and other people. At best, we are able to intellectualize these understandings and traduce them to our own conceptual frameworks. In the process, we kill the living/experiential element of knowledge, and the existential essence is left only as an empty vessel. From our perspective, we, Westerners, must acknowledge our epistemological grounds and heritage if we want to enter into honest and fair dialogue with other cultures. In this vein, in this chapter, we will build on the work of Gilles Deleuze, Felix Guattari, and Johann Wolfgang von Goethe (1749–1832) to construct a research perspective that can prove useful to foster dialogue with indigenous wisdom and to recover a sense of unity with the outside world. This is the way we will follow to escape the modern epistemological gridlock that have chained our creativity for centuries.

In the next section, we will depict the epistemological basis of the modern scientific outlook and the responses from post-structuralists and postmodernists. In the third section, we examine the epistemological potential that holds ancient ways of knowing and the difficulties of current critical social science in acknowledging the intangible elements they contain. In the fourth section, we use the work of Deleuze, Guattari, and Goethe to construct a relational, participatory, and intercultural approach to knowledge production that can enliven the joint production of comprehensive knowledge. In the concluding section, we look ahead by stressing the ethical and political dimensions of the perspective here constructed.

Modern knowledge and the poststructuralist/postmodern response

The intervention of the thinkers of the intellectual renaissance set into motion a way of thinking that seeks knowledge by means of an intellectual reflection about an external world that remains separated from human consciousness. Thus, Francis Bacon starts from the scientific observation of the visible world and through mathematical abstraction tries to infer the general laws that govern natural phenomena (Bacon 2009, see aphorisms 3 and 71). Rene Descartes, meanwhile, with his *cogito ergo sum*, gives analytical primacy to the rational and transcendental subject that thinks (Descartes 2010) and Immanuel Kant warns that "the things in themselves" are inaccessible to our understanding, consolidating a methodological dualism and limiting the possibilities of human knowledge to know the essence of the world (Kant 2002). Mathew (2006:89–90) argues that modernity can be identified in terms of a unique and historically unprecedented metaphysical premise, namely, materialism, which transcends an era dominated by mystical revelation.

Here we want to simply characterize, broadly, some key elements of the modern way of thinking that is naturalized and subtly permeates our contemporary lives: the world is outside, and I observe it from a marginal point of the universe; to know is to create an inner representation of an external world that is foreign to me (see Steiner 1981).

Beyond the epistemological discussions, let us observe the methodological and existential solitude to which this way of thinking relegates us. It inhabits us and colonizes us stealthily: we perceive the external world, weigh it, measure it, and count it; from there we derive, intellectually, abstract laws. To *predict is to know*, said Auguste Comte in 1844 (1965:15), transforming knowledge into a soulless device that makes possible to understand the world mechanically and devising directives either to intervene beneficially in the world or to destroy it.

In the last 50 years, responses to the supremacy of modern representational thought have been broad, diverse, and powerful. Postmodernity is born with a response to the crisis of modernity. With the student revolts of 1968, a stage for humanity marked by Nihilism and by the denial of any belief or moral, religious, political, or social principle is inaugurated (see Carabante 2018; García-Flores 2008; Nietzsche 2012; Pinillos 1996). Thus, postmodernity is presented as a project that demolishes the dogma of absolute reason, of the modern subject and of the idea of progressive development toward a state of greater well-being (Lyotard 1979).

Associated with postmodern times, poststructuralism proposes that social reality is constructed discursively, and subjects appear as products of historical conditions and power devices that determine them. Knowledge acquires a dynamic and always provisional character (Restrepo and Rojas 2010), admitting only partial, circumstantial, or ephemeral truths that change, or are discarded based on their own dynamics (Lipovetsky 1986; Vattimo 1986). Foucault (1988a:32) points out the discursive practices, understood as a way of acting and thinking, that provide the key to intelligibility for the correlative construction of the subject and the object. Haraway (1995), in turn, questions the neutrality and universality of science, postulating that knowledge is always partial and situated, involving relationships between human and non-human entities. For Deleuze (1994), meanwhile, the modern regime of knowledge uncritically assumes that thought is a natural faculty aimed at truth and argues that representation – typical of Cartesianism – conceals difference, under a cloak of generality and transcendental illusion.

Ancient ways of knowing and the (problematic) reimagining of our research practice

Since the "intellectual rebirth" of the 16th and 17th centuries, the wisdom of non-Western and ancient thinkers has become "retrograded beliefs". In 1620, Bacon was particularly critical of the old idlers who created and sustained sects and philosophical heresies (Bacon 2009). Two centuries later, Comte vehemently rejected the vein claims of a retrograde, increasingly degenerate theology (Comte 1965:39–40). The ancient world crumbled, and modern science took its place as the new religion. The living Mother Earth, venerated by ancient wisdom, is replaced by material relationships. We began to think of the earth as a giant ecosystem, a marvelous clockwork.

From Latin America, the decolonial perspective argues that modern scientific knowledge has been placed at the service of the European colonial powers, relegating and discrediting other forms of knowledge (Mignolo 2002). According to decolonialism, there is a first modernity that remains hidden for many European post-structuralists, and that precedes the second modernity of the Enlightenment (Mignolo 2003). This first modernity began with the discovery of America in 1492, which marks the starting point of the epistemological oppression suffered by Latin American peoples (Escobar 2005, 2011, 2015; Walsh 2009). The Latin American decolonial approach proposes the implementation of methodologies that allow valuing and considering "other knowledge"; through this it encourages a political project that seeks to change the terms and conditions of conversation, between European knowledge and subaltern knowledge (Mignolo 2002, 2003; Restrepo and Rojas 2010; Robbins 2006).

Yet, in many contributions that seek to rescue ancestral wisdom, attention is placed mainly on the utility of indigenous knowledge to advance development goals, for example, sustainability, reducing the expression of indigenous knowledge to an ancestral legacy, useful to inform a particular foreign ethic of nature (Aikenhead and Ogawa 2007; Agrawal 1995; Jones and Cloke 2002). Thus, indigenous researchers have expressed their concern about the insufficient recognition of the ontological and epistemological basis of indigenous knowledge, promoting claims for cognitive self-determination and epistemological sovereignty (Cajete 2004). We will briefly characterize some elements of this knowledge and its production.

For indigenous peoples, knowledge is revealed through an inner journey in which the person relates experientially and empirically with the living territory and to the whole cosmos. This inner journey toward knowledge is part of a collective identity that is articulated around the living earth (Ñuke Mapu for the Mapuche people in Chile). It allows the recognition of spiritual forces or energies that govern space and grant powers to the elements contained in it (Neira et al. 2012). Following Cajete (2004), indigenous knowledge about nature is related to a direct phenomenological experience of a lived space, a practice from which cosmologies are derived that connect human beings with the universe, with humans being active, and creative participants in that totality. Similarly, for Ñanculef Huaiquinao (2016), the Mapuche people uses phenomenological observation to apprehend the world and grasp their connection with the natural world. This human–nature liaison echoes in their language: Mapuzugun is the onomatopoeia of the primordial sound that emerges from nature and that communicates its essence and state. *Nature speaks to the human being.*

In the examination of scientific problems, researchers trained in western universities develop a style of thinking that directs the attention only to what is external and has location and form, disregarding sensations and feelings also arising as inner experience in the encounter with the external world. For us, it is certainly difficult to grasp and internalize forms of knowledge production that are present in many indigenous people around the globe. Although we could rely on our abstract thinking to form a conceptual image, it appears difficult to vividly experience the connections with the surrounding natural and social world. Further, how can we possibly imagine inner revelations as facts that present themselves inside our being, and not as events happening outside? Although decolonial and poststructuralist literature is essential to reveal the interwoven power relations in the historical evolution of people and societies, these approaches remain ill-prepared to understand the communion that people from the indigenous world establish with spiritual forces that operate at cosmic and planetary scales. This raises the question: In what ways can postmodern methodologies enter more fully in dialogue with ancient knowledge? Here, we trace a path of reflection based on the work of Gilles Deleuze, Félix Guattari, and Johann Wolfgang von Goethe (1749–1832).

Immanence and formative forces: Deleuze, Guattari, and Goethe

In *Difference and Repetition*, the postmodern thinker Gilles Deleuze (1994) argues that representational thinking highlights difference and cover up reality in a mantle of transcendental illusion. He invites us to consider difference as the sustenance of knowledge. From a Deleuzian perspective, thinking operates as a *force* that manifests in a constant flow of concepts and ideas. Deleuze proposes us to imagine a place where concepts are in "pure movement", creating a dynamic and ever-changing fabric. When we observe the world, there is "something" out there that forces us to think and create concepts to interpret it. To understand that "something", Deleuze and Guattari (1993) introduce the notion of the *plane of immanence*, conceived of as a "sea of concepts", where movement captures everything in an incessant flow across an infinite plane. This plane of immanence does not belong to any subject or object and represents the origin of all our concepts and ideas about things and about ourselves. For Deleuze and Guattari, the plane of immanence interweaves natural and thinking forces:

> The plane of immanence has two facets, as thought and as nature, as Physis and as Nous. This is why there are always many infinite movements intertwined one inside the other, folded one inside the other, to the extent that the return of one instantaneously triggers another, in such a way that the plane of immanence does not stop weaving itself.
>
> (Deleuze and Guattari 1993:41–42)

The relationship that Deleuze and Guattari establish leads us to visualize the foundation of all the things perceived by our senses as an interweaving of human thoughts and natural forces; the human represents a creative power that weaves the world in harmony with nature; they establish a product–producer relationship in which the subject cannot be separated from the object. The notion of immanence avoids transcendence and a Kantian (and Platonic) dualism, establishing the existence of an invisible plane that is constantly updated (manifested) in objects and subjects. In a way, the transcendental sphere descends from an abstract plane to our current experience of the surrounding world, establishing a *field* in which material, natural, and social processes unfold.

The real and not merely formal integration between the natural and the human takes us back to the ideas that make up the knowledge of many ancient peoples of the world (see also Fornssler et al. 2014). In fact, for Deleuze and Guattari (1993:47), the pre-Socratic Greeks were the first to think in terms of a plane of immanence, which they called logos: the cosmic intelligence that penetrates everything. The universal cosmic force is also present in the idea of Pachamama, mother nature, venerated by ancestral peoples of our continent (Fals Borda 2009; Zaffaroni 2011); it also refers to the living Prakriti, the source of universal consciousness that inhabits thought in eastern traditions (Shiva 1995, 2001). In one of his latest works, Deleuze thinks of immanence as "a life", as a force that is everywhere and is the carrier of events (singularities) that manifest (actualize) in objects or subjects (Deleuze 2007). It is important to point out that the contributions of Deleuze and Guattari do not necessarily lead toward a mythical–religious vision: although some authors use their approach to conceive a plane formed by conscious spiritual entities (Dahlin 2013), most contemporary scientists interpret it from a materialist perspective, relating the movements of the plane of immanence with the material processes intrinsic to life (Herzogenrath 2009; Protevi 2012).

Deleuze and Guattari (1993, 2009) understand thought as a living/natural force that incessantly creates new concepts on the plane of immanence. Thinking activity weaves the plane of immanence and originates a form in movement composed of interwoven concepts along an infinite plane. In the book *Capitalism and Schizophrenia* (Deleuze and Guattari 2009), they use the natural metaphor "arborescent" to describe the hierarchical and linear form that modern thought takes on the plane of immanence. In contrast, "rhizomatic" thought originates a complex form in movement, a rhizome, which allows connecting a point with any other point, adapting to the living character of nature, and admitting connections that are excluded in more formalized methods (Fornssler et al. 2014). Thus, Deleuzian–Guattarian thought opens infinite possibilities for concepts to manifest (update) in space and time in moments of creative rupture, which marks the point of emergence of the new.

Goethe (1749–1832) also conceives of nature as a living entity that is related to human experience and thinking. This romantic poet, artist, and scientist recognizes in all the forms of nature the continuous action of life (Goethe 2007). For Goethe, it is in the form in movement, and not in matter, where we find the essence of life (Simms 2005:167). In natural processes, Goethe recognizes the manifestation of archetypical *formative forces* (Urpflaze) that are singularized in the unique forms of the phenomena. Goethe observes the manifestation of these forces through precise and scientific observation on natural events. He observes and reports, in minute details, the successive transformation that a plant undergoes, from germination to death. Goethe creates an image of the phenomenon at point t1 and then follows it to t2, t3, t4, etc. In this way, a later state can be deduced from an earlier one and from an earlier state we can deduce the form that the phenomenon will take in a later state. This is the metamorphosis of the plant, the process of transformation of life that expresses the relationship between what is perceived and its essence.

Thus, Goethe's methodology is aimed at observing the phenomenon from all points of view, situating the observer in a stage of "pure perception" connecting all possible manifestation with concepts rising from intuition. The concepts so emerging are not dead abstract concepts, but actively produced while engaging with the physical manifestation of the concept. This idea correlates with the Deluzian construction of the "sea of concepts" moving in an "infinite plane". While for Deleuze and Guattari concepts appear as a subjective creation, with Goethe we can connect the intuitively perceived concepts to something that is objective in the plant and manifests in all the forms it takes. Goethe enable us to enter into an objective relation with a world of phenomenon that presents to us as subjective inner experience, enabling us to grasp that space in which the objective–subjective dualism is replaced by an inner experience that corresponds to something real in the physical world.

Goethe's work was primarily directed toward nature, rather than toward the thinking subject. Contemporary Goethean researchers have extended Goethe's formulation, placing the attention on the relationship that the thinking subject establishes with the wholeness of the world through their practices.

For them, in a similar way that nature is bounded together by the formative forces, in society there is also a force, the social field, that binds all individuals together in a wholeness. From this angle, the social field is gradually configured and reconfigured from actual encounter among people, where they can face each other in ways that can, eventually, go beyond their purely individualistic perspective. Knowledge, the production of truth, becomes a social act steaming from active and conscious social encounters, where it is revealed in a creative act, as something new emerging (Brook 1998; Seamon 2012; Seamon and Zajonc 1998; Simms 2005; Stahel 2013; Wahl 2005).

For Deleuze, Guattari, and the contemporary Goetheanism, thinking appears as a force connected to the entire world. Indeed, the essential nature of the world is composed of thoughts that encounter, combine, and/or repel each other: two ideas come together and create a completely new idea, in an incessant movement of transformation guided by the meeting of thoughts activated by people holding contrasting perspectives. We will now move on to examine how these ideas can assist us in revisiting the abstraction that runs through modern thinking.

A relational and participatory perspective

With the insights provided by Deleuze, Guattari, and the contemporary followers of Goethe, we immerse ourselves in the middle of the phenomenon itself. We observe the phenomenon from within and open the space to experience the incessant movement of different interwoven concepts that run through the social field. This field has no subjects or objects and contains all possible manifestations in potential form. Any human encounter is permeated by differences, concepts, and ideas that are in constant movement. Fixed, standardized relations dissolves into plastic, movable relations, generating a creative space that contains infinite possibilities of expression and realization. Thus, the field can only be experienced in and through practices: we must transform ourselves into the phenomenon itself.

In an interview, Deleuze (1995) invites us to think of a surfer riding the waves. In his practice, the surfer does not strive to control his movements and determine a direction, he lets himself be carried away by pure movement, the subject becomes a wave. This example offers us an intuitive idea about the way in which what is fixed, dissolves in pure movement, which we can also relate to those practices in the social field that form a collective wave and lead the different subjects in an upward spiral that ends in a creative act. Of course, the wave will break, and a new wave will come: all knowledge will be unstable, provisional, and situated. There is a point in the intermediate, in the "between-time", at that point when the wave is about to burst, in which the moment is charged with creative power, where there are no subjects or objects, everything is pure movement, and an infinite plane of possibilities opens up.

As poststructuralist rightly claim, power relations are at play in social life (see, for instance, Foucault 1988a, b). Certain interests prevail and colonize the matrix of thoughts, giving rise, for instance, to particular forms of social exclusion, imposed by those who possess the means to exert power. To get out of the gridlock, the only possible way out is to destabilize radically existing power configurations and engage in social interactions facing the other in an *equal and dynamic* relation. This does not mean to erase or to disregard power relations but rather to put them in movement, reconfiguring our relation to the other through active interaction in the continuity of social life. Is this even possible? Indeed, this is a central question. The answer must come from practice, not from abstract reflective thinking alone.

An approach to power of this kind has been advanced by post-phenomenologists who replace a structural–historical view of power by a conception of power as a relation that can operate and acquire reality only through practice itself, flowing as a capacity to act that is full of desire (Raffin 2008; Rose and Wylie 2006; Ruddick 2010). Thus, power relations are conceived of as constructions emerging from the encounter between different human beings (Ash and Simpson 2014). Thus, as power relations are produced and reproduced through social interaction, they can be reconfigured or replaced through an active social encounter. This insight opens the possibility of thinking and acting outside predefined categories, establishing an ever-changing social field in which the relations among people are reshaped in the continuity of life itself.

Post-phenomenologist rejects the preexisting transcendental ego contained in traditional phenomenology (see Rose and Wylie 2006). Contrary to the modern view of a rational subject that observes the external world as something alien, in this view we can understand humans as active co-creators of the world. Subjects are always trapped in the process of becoming different, always transforming themselves, coming out of themselves, emerging in events of rupture, creation, death, and chaos (Rodríguez 2016:237). Similarly, the community life can be grasped as a collective identity in permanent rearrangement, whose unity, always provisional, is constituted and reproduced over time through practices which mark the encounter of a variety of assemblages establishing relations between people (Goh 2006).

Thus, the subject appears as an instable construction that is continuously updated, as practices and encounters unfold within the field of social immanence. There is something in this social field that connect people, just as the formative forces to the cosmos hold together the different forms taken by the plant in metamorphosis. This something is not a stable entity like a belief system existing “out there”, but a co-created field that encompasses the diversity expressed by each individual and group of individuals. In this way, we can regain a sense of connection between human beings and their practice with the wholeness of the world. In this place, we are better equipped to understand experientially what ancient wisdom tells us from the four corners of the world: human beings are part of the wholeness of the world.

In this way, we can move from the production of knowledge as a prediction to a creative act that allows something new to emerge from the concrete richness of the sensible, from real events, not from an abstract generalization (Baugh 1992). People think thoughts within a social field, connecting creative human beings that shape their own existence and the world in the continuity of practice itself. This entails a performative understanding of existence and knowledge production (Ash and Simpson 2014; Haseman 2006). The performative impulse is also central for Goethe (2007), for whom artistic imagination and intuitive thinking are the sources of scientific knowledge, not abstract reflection. Contemporary Goetheans reveal the imaginations that emerge with interaction on the social field by means of various artistic techniques such as music, theater, dance, or photography (see Kaplan 2015).

In a participatory and relational approach, the moments of creatives ruptures are central in revealing the game of imaginations and visualize those moments of ruptures in which something new emerges (Baugh 1992). In participatory research, the researcher must gradually and sensibly lead the process of collective construction toward a creative breaking point, understood more as subtle microevents than as great paradigmatic moments. This is only a general methodological prescription that signals how can relational and participatory research can be put to work in concrete social settings. A more detailed explanation and formalization of a methodology is a work in progress that is beyond the scope of this chapter, which is intended mainly at recognizing conceptually the veil and one possible road to enliven academic research.

Through the veil: ethical and political dimensions

Contrary to the knowledge generated through the representation of an external world, the preceding considerations lead us to consider the production of knowledge as a creative act in which something new emerges from practice and social encounters. Deleuze, Guattari, and the contemporary Goethean researchers offer indications on how to think differently, inviting researchers to reveal the creative force that unfolds in social interaction and to shrink the separation between the researcher and what is being researched.

This attitude entails reaching a point where the relations among agents acquire a quality of "plasticity": each participant maintains a capacity to metamorphose his/her feeling/thinking according to the flow of events, remaining open to the wholeness of what is unfolding. This inner attitude will eventually lead to experience and enter into relation with the plane of immanence operating behind events. This is not, of course, an easy endeavor. We are well aware that a researcher with a conscious or unconscious modern outlook may consider this a fantasy. Yet, no one said it is easy to think differently.

Logical reasoning here doesn't suffice, and it is necessary to *experience the experience*. At least make the effort to do so. At present, intensive field work is conducted in the south of Chile bringing together activist, indigenous people, and local politicians. Although it is beyond the scope of this chapter to enter into methodological considerations or concrete outcomes, it will suffice to say here that promising results have been reached at a local and intimidate scale, still remaining the challenge to expands the reach of the work to broader scales.

Among the number of objections that can be raised, possibly the most pervasive is the neglect of the power relations, engrained in society. Indeed, there are several unbalanced social arrangements that exhibit an apparent solidity that makes them appear as invulnerable to change. However, following recent advances in critical post-phenomenology, we argue that power relations own their existence and realty to practice, and can be transformed through practice itself. This means people, participating in research conducted under the tenants here presented, must be ethically active to attain this plasticity. In the last instance, the outcomes of the research cannot be predefined aprioristically, but rather they will depend on the ethical conduct of the participants.

In the introductory section, we refer to a global bureaucracy that today reigns in universities and other institutions. Here we are not speaking in the abstract, but rather guiding the reader to take hold of his/her own inner experience. This social arrangement takes as basis for action a system of knowledge that conceive of society as composed of isolated beings attempting to understand and live in an external world that is foreign to their consciousness. To be aware of this powerful system of knowledge, imposed on humanity with authoritative grip, represents a political necessity, a first step to reimagine alternatives and to emancipate us from their hold. This chapter challenges researchers on sustainability to face problems from a radically different perceptive and to be politically active in building new emancipatory grounds.

To advance in revising the way we produce knowledge along the lines here suggested demands piercing the epistemological veil constructed by science since the 15th century. Academic courage will be needed to think practically, and not only in abstract terms, about the relation between the invisible and the visible realms. Mainstream sciences largely disregard the former, and focuses the attention on the latter, adding, in turn, a self-constructed system of thought to what is seen by our eyes.

This general attitude toward knowledge production is still engrained in much of the critical literature on sustainability and on other related fields of knowledge. Traversing the veil of modern science means to permeate the visible and reach the complex of forces vibrating behind what our senses perceive. We can experience this complex of forces only inwardly and in real time, in the continuity of social practices themselves, dwelling inside the phenomenon, not outside it.

References

Agrawal, A., 1995. Dismantling the divide between indigenous and scientific knowledge. *Development and Change*, 26(3), 413–439.

Aikenhead, G., y Ogawa, M., 2007. Indigenous knowledge and science revisited. *Cultural Studies of Science Education*, 2(3), 539–620.

Ash, J., y Simpson, P., 2014. Geography and post-phenomenology. *Progress in Human Geography*, 40(1), 48–66.

Bacon, F., 2009 [1620]. *Novum Organum. Aforismos sobre la interpretación de la naturaleza y el reino del hombre.* Madrid: Tecnos.

Baugh, B., 1992. Transcendental empiricism: Deleuze's response to Hegel. *Man and World*, 25, 133–148.

Brook, I. (1998). Goethean science as a way to read landscape. *Landscape Research*, 23(1), 51–69.

Cajete, G. 2004. Philosophy of native science. En: Waters, A., ed. *American Indian Thought*. London and New York: Routledge, 45–57.

Carabante, J., 2018. *Mayo del 68: claves filosóficas de una revuelta posmoderna*. Madrid: Ediciones Rialp.

Comte, A., 1965 [1844]. *Discurso sobre el espíritu positivo*. Madrid: Alianza Editorial.

Dahlin, B., 2013. Gloves of ice or free hands? A nomadic reading of Rudolf Steiner and Bergson and Deleuze and others on knowledge as nonrepresentational and the importance of aesthesis. Other education. *The Journal of Educational Alternatives*, 2(2), 67–89.

De La Cadena, M., 2010. Indigenous cosmopolitics in the Andes. Conceptual reflections beyond "politics". *Cultural Anthropology*, 25(2), 334–370.

Deleuze, G., 1994. *Difference and Repetition*. New York: Columbia University Press.

Deleuze, G., 1995. *Negotiations Gilles Deleuze*. New York: Columbia University Press.

Deleuze, G., 2007. La inmanencia: una vida. In: Giorgi, G., y Rodríguez, F., eds. *Ensayos sobre biopolítica Excesos de vida*. Buenos Aires, Argentina: Paidos, 35–40.

Deleuze, G., y Guattari, F., 1993. *¿Qué es la filosofía?* Barcelona, España: Editorial Anagrama.

Deleuze, G., y Guattari, F., 2009. *Anti-Oedipus: Capitalism and Schizophrenia*. Minnesota: University of Minnesota Press.

Descartes, R., 2010 [1637]. *Discurso del método*. Quito, Ecuador: Editorial Bolgspot.

De Sousa, M. L., 2012. Challenging heteronomous power in a globalized world. In: Krätke, S., Wildner, K., and Lanz, S., eds. *Transnationalism and Urbanism*. New York: Routledge, 172–196.

Escobar, A., 2005. El "postdesarrollo" como concepto y práctica social. In: Mato, D., ed. *Políticas de economía, ambiente y sociedad en tiempos de globalización*. Caracas, Venezuela: Universidad Central de Venezuela, 17–31.

Escobar, A., 2011. ¿"Pachamámicos" versus "modernicos"? *Tabula Rasa*, 15, 265–273.

Escobar, A., 2015. *Sentipensar* con la Tierra: Las Luchas Territoriales y la Dimensión Ontológica de las Epistemologías del Sur. *Revista de Antropología Iberoamericana*, 11(1), 11–32.

Fals Borda, O., 2009. *Una sociología sentipensante para América Latina*. Buenos Aires, Argentina: Consejo Latinoamericano de Ciencias Sociales (Clacso).

Fornssler, B., McKenzie, H. A., Dell, C. A., Laliberte, L., and Hopkins, C., 2014. I got to know them in a new way rela (y/t) ing rhizomes and community-based knowledge (Brokers') transformation of western and indigenous knowledge. *Cultural Studies ↔ Critical Methodologies*, 14(2), 179–193.

Foucault, M., 1988a. *Tecnologías del Yo y otros textos afines*. Buenos Aires, Argentina: Paidos.

Foucault, M., 1988b. El sujeto y el poder. *Revista Mexicana de Sociología*, 50(3), 3–20.

García-Flores, J., 2008. La Problemática del horizonte de sentido entre la modernidad y la postmodernidad. *Temas de Ciencia y Tecnología*, 12(34), 57–70.

Goethe, J. W., 2007. *Teoría de la Naturaleza. Estudio preliminar, traducción y notas de Diego Sánchez Meca*. Madrid: Tecnos.

Goh, I., 2006. The question of community in Deleuze and Guattari: Anti-community. *Symplokē*, 14(1/2), 216–231.

Gudynas, E., 2011. Buen vivir: Germinando alternativas al desarrollo. *América Latina en Movimiento, ALAI*, 462, 1–20.

Haraway, D., 1995. *Ciencia, Cyborgs y Mujeres: La Reinvención de la Naturaleza*. Madrid: Ediciones Cátedra.

Haseman, B., 2006. A manifesto for performative research. *Culture and Policy*, 118, 98–106.

Herzogenrath, B., 2009. Nature|geophilosophy|machinics|cosophy. In: Herzogenrath, B., ed. *Deleuze|Guattari & Ecology*. Hampshire: Palgrave Macmillan, 1–23.

Jones, O., y Cloke, P., 2002. *Tree Cultures: The Place of Trees and Trees in their Place*. Oxford: Berg Publisher.

Kant, I., 2002. *Crítica de la razón pura*. Madrid: Tecnos.

Kaplan, A., 2015. *Artistas de lo invisible*. Buenos Aires, Argentina: Editorial Antroposófica.

Leff, E., 2004. *Racionalidad ambiental la reapropiación social de la naturaleza*. Buenos Aires, Argentina: Siglo XXI editores.

Lipovetsky, G., 1986. *La era del vacío. Ensayo sobre el individualismo contemporáneo*. Barcelona: Anagrama.

Lyotard, J. F., 1979. *La condición postmoderna Informe sobre el saber*. Buenos Aires, Argentina: Editorial Rei.

Mathew, F., 2006. Beyond modernity and tradition: A third way for development. *Ethics & the Environment*, 11(2), 85–113.

Mignolo, W., 2002. El potencial epistemológico de la historia oral: algunas contribuciones de Silvia Rivera Cusicanqui. In: Mato, D., ed. *Estudios y otras prácticas intelectuales latinoamericanas en cultura y poder*. Caracas, Venezuela: CLACSO, Consejo Latinoamericano de Ciencias Sociales, 14–32.

Mignolo, W., 2003. *Historias locales/diseños globales: colonialidad, conocimientos subalternos y pensamiento fronterizo*. Madrid: Akal.

Ñanculef Huaiquinao, J., 2016. *Tayiñ Mapuche Kimün. Epistemología mapuche – Sabiduría y conocimiento*. Santiago: Universidad de Chile, Facultad de Ciencias Sociales.

Neira, Z., Alarcón, A., Jelves, I., Ovalle, P., Conejeros, A., y Verdugo, V., 2012. Espacios ecológicos-culturales en un territorio mapuche de la Región de la Araucania en Chile. Chungará. *Revista de Antropología Chilena*, 44(2), 313–323.

Nietzsche, F., 2012. *Así Hablaba Zaratustra*. Madrid: Brontes.

Pinillos, J. L., 1996. La mentalidad postmoderna. *Psicothema*, 8(1), 229–240.

Protevi, J., 2012. Deleuze and life. En: Smith, D., and Somers-Hall, H., eds. *The Cambridge Companion to Deleuze*. Cambridge: Cambridge University Press, 239–264.

Raffin, M., 2008. El pensamiento de Gilles Deleuze y Michel Foucault en cuestión: las ideas en torno del poder, el sujeto y la verdad. *Lecciones y Ensayos*, 85, 17–44.

Restrepo, E., y Rojas, A., 2010. *Inflexión decolonial: fuentes, conceptos y cuestionamientos*. Popayán, Colombia: Editorial Universidad del Cauca.

Robbins, P., 2006. Research in theft: Environmental inquiry in a postcolonial world. In: Aitken, S., and Valentine, G., eds. *Approaches to Human Geography*. New York: Sage, 311–324.

Rodríguez, S., 2016. Toward a methodology of death: Deleuze's "event" as method for critical ethnography. *Critical Questions in Education*, 7(3), 232–248.

Rose, M., y Wylie, J., 2006. Animating landscape. *Environment and Planning D: Society and Space*, 24, 475–479.
Ruddick, S., 2010. The politics of affect. Spinoza in the work of Negri and Deleuze. *Theory, Culture & Society*, 27(4), 21–45.
Seamon, D., 2012. Place, identity and phenomenology: A triadic interpretation. In: Caskin, H., Bernardo, F., Dubai, A., and Bentham, A. E., eds. *The Role of Place Identity in the Perception, Understanding, and Design of Built Environments*. Sharjah. U.A.E.: Bentham Books, 3–22.
Seamon, D., and Zajonc, A., 1998. *Goethe's Way of Science: A Phenomenology of Nature*. Albany, NY: State University of New York Press.
Shiva, V., 1995. *Abrazar la vida. Mujer, ecología y desarrollo*. Madrid, España: Editorial Horas y Horas.
Shiva, V., 2001. *Biopiratería. El saqueo de la naturaleza y del conocimiento*. Valencia, España: Icaria Editorial.
Simms, E. M., 2005. Goethe, Husserl, and the crisis of the European sciences. *Janus Head*, 8(1), 160–172.
Stahel, A., 2013. Alternativas a la crisis desde una perspectiva fenomenológica y antroposófica. *Càtedra UNESCO de Sostenibilitat*, 14, 25–44.
Steiner, R., 1981. *Thruth and Knowledge*. New York: SteinerBooks.
Vattimo, G., 1986. *El fin de la Modernidad*. Barcelona: Gedisa.
Wahl, D., 2005. "Zarte Empirie": Goethean science as a way of knowing. *Janus Head*, 8(1), 58–76.
Walsh, C., 2009. *Interculturalidad, estado, sociedad: luchas (de)coloniales de nuestra época*. Quito: Universidad Andina Simón Bolívar.
Walsh, C., 2010. Development as Buen Vivir: Institutional arrangements and (de) colonial entanglements. *Development*, 53(1), 15–21.
Walsh, K., 2007. ¿Son posibles unas ciencias sociales/culturales otras? Reflexiones en torno a las epistemologías decoloniales. *Nómadas*, 26, 102–113.
Zaffaroni, E., 2011. *La Pachamama y lo humano*. Buenos Aires, Argentina: Ediciones Madres de la Plaza de Mayo.

13 Goethe's scientific method

The road not taken[1]

Isis Brook

Introduction

In many debates on the environmental situation in which we find ourselves, there have been attempts to identify the source of the problem in ourselves – our attitudes and approaches to nature – and in the wider related enlightenment project of understanding the world through a broadly dualist and then reductionist scientific approach. Setting ourselves apart from the physical world leads to an alienation of human beings from nature and a resulting attitude where nature is viewed as a commodity (Gómez-Baggethun et al. 2011). Thus, we can speak of nature as resources, standing reserve, natural capital or ecosystem services, and so on (Washington 2020; Deliège and Neuteleers 2015). The task facing us then becomes one of ensuring human survival by enrolling nature into the human-constructed, mainly capitalist, structures for our use. This use can include beneficent goals such as equitable distribution or the assurance of nature's continuance (Davidson 2014), but note that this is still in order for it to serve humans.

The intertwining of cultural attitudes of dominance over nature and a mainstream science that prioritized physics as the guiding discipline has led to a perception of the world as inert matter (Merchant 1980; Sheldrake 1990). That approach has led to astounding achievements alongside potential devastation. In fact, we would not know the depth of the problems we face without some of those achievements, for example, climate modeling and ecosystem monitoring.

I want to explore a path in the development of the scientific method that was not the one taken. This is the scientific work of the 18th-century German poet Johann Wolfgang von Goethe. To do this, I will first place him in his historical context to see in what ways his thinking was diverging from the way science was developing and the trajectory it took. Then I will set out his

1 Acknowledgement: Material in this chapter draws on Brook I. 2021 'A New Science from a Historical Figure: Goethe as Holistic Scientist' and 'Engaging in the Goethean Method. An Approach for Understanding the Farm?' Chapters 6 and 20 in Wright, J. (ed) *Subtle Agroecologies: Farming with the Hidden Half of Nature*, Abingdon: CRC Press.

DOI: 10.4324/9781003043577-13

method of working in such a way that the reader could explore this approach firsthand. Whether this path would have not led to the crisis that faces us we cannot know but engaging with this method does seem to place us in a different relationship to nature: a relationship that is more relational and collaborative than dominating.

Goethe's world

Johann Wolfgang von Goethe's dates are 1749–1832, which placed him chronologically at the watershed between the study of what was then called natural history and the new science of biology. His scientific approach countered both a form of vitalism (that in Goethe's time was characterized by belief that organic entities were infused by a mystical/imperceptible life force or spirit) and mechanistic ways of seeing the world that were coming to dominate the science of his time (Steiner 1985:92).

Goethe's scientific work included the areas of geology, meteorology, osteology, and botany and the study of color. Working across a wide range of subjects seems strange to our ever-more specialized contemporary approach to science. Why would someone hop from one realm to another, and risk being labeled a dilettante? Even in the late 18th century, this would have seemed to be spreading oneself too thinly to make progress in any field of science. To understand this and to make sense of Goethe's scientific writings and discoveries, we have to understand him as embracing a form of holism, of seeing nature as one, and investigating sides of that whole in order to see it in its fullness. Moreover, his individual findings were only part of what he was doing. The core of Goethe's work was to address the nature of science itself. For example, he was also an early exponent of the history of science as a discipline in itself (Fink 1991:70). Attending to the cultural milieu from which any scientific work was emerging he believed would direct our attention to why it takes the shape it does and the ways it can be mistaken. Goethe's example of such a problem in science was the tendency to assume that everything is created for a purpose beyond itself, namely, for us. He says:

> When a science falters and comes to a standstill despite the best efforts of many researchers, it can often be seen that the fault lies in certain traditional concepts of things, a conventional terminology, which the great majority accepts and follows unconditionally. . . . The progress of natural philosophy has been obstructed for many centuries by the conception that a living being is created for certain external purposes and that its form is so determined by an intentional primal force. This idea still holds us back, although some have voiced vehement opposition to it and drawn attention to the stumbling blocks it creates.
>
> (Goethe 1995:53)

And he proceeds to dig further into the psychological roots of this problem:

> Man [sic] is in the habit of valuing things according to how well they serve his purposes. It lies in the nature of the human condition that man must think himself as the last stage of creation. Why then should he not also believe that he is its ultimate purpose? . . . Given his need for objects and his use of them, he draws the conclusion that they have been created to serve him. Why should he not resolve the inner contradictions here with a fiction rather than abandon the claims he holds so dear?
>
> (Goethe 1995:53)

To outline how Goethe's approach was different, I will focus on three aspects of Goethe's science: the rejection of an overreliance on theory; the grasping of nature as being in flux; and the role of human faculties in understanding nature. In all of these aspects, he was neither representative of his own time nor foreshadowing the direction that science was to follow. Furthermore, although he did have an early fascination with alchemy, and draws on it in his literary work (Gray 1952), what he was doing in his scientific work and reflection on science as a discipline was something entirely new.

Rejection of overreliance on theory

At the time that Goethe was writing, mechanistic, reductionist ways of thinking were becoming prevalent in society and science, and these were leading to impressive discoveries and inventions. However, for Goethe, these discoveries were often misunderstandings and science itself was being driven up a blind alley. One aspect of the practice of science on which Goethe wrote extensively was the movement toward an overreliance on theory. His final extended scientific study and most controversial work was an exploration of color. It included an extended and, admittedly, polemical attack on Newton's approach to color as theory laden and reductionistic (Sepper 1988). Goethe believed that this approach leads us to no longer see the phenomenon we seek to understand; we would see only our own construct. As Goethe argued:

> Someday someone will write a pathology of experimental physics and bring to light all those swindles which subvert our reason, beguile our judgement and, what is worse, stand in the way of any practical progress. The phenomena must be freed once and for all from their grim torture chamber of empiricism, mechanism, and dogmatism; they must be brought before the jury of man's common sense.
>
> (Goethe 1995:309)

Goethe was not saying that theory or hypothesis had no role in scientific investigation; he did, however, want to rein it in and saw it as a tool that needed very careful handling. Goethe rejects both the overreliance on theory to determine what to look for and the habits of mind that construct phenomena in the way suggested by the preconceived theory. A strong example of this criticism can be seen in the following quotation:

> To rid the human mind of a hypothesis that has unduly restricted it, forcing it to observe erroneously and to combine falsely, to muse instead of seeing, to sophisticate instead of judging, is already to render it an inestimable service. Henceforth it sees the phenomena with greater openness of mind, in other relations and interconnections, orders them after its own manner, and once more gets the chance to err after its own manner, a chance that is invaluable if it soon succeeds in perceiving its error.
> (Goethe cited in Heinemann 1934:68)

The method by which one is to "rid the mind" of constricting hypotheses is to approach the phenomenon from all directions. One of these directions will be prior hypotheses, but that is only one direction among many diverse approaches. Apart from the importance of a full investigation of the phenomenon as experienced by the scientist, there is also, as the latter part of the quotation reveals, the conscious attention to the action of theory, hypotheses, and opinion on perception itself. Thus, the scientist has to investigate their own mind as well as nature to ensure that the necessary openness is maintained and to let the mental faculties be guided by nature.

In considering Goethe within the context of a contemporary search for a holistic science, it is interesting to note that Goethe draws on ideas from Francis Bacon: the *bête noire* of nature respecting holistic approaches (Merchant 1980). However, his use of Bacon's "idols of the mind" was to accept the need for self-criticism and to always return to the phenomenon being studied. And again, when he used Bacon's concept of experiment to criticize Newton, it was because Newton had unduly narrowed down the phenomenon of color.

For Bacon, an experiment is not to confirm a hypothesis but to rule out some hypotheses. It should place the experimenter at a crossroad and the experiment should rule out one or more of the possible hypotheses and perhaps suggest new avenues to follow. Newton's presentation of his prismatic experiment as *experimentum crucis* shifts its meaning and with it the more open context of Baconian induction. As Sepper says:

> [H]e [Newton] accomplished a significant transformation of the Baconian method and placed the new concept of the crucial experiment at the centre of natural philosophy, at the centre of science. . . . Newton's Letter, though quasi-Baconian in appearance, proceeds more positively than negatively in that it judges what is wrong from the perspective of what the theory requires; also the phenomena are selected to test the theories, rather than to provide a comprehensive array.
> (Sepper 1988:134)

The way in which Newton had distanced the scientist from the phenomena as they arise – in their interacting multiplicity – was the mainstay of Goethe's extended attack on Newtonian optics. To isolate a single complex experiment in order to prove a theory was for Goethe poor science. However, Goethe was by no means uncritical of Bacon. In Bacon's radical realigning of science with the world as experienced and his assumption that an entirely theory free observation was possible, Goethe believed that Bacon had "thrown the baby out with the bath water" (Goethe cited in Nisbet 1972:30). The idea that an objective view from nowhere is possible denies our inevitable engagement and relationship with the thing studied and the possibility of using that relationship.

Thus, theory had a role to play but to understand Goethe's way of working it is more helpful to think in terms of the principles that open, respectful, diligent observation can find in nature. Or perhaps, in Goethean style, one should say principles that one is led to by nature herself.

Through the many scientific works by Goethe across a range of fields, he found certain principles that are evident in nature and these, for him, helped to explain its endless creativity (Tantillo 2002). Rather than having strict laws circumscribing what is possible, these four principles are better seen as ways of capturing the way nature does what it does. They are as follows:

Polarity:	Goethe's term for the way nature is creative through the juxtaposing of opposites; it remains in flux by breaking apart and rejoining.
Intensification (*Steigerung*):	Goethe's term for the way all of nature seems to be striving or overcoming itself through increasing complexity.
Compensation:	This recognizes that nature creates within limits; forms can develop and change, but they are always sacrificing something to develop something else; for example, adaptability is sacrificed for specialization.
Generativity:	This is the principle Goethe recognized to explain the way an organism's parts can work together or even compete, in some sense, to grow. For example, a plant can produce more of itself vegetatively or sexually and these two impulses are both working generatively and express an inner vitality in the plant. Goethe considers that an abundance of generativity suggests a radical freedom at work, as opposed to a hierarchy where certain parts direct development.

These principles operated for Goethe in the way that a theory or hypothesis operates, although they were, he maintained, found in nature, not created by scientists for their convenience nor to constrain or shape what we are seeing. If we approach nature as a machine, we will find machine-like aspects, not because nature is like a machine but because that is the lens we look through.

These principles were for Goethe a means to capture something of nature for the human mind to grasp, but they are, like nature itself, ever malleable and undergoing change. One way to understand these principles is as guides for reasonable ways of adding the non-physical meaning aspects of a phenomenon, through contemplation of the phenomenon in the light of these principles.

Moving with the aliveness of nature: all is flux

An aspect of Goethe's science that seems unusual for the time he was writing was his appreciation of nature as continuously in flux. The project of biology in the 18th and early 19th centuries was to catalogue and order nature. Nature was understood as created by God, and the human task was to know what was there and to order its objects into static tables. Taxonomy was, in a sense, just the task of putting things in the right boxes. Goethe, as we see from the principles mentioned earlier, is more interested in nature as a whole and how it generates itself.

Indeed, his first published scientific discovery in 1786 of the intermaxillary bone in humans upsets one of the most fundamental divisions: that between humans and other animals. This bone in humans is fused with the maxillary bone, whereas in other animals (even apes) it is freer. Goethe saw this fusing, as opposed to seeing one discrete bone because his scientific question was not "which box does it fit into?" but rather, how does dynamic nature shape bones? He was seeing the hard material of bones more fluidly without preconceptions driven by a theory that must separate humans. This fluidity was so new as a concept that Goethe needed to introduce a new term for it, and in 1796, he used the term "morphology" for the study of the transformation of organisms (Jensen 2019).

Morphology also suggests the need for a kind of lively apprehending, not only to see the whole form, but to understand it as a whole. The organic, for Goethe, had no fixed form (Goethe 1995:64). Thus, the organism seen in everyday consciousness is not the organism as a living thing, to see this we need to bring to it the sense of flux that it is engaging in.

With plants, the picture is even clearer. Unlike his contemporaries, Goethe does not stick to studying plants as dry preserved specimens in herbarium samples. As he discovers, particularly on his extensive and life-transforming trip to Italy (Goethe 1989), the way a plant grows reflects aspects of its environment such that there can be no stable sample of a plant – its size and development are shaped by where it grows. Alexander von Humboldt dedicated the German edition of his groundbreaking 1807 *Essay on the Geography of Plants* to Goethe (Magee 2007:227). This essay presents an entirely new approach and was the forerunner of biogeography.

In *The Metamorphosis of Plants*, Goethe set out his related insight that the plant develops by transforming its material through time by shaping and reshaping itself. It does not unfold to a strictly determined plan, but it does express a vitality and a drive to develop and reach a point of fruition, when circumstances allow. His cryptic statement "all is leaf" means that the plant substance, which might now be expressed as leaf, undergoes a transformation (akin to reproduction) to become the other plant parts: sepal, petal, carpal, stamen, etc. Nature's creative process is visible to us once we shift our attention from static form to fluid process. This insight he attributes to encountering the discontinuous metamorphosis in the shape of leaves, which he first noticed with the dwarf palm (*Chamaerops humilis* L.) in the botanical gardens in Padua in 1786 (Arber 1950:42). Once observed with that plant, he could then see the way many annuals produce at first simple leaves on petioles, then more elaborated leaves, and then back to simple shapes that hug the central stem as a process of metamorphosis, not of the individual leaves, but of the plant. By engaging his imagination enough to follow that story, he then saw that the flower parts are all also a continuation of that same process. The plant's creative process is visible and laid out in front of us if we shift our attention from static form to fluid process.

Human capacities and their role in science

The third aspect of Goethe's approach to science that needs some explication is the role the human being plays in scientific investigation. In the science of his time, the aim of distancing the inevitably fallible subjectivity of the human being to leave, a view from nowhere, objectivity was growing (Bortoft 1996). This aim was reconceptualized by Goethe as the need to train human subjectivity through immersion in the phenomenon studied in order for the full capacities of the human being to be available. Here we can see that the prevailing subject–object dualism is radically overturned, and science becomes something relational. This is connected to both the preceding points about how we get to retain the baby but not the bath water and about nature in flux.

Goethe said, to understand the whole as a metamorphosing possibility we need to "remain as quick and flexible as nature and follow the example she gives" (Goethe 1995:64). What does he mean by this exactly? The bridge that Goethe used to make the sequential connection from one instance to another, from one leaf to another, from one plant to another, etc., was that of the human imagination. Rather than seeing human faculties such as imagination, or indeed intuition, as impairing scientific objectivity, Goethe saw these human faculties as the means to really understand nature. By allowing nature to work with the human faculties, he thought we would begin to see with the mind's eye and thus make connections and reveal the inner life of nature that was not initially or easily apparent. He was resistant to the notion of a free-floating idea or concept that was not tied to, or indeed given by, the sense perceptible to the human mind through what seems like a shared contemplation (Stephenson 1995:113). The observation and identification of key aspects of nature and an open contemplation of them helps to reveal nature as a whole, and this activity, in Goethe's view, was crucial for the progress of science. His use of the term *Anschauung* captures this well, although it is not directly translatable. The English botanist Agnes Arber, commenting on Goethe's work, renders it as "intuitive knowledge gained directly through contemplation of the visible aspect" (Arber 1954:122).

Goethe's lively discussion of philosophical problems with the activity of scientific investigation helps to fill out the picture of how he worked and what he thought was possible. Accounts of Goethe's own perceptual abilities abound (Amrine et al. 1987:379), and he acknowledges that he was naturally very perceptive, but he also believed in the development of one's faculties. It is clear that he thought the human sensorium was capable of improvement when he discussed the idea of scientific work opening "new organs of perception". As he said: "The human being knows himself only insofar as he knows the world; he perceives the world only in himself, and himself only in the world. Every new object clearly seen, opens up a new organ of perception in us" (Goethe 1995:39). The means to improvement was the exercise of looking at, and building a relationship with, nature. He seemed to invoke a cycle of looking at nature, examining oneself, looking at nature again, and so on. The procedure then becomes a spiral of enhancing capabilities through experience and closer and closer relationship with the aspect of nature studied. For Goethe, this was the activity of science, he said: "Insofar as he makes use of his healthy senses, man himself is the most exact scientific instrument possible" (Goethe 1995:331). That sense of having to live into, to merge with, the thing studied comes through in his writing.

Thus, we have a scientist very alive to the problem of presuppositions and human tendencies to shape the world to themselves and yet endorsing faculties such as imagination and intuition to arrive at insights about the workings (or rather the being) of nature. To highlight that seeming conundrum, he talked of the kind of approach that was needed as a "delicate empiricism". He said, "There is a delicate empiricism which makes itself utterly identical with the object, thereby becoming true theory. But this enhancement of our mental powers belongs to a highly evolved age" (Goethe 1995:307). This delicate empiricism is *Anschauung:* where contemplation leads to intuitive insights by allowing nature to speak because we have placed our faculties at her disposal. Such an enhancement would perhaps allow us to become, for a moment, that which we study, for example, to experience vegetative growth or even photosynthesis. Moreover, it allows us to bring the wider picture of relationships and forces in nature to bear on our perception.

It is hard to imagine exactly how Western culture, and the way we have treated nature, would be different if the development of science had adopted something of this relational quality. To begin with, we would be different people. The shifts in consciousness that Goethe believed were necessary to understand nature would have instilled different abilities and, along with them, different values. What comes across from the bulk of Goethe's scientific work is a heartfelt reverence for nature. Had science taken that direction, I suspect that now (to play on John F. Kennedy's inaugural address) we would not be asking what nature can do for us but what we can do for nature.

Practicing Goethe's method

If Goethe's method is an approach to science that was not taken, can it be found again or reinvigorated? In some ways it was never lost, not just because the written work has always been available but also because there are examples of Western scientists, who do not quite fit the mainstream being inspired by Goethe. For example, the English botanist Agnes Arber whose work on the mutability of plant forms involved the necessity of avoiding hard and fast distinctions. This is an approach in botany now termed "fuzzy Arberian morphology" (Rutishauser and Isler 2001:1179) and shows definite aspects of working in a Goethean way. However, the major impetus of keeping alive something of Goethe's method has to be credited to Rudolf Steiner. From his work in editing Goethe's scientific writings for the *Deutsche National-Literatur*, Steiner could see that what Goethe was doing was not just outlining certain discoveries, but is also setting out how different states of consciousness are needed to bring out a fully rounded appreciation of any phenomenon. Moreover, in Goethe's scientific writing, it is clear that these states of consciousness could be schooled and used with the kind of rigor expected in science (Steiner 1978).

Some historians, such as Karl Fink, see in Steiner's work on Goethe a one-sided or overemphasized view that does not equate with the bulk of Goethe's scientific work (Fink 1991). Steiner's work certainly developed Goethe's scientific insights, but features, such as a living richness, a clear divergence from the mechanistic/reductionistic orthodoxy, plus a new working method that includes more of the specifically human faculties, were certainly already there to be developed.

Steiner's interpretative work on Goethe's science is a key inspiration for contemporary Goethean scientists, such as Craig Holdrege (2013), Jochen Bockemühl (1985), and Arthur Zajonc (Seamon and Zajonc 1998). Goethe's science, in particular his way of approaching phenomena with the human faculties operating in a disciplined "objective" way – a form of schooled subjectivity – is emphasized repeatedly in the various fields developed out of Steiner's spiritual science, which he called anthroposophy, including Waldorf education, anthroposophical medicine, and (relevant to questions of sustainable land use) biodynamic agriculture.

To bring out the sense of working in the way that Goethe was suggesting, I want to now lay out his method as a practical approach with suggestions and guidelines to follow. The aim here is to give the steps that can lead to the development of new organs of perception. Why this might be worth trying out is not just to better understand this historical figure but to experience a different way of engaging with the world. As already described, this method involves developing and utilizing human faculties, such as imagination and intuition, in a disciplined way that allows them to play an insightful role in a holistic understanding of nature and ourselves as part of nature. The four elements of earth, water, air, and fire are used here to assist with feeling our way into each of the shifts in consciousness that are involved in the Goethean approach. This journey challenges the reader to step outside their usual preconceptions and, through a type of phenomenology, allows them to experience subjectivity in a new way. The experience sheds light on Goethe's idea of delicate empiricism as an approach to scientific understanding. For addressing the problems we face and engaging with solutions such as living more lightly on the land, and of collaborating with nature's generative powers, this method can give a fresh view of what true sustainability could mean.

Some aspects of this method will seem mundane and others strange, but it is clear that we cannot approach it with our everyday habitual dualistic thinking in place. To set the scene for a radical change, I want to begin with a suitably enigmatic Sufi story:

> Once upon a time, a shopkeeper had an assistant who looked at everything cross-eyed, so he always saw double. One day the shopkeeper asked the assistant to go into the storeroom and fetch a large jar of oil to bring to the front. The assistant returned a moment later and asked, "There are two there. Which one should I bring?" The shopkeeper sighed, since she was used to these questions. So she decided to try a new tack. "Just break one of the jars and bring the other!" So the assistant did it, but when one jar broke so did the other.
>
> (Douglas-Klotz 2005:72)

Goethean observation as a careful process

Goethean observation, as a means to come to know a phenomenon, is widely recognized as having four steps (Bockemühl 1985; Hoffmann 2007; Holdrege 2005; Seamon and Zajonc 1998). My elucidation of it draws from the training given by evolutionary biologist Dr. Margaret Colquhoun and others through the Life Science Trust. Also from my subsequent working with and studying this approach, to inform what is set out here. The Life Science Trust organized a remarkable series of seminars through residential workshops (some 7 or 9 days and some 3 weeks), each focused on a particular realm of nature, for example, rocks, plants, color, animals or landscape. They were held over a number of years (from the 1990s to 2000s) at various venues in the UK and, latterly, at the Life Science Centre, Pishwanton, in Scotland. Participants generally attended at least four workshops to experience this method with a range of phenomena).

The four steps are usually preceded by a preliminary stage of recognizing one's habitual responses. In my own practice, and my own teaching of this process, I place more emphasis on this pre-stage than others do. Perhaps, this is because I am a philosopher by training and thus am aware of the epistemologically controversial nature of claims such as "being one with the phenomenon studied". Moreover, this pre-stage is useful for highlighting that Goethean observation requires a challenging switch from our culturally habitual dualistic thinking to a more holistic thinking.

A real strength of the Goethean approach is that it lays out a formal stepped process. Skilled practitioners will move between and meld these steps but, to begin with, it helps to explore and carry them out in a systematic way. This might seem and, at times, feel slow and pedantic, but the point is that we can follow consciously and rigorously what is happening in the world and in ourselves as the practice progresses. Experiencing this process can sometimes be unnerving or strange; for example, we may become aware for the first time of odd physical responses, insights, or mystical states. However, engaging in the method produces a careful record and thus a fuller understanding of this shifting relationship between what we usually think of as our external and our internal worlds. Then we can move into and out of these feelings and responses by retracing our own journey.

Choosing a plant to study

When learning this approach, it is easiest and most instructive to begin with some aspect of the plant realm even if, later on, the intention is to apply the method to an animal or ecosystem or landscape or something from the human realm such as an organization or community. The plant, as we saw earlier, was foundational for Goethe's understanding of his method of investigation. In a sense, it led him into the necessary shift of consciousness to properly experience the plant as alive. Practice with a plant provides benefits later on with more complex organisms, such as animals, and less responsive entities such as rocks. With a plant, it is easy to reach an understanding of what these steps feel like and so be more comfortable and assured about where one is in the process.

As with any research question or journey of inquiry, we need to have a focus – what is it that will be studied in this way? For early practice with Goethean study, it is good to work with something that draws the attention in some way; not necessarily something that is well-known or that serves a practical purpose – that can come later when one is familiar with the method. Being able to find the thing that would be fruitful to study is as much about waiting for it to call to or stand out to the observer. This requires patience and a childlike receptivity to hear what it is about the world that one is particularly suited to explore at the time. Being drawn to something could be an attraction or curiosity or it could be a feeling of revulsion or challenge – a troublesome "weed" plant, for example. As can be seen right at the beginning of this process, the mindsets of "what can this plant offer me", "what would be useful", "can I multi-task this to save time", and so on have to be set aside.

I will be using italics when giving direct suggestions of ways of working in the preliminary stages and each of the steps and reminders of what needs to have been completed at the end of each step. (Note that taking on this process and working with it is something that requires a personal impulse to work in this way and should not be imposed.)

> *To begin adopt that child-like receptivity and just wander around and let something strike you and come to (what you might later see as) a mutual decision to enter a relationship. Before reading further, you need to have selected – or perhaps we can say you need to have been selected by – your plant.*

Before the four steps: clearing the workspace

As human beings, our ways of thinking, feeling, moving, responding, and simply being in the world are shaped by our physical environment, culture, personal history, and a whole web of interactions. In the West, many of these are permeated by the dominant tradition of mind–body dualism, scientific reductionism, and (perfectly justifiable) practical concerns of survival. It is one thing to acknowledge this, but altogether a more difficult matter to escape it. Mind–body dualism, for example, shapes our language so that even to express how something could be different involves using the language that has developed to embody those cultural assumptions. An obvious example is when we talk of "the environment", the use of "the" separates us and places us outside of that in which we are actually environed. The process of Goethean observation is a honing of the human being as a scientific instrument. If this is what we are going to be engaged in, understanding the starting point – that is, our normal way of thinking and being – is going to be essential. There is self-reflection involved here and this can bring disquieting insights into oneself. For example, attitudes that we imagined were long outgrown can be unearthed.

What should become apparent is just how much the Goethean method involves self-examination and critical reflection. It is a qualitative approach to the world, but not one that revels in undisciplined subjectivity.

Guidelines for clearing the workspace

To clear the workspace take a preliminary look at what is actually there and how it usually works. Return to your chosen plant in a normal, everyday way and then set down your first impressions in whatever form they may take. These might be: your habitual likes and dislikes, how you might use the plant, feelings of boredom or anger, snippets of information, inspiring ideas or urges to put something right and so on. Instead of acting on these impressions or continuing them with further thoughts or day-dreams, you need to consciously lay them out for inspection and then set them aside. Keep a record of these thoughts and feelings. By naming attitudes and presuppositions that are in the background, you can spot when they might creep into your work in the later steps of the process. That said, a first impression may also be very perceptive and its veracity may re-emerge later on.

Before moving on, make sure you have made some notes about your usual thoughts and feelings about your phenomenon and any prompts to action that might have come up. A small journey into speculation about the origins of those attitudes can be helpful, but avoid holding up the next steps of the process that need to be worked through far more slowly. Your plant is waiting.

Step one: exact sense perception

This first step is characterized by standing away from that very personal first encounter and observing the plant afresh. This is a process of meeting a being and, as "clearing the workspace" emphasized, this needs to be done on the plant's own terms, that is, without overlaying it with personal preconceptions or normal ways of thinking. Not only personal feelings, but also any known theories about a phenomenon need to be held back in order to let the "facts" speak for themselves. An example of this very different approach can be seen in Goethe's study of color as a phenomenon. Rather than draw hasty hypotheses or work from an existing theory, such as Newton's, his painstaking investigations followed every conceivable avenue of experimentation (Sepper 1988). Thus, for Goethe, finding out about the nature of color involved aspects such as complimentary color after images, how artists use color, how dyers use reagents, and so on, rather than just the latest idea on color from physics.

Recording observations can be done in a number of ways such as writing detailed descriptions. However, drawing the phenomenon is one of the best ways to focus attention on the hitherto unnoticed detail and the relationships between parts. If the aim is to really see a particular tree that happens to be an oak, drawing can be very helpful to prevent the observer slipping into their usual "seeing oak trees" mode of perception. Artists have the additional problem of having to avoid their personal "drawing style" influenced mode of perception. The categorized artifact that one's usual mode of perception creates must be ignored in order to let the observer see the oak tree as if it had never been seen before. A number of drawing exercises are helpfully detailed in Margaret Colquhoun and Axel Ewald's 1996 book *New Eyes for Plants: A Workbook for Observing and Drawing Plants.*

Another tool that is used to help in this first step is to ignore preexisting knowledge, for example, the names of things. Instead, we need to see and describe them outside of learned classifications. This restriction on nomenclature is helpful when sharing observations in a group as the observers have to struggle to find descriptive language that makes sense to others but avoids the readymade and therefore potentially occluding terminology. It is interesting to observe in general how people often want to name, for example, wildflowers and seem disquieted by not knowing the name. For this process, it is helpful to recognize that disquiet and then turn it into a more open curiosity.

Jochen Bockemühl's work on phenomenology (1985) uses the four elements (earth, water, air, fire) as a helpful way of characterizing each of the steps. This first step has an "earth" quality: the solid facts are gathered, and the feel of the process here is one of careful exactitude. Although fascination with detail is appropriate, one should not get carried away. For some, this step is experienced as being rather tedious, yet for others it is very satisfying, and such differences in themselves show something of our personalities. In the training mentioned earlier, we spent several days on each step with our chosen phenomenon. With the plant, it can be hard to stay in this mode, for example, one will be tempted to describe things as "growing" or "wilting" and generally anticipating future states or imagining past ones that we can recall; it is quite a discipline to hold those imaginings back and stick with the present. *Make notes about this when it occurs and how it feels to pull back from it.*

Secondary sources can also form part of this first step as a part of the information gathering. It can be helpful to know what other people have discovered. For a plant, this could be some botanical knowledge, or the relevance of the plant in agriculture, herbalism, myth, or even the language of flowers in mediaeval paintings. The important point about using secondary sources is to avoid them to begin with (apart from checking in regard to poison) and then have a skeptical eye and always return to the plant itself for verification. With secondary sources, one is gathering what other people have said and they could be working from presuppositions and not from the plant itself. However, secondary sources might suggest new ideas or other forms of access to the phenomenon or questions to pose for further investigation. On the other hand, it also gives more material that has to be set aside in order to really see the plant.

Guidelines for carrying out step one

Try to experience your chosen plant as if you had never seen it before. From that perception, begin to record all that you can about it. Recording as you go is important: have a notebook and drawing materials to hand. It is important to remember that you have multiple senses, so don't just concentrate on sight. With plants, smell and texture are obviously important as is, for example, response to the wind. Taste can also be explored – with caution (do check that your plant is not known to be poisonous before tasting it).

When writing descriptions try, as explained above, finding a word that expresses what you are seeing rather than taking the ready-made ones such as stalk, leaf, flower etc. This will prompt more looking and thus more potential to see fresh relationships. Also, you could try for a moment to ignore assumptions about where your plant begins and ends – can a caterpillar on the leaf be described as part of the plant and soil around the roots be described as part of the plant? Ensure you know the details such as: the number of flower parts, how the leaves are arranged around the stem, is the stem ridged or round and is it the same color all the way along, and so on.

Try drawing exercises as: drawing the outline of a plant without looking at the paper, using shading for depth with no regard for actual shades or shadows, or creating the form by shading the outside area of the paper as if chipping the form from a block. You could also use watercolor paint, pencils or pastels to mix the exact color of different parts. One of the most useful drawing exercises, and this one should never be left out or cut short, is to draw from memory. You may think that you know everything about the appearance of a plant, only to have that assumed knowledge disappear the moment the plant is hidden from view. Drawing from memory by first closing your eyes and building the plant in your imagination as you have come to understand it, is extremely helpful in trying to build the bridge – so crucial to Goethean science – between the phenomenon and the human being as a scientific instrument. It is important to set aside any personal concerns about your ability to draw; the point is not to produce a beautiful picture, but to train your perceptions.

Do not move on until you have a collection of drawings and descriptions and a solid feeling of getting to know the "what is" of your plant. This will provide your anchor for the work ahead.

It is impossible to continue in exact sense perception indefinitely. To register all the great amount of variety and detail would be, as Goethe said, "like trying to drink the sea dry" (1995:24). Simply amassing facts about the phenomenon as a static object at the moment at which it is observed will not allow us to really see what the thing is nor come to any firm idea of it. Exact sense perception is only the foundation on which the following steps rest and to which they return, when necessary, to compare conclusions reached by other means.

Step two: exact sensorial imagination

The previous step was an attempt to capture what the phenomenon is presenting right there in the present. The activities involved are exact about what is seen, heard, felt, or smelled etc., but the entity studied cannot really be captured in a frozen present. It exists as a process, and so to get to what it really is, one has to live into that process so that we can begin to accompany it in its being. We do this by using the human faculty of imagination. But not imagination as we often think of it in the human realm as distanced from reality. Imaginative activity in Goethe's sense is called by him "exact sensorial imagination" (Bockemühl 1985) and it builds on the rigor of the previous step, but now it is set in motion. The aim of this activity is to perceive the phenomenon as a dynamic entity. Just as the previous step required a certain policing of one's usual ways of thinking, our imaginations also need some schooling to allow us to stick with the phenomenon, not as we have come to know it as an entity frozen in time, but as a being in process. It is the plant realm that gives us a clear picture of the need for a shift in our thinking if we are to understand a plant as a plant and not just as a marker in our system of classification or an entity to be utilized. This is where we can first detect that the quasi-empiricism, which in the first step seemed exacting and pedantic, is now on a path to what Goethe termed a delicate empiricism. Delicate because it does not impose a theory, and neither does it deny the human faculties their role in coming to know the world. Rather, the faculties have to treat the world with delicacy in order to find it, rather than just find humanness reflected in it.

There is something dreamlike about this step. However, because it follows laying aside our theories, our presuppositions, and undergoing a rigorous working in exact sense perception, our dreams are "in the style of" the phenomenon and not drawn from our own personal fancies or presuppositions. These experiences need to remain dreamlike, because any fixing of them will put us back into the first step.

One of the easiest ways into this type of imagination and seeing how it could lead to understanding something about a phenomenon is through Goethe's work on the metamorphosis of plants. It is here that we can see his use of exact sensorial imagination as a shift in consciousness that now connects with the phenomenon in a new, but nevertheless still rigorous, way. As we saw earlier, in the section on flux, discontinuous metamorphosis in a sequence of leaves gave Goethe the invitation to enter the process of the plant. Many plants produce a sequence of different leaves, often beginning with a simple shape, becoming more differentiated and then contracting to a more pointed form, and eventually there is a transition to the flower parts (Holdrege 2013:76).

Jochen Bockemühl, whose work makes extensive use of leaf sequences, explains the process and shows just how different this kind of watery perception is from the earthy style exact sense perception of step one.

> With the mode of observation corresponding to the watery element, it becomes possible to go beyond the single elements of form and reach a realm not directly accessible to sense perception; here the sequence of forms appears as formative movement, and the formative forces can be experienced. If something is observed as an object, it is always seen from the outside, it is seen separately and seemingly from all sides at once. There, one's own standpoint is unimportant. The object exists without me. If, however, one begins to become aware of the formative forces in the way described one's own inner activity (intentionality) and one's own position within the whole becomes significant.
>
> (Bockemühl 1985:21)

With any living entity, it is easy to move into this second step because the phenomenon seems to require it. We cannot capture the livingness of a plant if we stick with exact sense perception. Our thinking in that mode is too static to live into the phenomenon and experience it as changing and growing. Something of the phenomenon has to live in us if we are to make a connection between, for example, the sapling and the tree. Our thinking has to be mobilized to grasp the becoming of nature and the way nature is constantly creating. Although we know, in ordinary consciousness, that plants grow and change, here we experience it afresh and understand as if from the inside.

It is imagination that makes this mobilization of our thinking possible. In this mode of perception, we are living in the phenomenon as a process. We are imaginatively engaged in those same processes. We need to enter this not by bringing in human meanings but by living in the phenomenon as the being it is. Although we cannot leave our humanity out of the picture, as this is the source of our imaginative ability, in this and the steps that follow we are placing our faculties *in the service of the phenomenon*. Exact sensorial imagination leads us to a holistic apprehending where we understand, for example, the plant as a metamorphosing possibility, connected to its place and to the plant realm.

Guidelines for carrying out exact sensorial imagination

Working with a plant will draw you into this mode (observing growing, wilting, etc.), so you may have already ventured there and had to hold back. How does it feel to now open that possibility?

An easy experience of moving a plant imaginatively is through observing your plant at different stages of growth by finding others of the same type. For example, you could be studying a particular wild carrot (Daucus corota) but around it are others at earlier and later stages of growth or flowering or fading. See if you can imaginatively take your wild carrot backwards or forward in time using these others as indications of your plant's process. You could also enter into the plant as a process the same way as Goethe through its discontinuous metamorphosis. Try to move through the sequence of leaves on one stem as if you are inside the plant to exercise and school your imagination. In your sketch pad try to imaginatively produce leaf forms that could appear in between those that are evident in your plant. This helps you to experience the plant as a dynamic process of metamorphosis as opposed to recording only its visual form. (If your chosen plant does not show any obvious aspect of this discontinuous leaf metamorphosis, you could practice with one that does, such as mustard, buttercup or groundsel, to get a feel for these changes before returning to your plant.)

Other exact sensorial imagination exercises to help here include imaginatively growing your plant through a whole life cycle or imaginatively taking your plant through the seasons. The richer the pictures you can build, for example, the changing light, the changing insect visitors and so on, the more you will be sensing into the plant. Growing the plant in your imagination before sleep can be very helpful for continuing your study the following day. This activity seems to connect you with the dreamlike realm that is needed here, and it facilitates the later steps.

You could also try, with care, "imagining it otherwise" (Brook 1998:55). This is where you purposely misuse your imaginative faculty to impose something on the plant, for example, imagining an ash tree (Fraxinus excelsior) is evergreen. The purpose here is to feel for a response. By now, you will be in a relationship with your plant and will feel (perhaps viscerally or emotionally) the wrongness of your imaginative fancy. Such a response kick helps to get you listening in to what is really there and tuning into where insightful responses might come from in your own body.

Ensure you have felt the shift in consciousness into this dreamy, watery style of thinking and feeling before moving on. You should have things such as a leaf sequence with some imaginatively realized intermediate stages that do not actually exist but could have. You should also have drawings entirely from imagination of your plant in past and future stages in your sketch book and notes about feelings and responses. You should now be able to move freely into and out of the plant and carry something of its being in your consciousness.

Step three: seeing-in-beholding

This third step goes beyond the fluid mode where the phenomenon could be built up imaginatively through its changing forms. However, that fluid mode is still somewhat attenuated. It provides the training of the imagination, but what is now necessary is to tune in to the phenomenon as it is, as a living whole, rather than as parts brought together imaginatively. The element here is air. This takes the human being even further back from our ordinary way of engaging with the world. Here what is needed is to quiet the imaginative activity in order to make space for the phenomenon to present itself.

This needs to be like air, not flowing through the forms like water, but making them visible. Using elemental language to describe these steps does not mean that we are looking for particular characteristics in a phenomenon. We are not looking for airy aspects of the phenomenon; instead, what is required is a shift into an airiness in one's consciousness. Bockemühl describes the way the observer must be:

> It is characteristic of air to expand in all directions, offering its own being and activity in order that the being and activity of another can appear. Insofar as we move inwardly in accordance with this image of air, we reach the cognitional attitude corresponding to the air element. An inner readiness is thus created for that which manifests in the world to reveal itself in us, as an image which discloses a being.
>
> (Bockemühl 1985:26)

It is the human faculty of inspiration that now offers itself to the phenomenon. Through stillness, the phenomenon can present its real self, and this is often felt as a particular gesture, a gesture that somehow speaks or presents that phenomenon. The insights that arrive in this third step can seem strange to any normal ways of thinking. It can be exciting or emotionally moving, and because what arrives seems foreign to oneself, it enhances the sense that it is given. It should feel like something received rather than made. These received indications can be captured and explored in some form of artistic representation where gesture and meaning are brought out and the inner mood is expressed.

Guidelines for carrying out seeing-in-beholding

> *Try to adopt a state similar to some meditative practices where your habitual self is quieted and you just are the contemplation of your plant. For this inspirational step don't be concerned about expressing what is given in emotional language, this is just how it often comes. Emotional language might seem paradoxical as this step is far from normal, self-absorbed emotional subjectivity. Working artistically with color can help to deepen the indications that arrive. For example, aim for the mood color of the plant, which*

might be very different from its physical color. What should be expressed is the gesture of the phenomenon, something that points to its essential nature. Again, do not worry about any lack of familiarity with, for example, writing poetically or painting, it is the process of attempting that is important.

Make sure you have captured any insights into the gesture of your phenomenon. This can be hard to record in language or image but try to get something down. The process of writing, painting, moving or singing etc. is not just the representation of the insight, the activity itself helps you to live into the insight and to feel it more deeply. In your sketch book you should have attempts at capturing the atmosphere/mood and gesture of the plant. You might just have a list of words, try for a freely rendered haiku from those words as a way of really honing down and capturing the gesture of the plant.

Step one gave us the solid facts that anchored our imagination to the phenomenon and allowed us, in step two, to enter into the livingness of the phenomenon, which in turn gave us enough familiarity with it to see through to and express its inner/fundamental gesture. This gesture of the whole can push us into the fourth step of being-one-with the object (Goethe 1995:75).

Step four: being one with the object

The first three steps of the Goethean method involve different activities and ways of thinking, and these could be characterized as first using *perception* to see the form; second, using *imagination* to perceive its mutability and, third, inviting *inspiration* to reveal the gesture. The fourth step uses *intuition* to both combine and go beyond the previous steps. Here we experience the "what it is" of the phenomenon in its full power and potentiality. It is here that the phenomenon can be understood, and it presents itself to the human being as an idea or even a theory. Thus, in the Goethean process, we do not start with theory and overwrite the phenomenon with our own thinking, instead, we place our human thinking and theorizing capacities at the service of the phenomenon.

In terms of the elements, the shift in consciousness we now need is to fire. This fourth step is also the most physically abstract, having the least connection to the outer appearance of the phenomenon. However, that abstraction from the specific allows the perception of what is essential to the inner nature of the thing. Bockemühl calls this an experience of the "being's beingness". He expresses the fire step thus:

> We are here at the limit of what can be called a mode of observation. The warmth enters us – our inner activity itself becomes an organ. We do not experience the outer expression of a being, we become aware of its inner impulse. At these moments of inner identity, all outer manifestations disappear. They are "burned up".
>
> (Bockemühl 1985:30)

Fire or warmth is also suggestive of another feature of this step: the connection to the inner impulse of the thing comes about through our own inner impulse to act. At this point, we are prompted to action, not only in the sense of wanting to express something of the being, as in the third step, but to *do something about it*. We feel "fired up". Because of the journey that precedes it, this is not a subjective expression of one's own personal will. The intention is to combine the being of the phenomenon with the human ability to both think and to act in the world.

In teaching this method with groups, particularly with regard to landscape, it is interesting to note the difference between a first impression impulse to do something – make a change, clear an area, develop a path – and an impulse that arrives out of combining one's thinking with the phenomenon. Clearing the workspace and the four steps are intended to bring us to a point of understanding and collaboration with the phenomenon. Now we are in a position to move forward, to act collaboratively out of that relationship.

Guidelines for carrying out being-one-with the object

> *How you work to bring about this is step harder to explain, in my experience it just arrives suddenly when working in step three. It is as if from the acquiescent air process, where our usual thinking is held back, concept or idea or determination suddenly arrives, but with a strange clarity, a kind of shininess, that is unlike one's normal thought processes or perceptions. As a process I suggest continuing with the activities of the third step but with even less self-intention. Free yourself from having to express the gesture and just wait for an intuitive impulse – a thought and its attendant action. This will most likely come when you give up. It feels even more distanced from your normal self and something of a shock – an intuition can feel like a thought or concept without the thinking of it. Write it down as best you can and just live with it soulfully before trying to analyses what it means.*

Summary of the method

The steps with their indicative elements and human faculties emphasized can be summarized in Table 13.1 for ease of reference.

Table 13.1 The Four Steps of the Goethean Approach and Related Human Faculties and Elements

Steps		*Human Faculty*	*Elemental Feel*
1	Exact sense perception	Perception	Earth – solidity
2	Exact sensorial imagination	Imagination	Water – flowing
3	Seeing in beholding	Inspiration	Air – surrounding but ephemeral
4	Being one with the object	Intuition	Fire – fast, concentrated

Once you are familiar with the shifts in consciousness of these four steps by experiencing them through a journey or two with something from the plant realm, it is possible to branch out to other realms and take a phenomenon such as a rock, or an animal. This approach is particularly helpful when considering actions involving whole landscapes, for example, in agroecology, conservation, restoration, or rewilding. Within a Goethean process, if you want to make changes to landscapes, social spaces, or green spaces, etc. inspired by insights from the fourth step, you need to move forward into a new set of phases that mirror the previous steps (3, 2, and 1). This is about testing out and double-checking a planned action before landing it (earthing it) in the world. A useful discussion of these mirrored steps with practical examples can be seen in Christopher Day's 2003 book *Consensus Design: Socially Inclusive Process.*

Conclusion

In this chapter, we have seen that Goethe was creating a different approach to science. One where we can approach nature with an objective subjectivity that also becomes a kind of practical reverence. Historically speaking, this is important as it avoided the errors of two conflicting approaches that were informing the science of his time: vitalism and the mechanistic approach. For Goethe, vitalism gave away the self-generative power of organisms and of nature herself to a vague mystical idea of some free-floating force; it was thus a kind of obfuscating mystery mongering. However, for him the mechanistic approach ignored the complexity of the intertwined relationships in nature and denied the generative power of organisms in their full majesty. Science, as it developed from the 18th century into the 19th century, rejected vitalism and fully embraced mechanism. This ignored the possibility that other options were available.

To address the problems we face in the 21st century, we need a way of working with nature, one that does not destroy it or replace it with some limited imaginings. Goethe's approach provides a means to begin the process of undoing our habitual dualistic patterns of thought. It nourishes new organs of perception that are receptive to, and respectful of, nature so that we might begin to glimpse what nature really is, what place we have within it, and how we may collaborate or coevolve with it. For example, imagine how such an approach might impact on preservation, conservation, agroecology, or teaching the ecology-related disciplines?

As the presentation of the steps and the whole process of the Goethean method has shown earlier, the roles of imagination, inspiration, and intuition involve both a personal engagement and a transformation of our thinking and our being. This heartfelt engagement means that in our humanity, we cannot conceptualize nature as simply a resource for ourselves, nor can we absent ourselves from nature, as if we are only a problem. Through this method, we can become alive to the way nature herself seems to be teaching the means to and inviting collaboration.

References

Amrine, F., Zucker, F., and Wheeler, H., eds., 1987. *Goethe and the Sciences: A Reappraisal.* Dordrecht: D. Reidel Publishing Co.

Arber, A., 1950. *The Natural History of Plant Form.* Cambridge: Cambridge University Press.

Arber, A., 1954. *The Eye and the Mind.* Cambridge: Cambridge University Press.

Bockemühl, J., 1985. Elements and ethers: Modes of observing the world. In: J. Bockemühl, ed. *Towards a Phenomenology of the Etheric World.* New York: Anthroposophic Press Inc, 1–67.

Bortoft, H., 1996. *The Wholeness of Nature: Goethe's Way of Science.* Edinburgh: Floris Books.

Brook, I., 1998. Goethean science as a way to read landscape. *Landscape Research*, 23(1), 51–69.

Colquhoun, M., and Ewald, A., 1996. *New Eyes for Plants.* Edinburgh: Hawthorn Press.

Davidson, M., 2014. Rights to ecosystem services. *Environmental Values*, 23(4), 465–483.

Day, C., 2003. *Consensus Design: Socially Inclusive Process.* Oxford: Architectural Press.

Deliège, G., and Neuteleers, S., 2015. Should biodiversity be useful? Scope and limits of ecosystem services as an argument for biodiversity conservation. *Environmental Values*, 24(2), 165–182.

Douglas-Klotz, N., 2005. *The Sufi Book of Life.* London: Penguin Compass.

Fink, K., 1991. *Goethe's History of Science.* Cambridge: Cambridge University Press.

Goethe, J. W. von., 1989. *Goethe: Italian Journey.* Edited by T. Saine and J. Sammons. Translated by R. Heitner. New York: Suhrkamp Publishers.

Goethe, J. W. von., 1995. *Goethe: Scientific studies.* Translated and ed by D. Miller. New York: Suhrkamp Publishers.

Gómez-Baggethun, E., and Ruiz-Pérez, M., 2011. Economic valuation and the commodification of ecosystem services. *Progress in Physical Geography*, 35(5), 613–628.

Gray, R., 1952 *Goethe the Alchemist: A Study of Alchemical Symbolism in Goethe's Literary and Scientific Works.* Cambridge: Cambridge University Press.

Heinemann, F., 1934. Goethe's phenomenological method. *Philosophy*, 9(33), 67–81.

Hoffmann, N., 2007. *Goethe's Science of Living Form.* New York: Adonis Press.

Holdrege, C., 2005. Doing Goethean science. *Janus Head*, 8, 1.

Holdrege, C., 2013. *Thinking Like a Plant: A Living Science for Life.* Great Barrington, MA: Lindisfarne Books.

Jensen, A., 2019. Johann Wolfgang von Goethe. *Internet Encyclopaedia of Philosophy.* www.iep.utm.edu/goethe/#H4 (Accessed 5.9.2022).

Magee, J., 2007. Alexander von Humboldt: A vision of the unity of nature. In: Huxley, R., ed. *The Great Naturalists.* London: Thames and Hudson, 224–232.

Merchant, C., 1980. *The Death of Nature: Women, Ecology and the Scientific Revolution.* New York: Harper and Row.

Nisbet, H. B., 1972. *Goethe and the Scientific Tradition.* London: Institute of Germanic Studies.

Rutishauser, R., and Isler, B., 2001. Developmental genetics and morphological evolution of flowering plants especially bladderworts (*Utricularia*): Fuzzy Arberian morphology complements classical morphology. *Annals of Botany*, 88, 1173–1202.

Seamon, D., and Zajonc, A., 1998. *Goethe's Way of Science: A Phenomenology of Nature.* New York: SUNY Press.

Sepper, D., 1988. *Goethe Contra Newton*. Cambridge: Cambridge University Press.
Sheldrake, R., 1990. *The Rebirth of Nature*. London: Century.
Steiner, R., 1978. *A Theory of Knowledge: Implicit in Goethe's World Conception*. Translation by O. Wannamaker. New York: The Anthroposophic Press.
Steiner, R., 1985. *Goethe's World View*. Translation by W. Linderman. New York: Mercury Press.
Stephenson, R. H., 1995. *Goethe's Conception of Knowledge and Science*. Edinburgh: Edinburgh University Press.
Tantillo, A., 2002. *The Will to Create: Goethe's Philosophy of Nature*. Pittsburgh: University of Pittsburgh Press.
Washington, H., 2020. Ecosystem services -A key step forward or anthropocentrism's 'Trojan Horse' in conservation?. In: H. Kopina, H. Washington, eds. *Conservation: Integrating Social and Ecological Justice*. Cham, Springer, 3–88

14 Sustainable design

A critique of the tripolar sustainability model, 15 years later

Alain Findeli

Introduction

The essay inserted hereafter was first presented as a paper at the 7th European Academy of Design Conference held in 2007 at Izmir (Turkey) on May 11–13, then published in 2008 in the *Design Journal* (XI, 3, 301–322). I am sincerely grateful to the editors of this book for considering the following text "as a landmark" and having kindly invited me to include it in their book in its original form, on account of its "highly constructive critique and alternative of mainstream sustainability narratives". Although a serious update of its argument, content, and references is indisputably necessary after 15 years, I have chosen to present it in its original form (only slight corrections have been made), with an epilogue intended to alleviate what may today necessarily appear outdated if not somehow naive in a still harshly controversial and extremely rapidly evolving field.

Original paper

Sustainable design: A critique of the current tripolar model

ABSTRACT *'Tripolar model' here refers to the model, often also called '3 P model: Profit/Planet/People' or '3 E model: Economics/Environment/Equity', which is very widely used as the reference model for any project in sustainability, including sustainable design. The purpose of the paper is to present a critique of this tripolar model as it is structured and used. The research is based on the empirical evidence that too many projects based on this model have run into serious difficulties or have failed. The critique of the model is carried out in two steps. The first concludes that there is a lack of complex intelligence in handling the systemic logic of the model. The second shows that the anthropological foundations of the model are fallacious. A corrected version of the model follows, with a renaming of the three poles: comprehensive economics, social equity and human creativity. The new model is then discussed with regard to the so-called 'social threefolding' principles of philosopher Rudolf Steiner (1866–1925), which refer*

DOI: 10.4324/9781003043577-14

to the three ideals of the French Revolution. In this perspective, Fraternity or Solidarity should be associated with the economic pole, Equality or Equity with the social, and Liberty or Freedom with the cultural/creative. The social threefolding model is currently used throughout the world as a basis for various initiatives, on scales ranging from local enterprise to national political economy. The paper concludes with the proposal that the model seems fruitful for the conduct of sustainable design projects.

KEYWORDS: *sustainable design, 3E or 3P model, social threefolding, Rudolf Steiner*

The holy trinity

In February 2007 I was reading the quarterly newsletter of the alumni of an engineering school in France, where successful project case studies involving graduates of the school were regularly analyzed and discussed. The specific case I was reading dealt with a large European building management company, which aimed to develop a more sustainable management process. "A building can be considered environmentally sustainable if it performs efficiently on the economic, environmental, and social aspects", writes one of the company's executive. The next day at lunch, I was browsing through the monthly newspaper published by the local regional administration (in 2007 France was still divided into twenty-two such administrative regions, reduced to thirteen in 2014), whose executive council just recently adopted the principles and methodology of Agenda 21 for its regional development. "Sustainable development must integrate economic, environmental, and social aspects", said the Vice President of the regional Environment and Sustainability Board. Such observations could indeed be multiplied since almost not one day passes without us listening to the radio, browsing on the net, watching on TV, or reading in a newspaper or magazine that the "aspects" of economics, environment, and society are the three pillars of sustainable development.

What's wrong with that, may we ask? What's the point? Well, the point is not that sustainable development should indeed address "economic, environmental, and social aspects" but the fact that this principle, and its incarnation as Agenda 21, has become mainstream and consequently unquestioned: it is spelled out like a gospel, a mantra, it is used in various contexts (political, educational, social, agricultural, and so forth) with exactly the same abstract formulation, with apparently little concern for what it actually could mean for action in a specific context. As if this mere incantation were sufficient to make sure that projects conducted under such a label will actually contribute to or yield a development (whose development?) that will be sustainable (for whom?)! As if the actors who will be involved in the corresponding projects, having been blessed by the holy word, knew from the start how to behave and properly manage their respective courses of action in a developmental and sustainable way!

What allows me to be so outrageously cynical is not a personal, irrepressible trait of character but empirical evidence of the difficulties, the obstacles, and even the failures that so many projects have met, despite the goodwill and noble intentions of their initiators and stakeholders. Here again, I must confess that my argument draws on common, daily experience rather than on specific cases reported in the literature. The difference, in this instance, is that the "common experience" refers to that of the design and design research communities rather than, as was the case above, the larger civil community, a community to which design researchers all belong.

We members of the design community all know that the case studies literature in project management does not contain an over abundance of examples of failure and mismanagement.[1] Success stories are much more popular in professional journals, and this is also the case in design. Nevertheless, we probably all know, from our teaching experience for instance, or at least we intuitively feel from our familiarity with the complexity of design projects, that carrying out a project along the lines of sustainable development is likely to meet with inextricable contradictions and difficulties. It is precisely the aim of this paper to make clearer this feeling and to try to find out why these difficulties arise.

Problem, purpose, method, and structure of the chapter

I admit that my diagnosis is not entirely based on the kind of evidence one would expect and should require in a respected scientific community as ours. The hypothesis is that such evidence would be easy to gather, however. My starting point therefore is that the management of design projects along the tripolar model – economics, environment, society – of sustainable development is consistently hitting upon strong contradictions followed by practical powerlessness and inefficiency, if not total failure. Careful phenomenological observation of such design projects, both in educational and professional contexts, leads me to the hypothesis that failures might not be due to an inadequate management of the projects[2] but rather to

1 One noticeable counterexample is the so-called Chaos Report published by the Standish Group. See, for instance, the 2004 Third Quarter Research Report available at www.standishgroup.com/sample research/chaos-1994_2.php (accessed 15 February 2007).

2 This issue is raised in similar terms, although in a different theoretical context, by Lalonde (2007). The argument goes as follows. The causes of failure in engineering projects are usually diagnosed with reference to the standard, instrumental and normative, model in project management (the so-called Cost, Time and Quality model): cost overrun, time overrun, and/

the inappropriateness of the theoretical and conceptual framework – that is, of the tripolar model governing the projects carried out under the heading of sustainable development. My hypothesis is further strengthened by the "holy trinity" phenomenon described above – by the apparent loss in content and substance that this pivotal model has ended up exhibiting in various situations. More precisely, the meaning given to the three poles of the model is so indeterminate and loose that all sorts of interpretation become possible. Whereas this is particularly true of the "society" pole, it also affects the two other poles. The abstract character of the terminology makes it almost impossible to find a consensus on what exactly is meant by "society", "environment", or even "economics". And here I mean a consensus among experts, not only among "common" citizens. The latter is a political, the former a scientific issue. It is as if the tripolar model which was originally meant to actually, that is, *pragmatically*, orient, guide and constraint action in a direction compatible with sustainable development, had become a mere *rhetorical* artifice. The catchphrase has become evocative and convincing enough to justify any enterprise whose aims are closer to political, economic, ecological, or personal ambitions than to sustainable developmental constraints.

Stated as it is, my hypothesis can be split into two questions or problems, along which the paper is structured. My first critique contends that the problems encountered in sustainable design projects may be due to an improper interpretation and implementation of the model. My second critique is more radical in that it considers the model itself to be faulty. My purpose is to develop both alternatives, although I will dwell longer on the second argument. The first critique is carried out from a systemic/complexity theory perspective, whereas the second draws on economic-, political-, and anthropological-philosophical arguments. My conclusion is preceded by a critical discussion of the new model I propose, followed by a brief and parenthetical metaphysical discussion of design thinking.[3]

The method adopted for this first phase of the research is clearly speculative or philosophical, if the latter implies argumentation mainly based on critical analysis and deductive logic. The paper therefore potentially

---

or poor quality delivery. Various evaluation guides and methods have been designed to assess and audit such projects. However, there are instances where, although successful with regard to standard evaluation, some projects are nevertheless considered as unsatisfactory or unsuccessful by their stakeholders and clients. The source of this lies in a too restrictive definition of what one considers as a good project – in the very central theoretical model adopted by project managers.

3 In the plenary chapter he delivered at the Wonderground DRS International Conference in Lisbon (November 2006), entitled "Worldviews for Design Theory", Per Galle urged design researchers to be more critically aware of the basic metaphysic assumptions on which, according to Peirce, design theory rests. In his abstract, Galle thus proposes "a method by which the philosophy of design may develop sound metaphysical foundations (worldviews) for design theory".

contributes to the fundamental aspect of design knowledge. However, the conclusions of the paper must be considered only as a first step in a research program that is still in progress. As I have claimed elsewhere,[4] project-grounded research in design requires the field of a design project as its terrain where speculatively devised hypothetical models must be put to work in order to be declared relevant and fruitful.[5] The opportunity to proceed to such "situated-projectness" (pardon the neologism) is still pending in the course of my professional and research life. This is but one drawback of the method and its underlying phenomenological stance: it needs a long time in order to unfold properly, i.e., to be epistemologically and methodologically sustainable.

Systemic unsustainability

What I call "tripolar model" is also often called "3P model" (Profit/Planet/People), "3E model" (Economics/Environment/Equity), or The Triple Bottom Line. It is usually pictured in the shape of a Venn diagram or of the familiar triangle that we have all come across in the literature, academic and otherwise (Figure 14.1). Its origin is well established: the basic concept emerged from the Brundtland Report (1987) and later crystallized at the Rio Conference (1992), together with the very concept of sustainable development and the 27 principles of Agenda 21.[6]

According to this model, the three aspects of economics, environment (or ecology), and society are supposed to be addressed in any project that claims to be sustainable. More precisely, these poles are to be considered as goals or forces driving the model, each according to its own logic: the force of economics (to produce more for a larger number and a good profit); the force of ecology (to respect the environment, to preserve nature and therefore the living conditions of future generations); the force of social equity (to fight poverty and distribute justly).

4 Findeli (2004).

5 One must be very careful with the choice of terminology in such a pragmatic framework. In project-grounded research, a "hypothesis" becomes a "problem" or, better, a "problematics", since the main purpose of such research is to improve design practice, not to increase the truthfulness of descriptive theories. Therefore, theoretical and practical endeavors must be closely interrelated in a highly dynamic dialectical process. This means (1) that the theoretical "models" mentioned in the text are not tested as in conventional experimental science, but "put at work" or situated in context; models are heuristic devices (see Figure 14.10); (2) that the validation criteria of these models are not truth and generalizability, but relevance and fruitfulness (for practice); and, finally, (3) that each new practical situation is an opportunity to devise (or improve) these theoretical models, so that the latter are not the result of a deductive but of an abductive process. The entire methodological procedure of project-grounded research is recursive, with a Möbius ribbon or an Escher-like morphology/topology.

6 Brunel (2004).

Since the mutual compatibility of these three forces is far from evident or natural, it means some intervention is necessary in order to hold the triangle together.

Their "natural" tendency is indeed to mutually conflict (Figure 14.2). A closer analysis of the genealogy of the model has shown that it is the result of a "cooperative negotiation" and that it may be understood as the outcome of a win-win worldwide game mediated by the UN.[7] The resulting consensus of the game was supposed to be a new emerging value in which all stakeholders could then recognize themselves and satisfy what they value most. What was at first a potential battlefield of a priori antagonist and conflicting values and rationalities could now become the springboard of future sustainable projects carried out under the banner of that consensual value. Unfortunately, just the opposite happened. Instead of the expected cooperative process, a "competitive appropriation" of the consensual value – namely, of the concept of sustainable development and its interpretation- developed, leading to the observed difficulties, dissatisfactions, and failures.[8] Why?

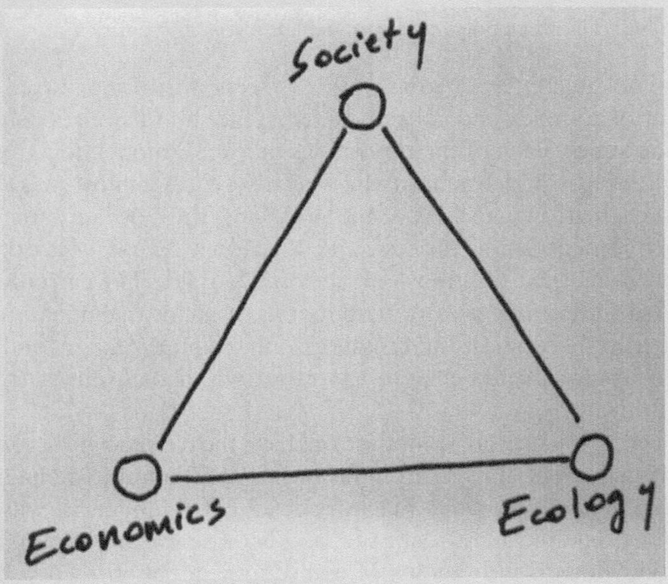

Figure 14.1 The "holy trinity" of sustainable projects.

7 This is a very short abstract of a long argument developed by Aurélien Boutaud (2006) in his doctoral dissertation.

8 Ibid.

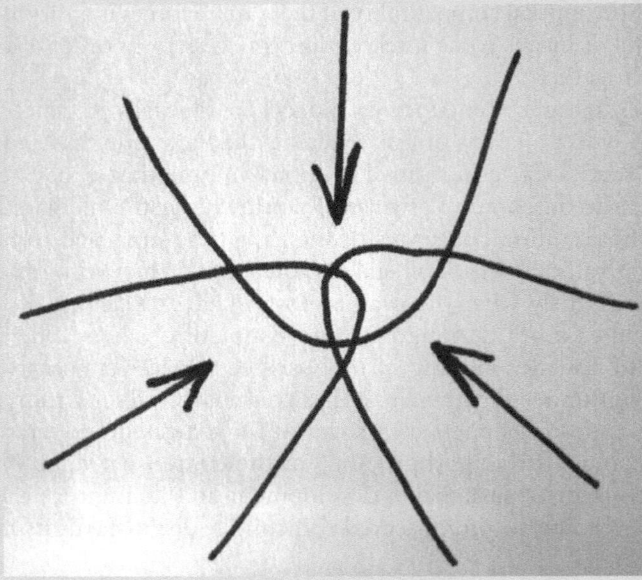

Figure 14.2 The three poles of the model are forces, not entities.

One possible interpretation of this aporetic situation is based on the concept of complex and collective intelligence and its relationship with systemic logics. Each of the three poles of the 3E model holds a specific worldview which determines the way its representative stakeholders construct their own analysis of the problems, their own argumentation during the negotiation process, and their own course of action when implementing their decisions into a specific project. The current state of our Western scientific culture, with its almost exclusive emphasis on analytical, chain-like logical and causal reasoning which is taught and valued by its educational and pedagogical institutions, is the primary source of this worldview.

As a consequence, an economist will have the tendency to consider environmental or social issues from his or her economical underlining epistemological and conceptual frameworks; an environmentalist will do the same for economic and social issues and likewise for the third pole. This situation, which is well known of course in negotiation theory, explains what we hear or read almost daily in the media and justifies the conclusions that can be and have actually been drawn from various experiences and experiments in the ten-year period between Rio (1992) and Johannesburg (2002). In the last decade or so, we have all been witnesses to a phenomenon, a tendency, a collective movement, a political party, a paradigm, a highly mediatized demonstration or even a mere fashion, which has tried to pull the triangle either in the all-economic (econocentric), all-ecological (ecocentric) or all-social (sociocentric) direction. The problem is due to the difficulty of a systemic dynamic whole, considering the triangle as a system rather than a mere assemblage (Figure 14.3) – that is the logic of which can only be understood by complex intelligence.[9]

I hope not to be misunderstood. My point is certainly not that all those projects that have been carried out under the banner of sustainability (or Agenda 21) with the methodological help of the tripolar model have failed. There are indeed many achievements, in all kinds of contexts

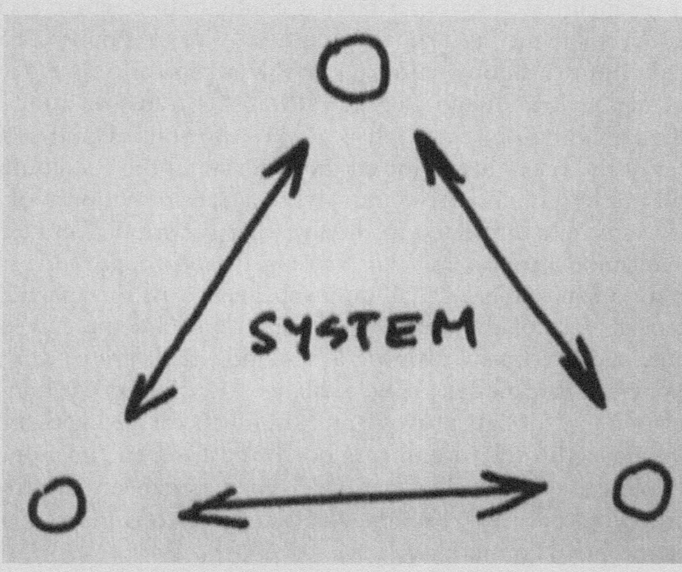

Figure 14.3 The model as a systemic dynamic whole.

9 The characteristics of complex intelligence cannot be developed in the framework of this chapter. It is an insult to the design research community to believe that these characteristics are unknown in the field of design. As a matter of fact, design thinking is a kind of complex intelligence in action. Which designer is not trying to come up with a beautiful, good, and inexpensive product? The epistemological status of a design outcome is indeed that of such an organic whole, since it is a "synthesis" of usually conflicting and mutually exclusive "variables". Throughout the design process moreover, the systemic structure of the object remains dynamic, until it is 'frozen' into a product. The necessity to develop complex intelligence in primary and secondary schools has been strongly advocated in France by sociologist Edgar Morin. He is also one of the main inspirations of the European research network built by Jean-Louis LeMoigne during the last 30 years (the other one being Herbert Simon) named Réseau européen pour la modélisation de la complexité (www.mcxapc.org).

and at various scales, which can be considered as success stories from a short-term perspective. In the long term however, there are numerous reports of problems eventually emerging. It is this observation that led me to question not only the logical structure but also the very substance of the tripolar model.[10]

Anthropological unsustainability

The historical and genealogical analysis of our tripolar model shows that it originally stems from a bipolar configuration where socioeconomic development (anthropic or cultural pole) is opposed to environmental preservation (natural pole). This bipolarity is indeed still pivotal in the Brundtland Report (1987). Its transformation into a tripolar configuration is strongly bound to the political and economic context of the years that followed the latter, until the Rio conference. We will not inquire into how and why this transformation took place, since we are only interested in the fact that the tripolar model is the dominant model today, and that there is a good chance of it remaining so for a while. What I shall do instead is discuss the analytical and theoretical bases of the model in order to unveil some core contradictions, and eventually propose a new, hopefully more fruitful and more operational, model. As we shall see, the theoretical framework or viewpoint of design proves quite useful to carry out this discussion, along with other arguments borrowed from the economic and socio-political spheres. The reasons I characterize my argument as "anthropological"[11] will become clear later; there is more to it than mere conceptual commodity.

The first problem with the model is the separation of the economic from the environmental or ecological domain. As their etymology indicates, economics and ecology belong to the same field and share the same logics: they both deal, very roughly speaking, with available resources, production,

10 From a different perspective, the Research Chair for Ethics at the University of Montreal (CREUM) came to the conclusion that the very concept of sustainability had become empty for having been subject to too large an extension. A fundamental philosophical and conceptual work on the meaning of the concept was therefore felt to be necessary. See the proceedings of the 2006 conference, "The Future of Sustainability?", available online on the Chair's website: www.creu m.ca.

11 "Anthropological" is meant here in its philosophical sense. In this respect, we are dealing with the different points of view under which the human being is considered, described, and represented ("theorized"). This appears somewhat in opposition with the conclusion of one of the participants to the conference mentioned in footnote 10, Catherine Carrère, who writes that the contradictions to which the diverging interpretations of the concept of sustainability lead us could only be overcome by changing our core question: "The issue . . . is not in the choice between humankind and nature, but in a reflection upon our concept of nature" (p. 17). As we shall see in this chapter, my conclusion is rather an invitation to reflect upon our concept of the human being.

transformation, exchange, consumption, value, and make all kinds of computation of one or the other and of their respective relationships. Roughly speaking, again, a science of economics that does not take into consideration the whole system involving these processes does not even deserve its name. Why, then, leave to a distinct science the task of figuring out the ecological impacts of the global handling of natural resources? If economics can indeed be defined as the science that deals with the production, distribution and consumption of goods, with the material, social, and symbolic aspects of exchange processes, and with the potential equilibrium between needs and their satisfaction, then the logical and structural separation of economics and ecology in a theoretical model meant to guide human agency is an epistemological and practical fallacy. In many cultures, especially those based on agriculture, such an idea of economics is plainly unthinkable as it proves totally unpractical according to the very guiding principles of sound agricultural practice. Logically speaking, there is no reason to consider that, since the archetype of economic exchange is a give-and-take or cost-benefit relationship, the "give" and the "take" aspects of the relationship should belong to different scientific domains. Obviously, my common-sense type of reasoning should be more qualified and discriminate. It actually could, and I am sure economists would have no trouble developing a conceptually more sophisticated and rigorous argument with regard to their discipline (they actually have, by founding the sub-discipline of ecological economics). What I am arguing here is not that there are some environmental aspects that should be taken into consideration by the science of economics. No, the statement is much stronger: since there is no radical difference between economic and ecological phenomena, economics and ecology should be merged into a single pole (Figure 14.4), namely, that of a more general and comprehensive economics (or Economics with capital E).

We are now left with a bipolar model again, except that it is not analogous to the previous one mentioned earlier. Both poles here deal with human agency. Their respective purpose is different however, if not divergent. (By the way, notice that this bipolar model is very similar to the central dialectics of Marxian philosophy in the 19th century. We will come back to this later.) Let us now consider the remaining pole, namely, social sustainability.

Thousands of pages have already been written on those human or social aspects and on the way they should be understood and dealt with in sustainable projects. I have no pretension – nor do I really have the competence – to add anything dramatically significant to this corpus in such a brief paper. I will try to show however that the point of view of design, seen as a discipline, the theoretical/empirical concerns and the subject matter of which are centered on the project, may constitute a contribution to what human and social sciences have already gathered on this rather vast topic. My field observations of

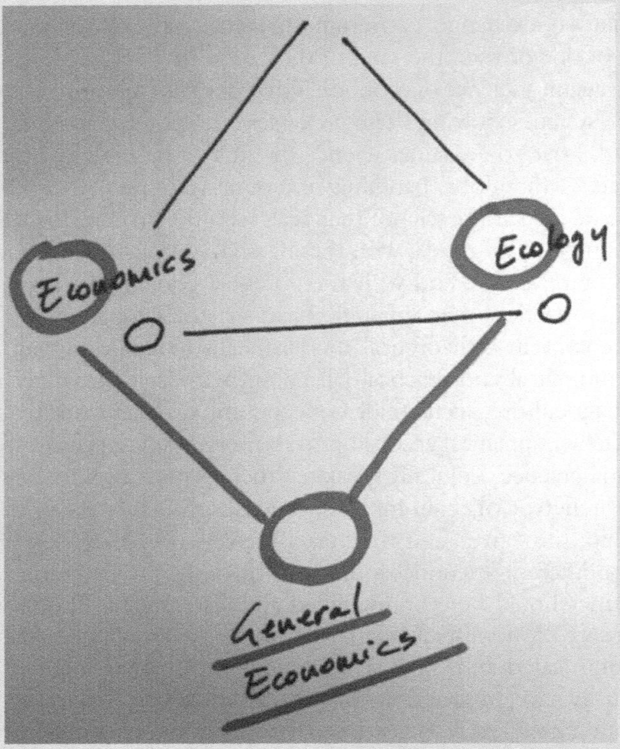

Figure 14.4 The merging of economics and ecology into single pole.

actual projects reveal that the third pole of the 3E or 3P model, corresponding to the social (or human?) aspects and consequences of sustainable projects, is often used as a "black bin" into which all aspects not directly related to the two other poles can be thrown, and this for lack of conceptual clarity.

The bin collects all "soft" aspects or variables of the analytic process, those which cannot really be quantified or rendered objective without losing their very meaning. As a consequence, the third pole is that which is subject to the widest and loosest interpretations. I have gathered a good deal of evidence of this fact from a number of ecodesigners and sustainable project managers. The situation is further beclouded by the fact that there is some confusion between logical and ethical arguments, between a hard deductive/objective reasoning based on figures and statistics (both in economics and ecology) and a more "fuzzy" or "philosophical" type of reasoning. In short, the third pole lacks adequate definition.

I believe that this indeed critical dimension of sustainable development should be treated in a more discriminate manner. To insist almost exclusively on the social dimension of human issues, as is the case in the majority of projects I have observed, is missing one half of the problem. A more comprehensive stance consists in realizing that any human phenomenon has two principal dimensions: the individual and the collective dimensions or qualities. Psychology and sociology in one part, anthropology and ethnology in another, are academic disciplines that have split and shared the corresponding knowledge in the two respective dimensions. If we then turn back to our model, we can derive the consequences of this almost banal but decisive distinction. Social equity is indeed an important aspect that deserves to be given priority in sustainable development. Thousands of relevant and convincing arguments have been presented on this topic. One should indeed expect that the wealth and riches created by the new Economic pole are fairly distributed, not only among the various social categories concerned by the project but also with respect to future generations. But then, the following questions remain: Who gives the impulse to produce wealth? Who is initiating the projects? Who is deciding how the riches are to be distributed? In other words: Where is the "projectual dimension" of the social pole? This is indeed a typical design question. If this dimension is not explicitly handled in the discussion, then we are left with a sort of Darwinian process, with an "invisible hand" phenomenon leading one to derive a kind of determinism from the model. As a matter of fact, the more scientific and sophisticated our theoretical models of sustainable development, the stronger the belief in their deterministic character. It could be contested whether the projectual dimension is naturally included in the economic pole. Such argument results from an epistemological confusion, of which design researchers are well aware as they have read Donald Schön: design is in no way the application of some pre-existing science, be it economics, sociology, semiotics or anything else; design has its own original epistemology. Contrary to most scientific disciplines, the task of which is to describe, analyze, and understand the world, design views the world not as an object but as a project. Being basically pragmatic (in the philosophical, Deweyan, sense), the discipline of design adopts a logics of goals, creativity and initiative. Basic courses in philosophy tell us to keep a distinction between the domain of facts and the domain of values. But every designer knows from experience that it is impossible to maintain these domains distinct in the course of a project, and furthermore that shifting from one domain to the other is not a Darwinian-like process of emergence: it takes human engagement to transform facts into action, that is into values.

For all these reasons and for the sake of logical clarity I believe the original "social" pole of the model should be split into the two separate poles of "social equity" and "social creativity" (Figure 14.5), leaving us with a new configuration: a systemic model consisting of three

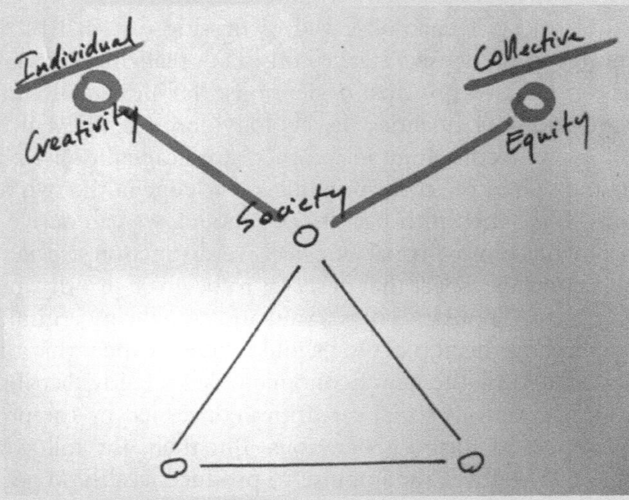

Figure 14.5 The splitting of the social into two poles.

dynamically interrelated dimensions corresponding to general economics (or Economics), social equity and human creativity (Figure 14.6).

A new "holy trinity"?

What is the cognitive gain of this transformation? What is the point of redesigning a tripolar model into another, inverted, tripolar model? Have we but "changed four quarters for a buck" as the saying goes? My answer is clearly negative. The first reason is that by including the design dimension in the model we have introduced a strong dynamic impulse that leaves us no choice but to address some very fundamental issues raised by human agency, especially when the latter takes the form of a project. The second reason has to do with the anthropological framework mentioned earlier.

The new proposed structure of our model happens to be analogous to the so-called "threefolding" of the social order proposed by philosopher Rudolf Steiner (1866–1925) at the turn of the 20th century (Figure 14.7). Steiner's first confrontation with social issues is mentioned in a series of short 1905 essays; it was revived during World War I, by first trying to change, unsuccessfully, German rulers' and military forces' course of action and then while lecturing at the Workers' University (Arbeiter-Bildungsschule) founded by Karl Liebknecht, in Berlin. Although his first article on this topic dates as early as 1905, his major publishing and

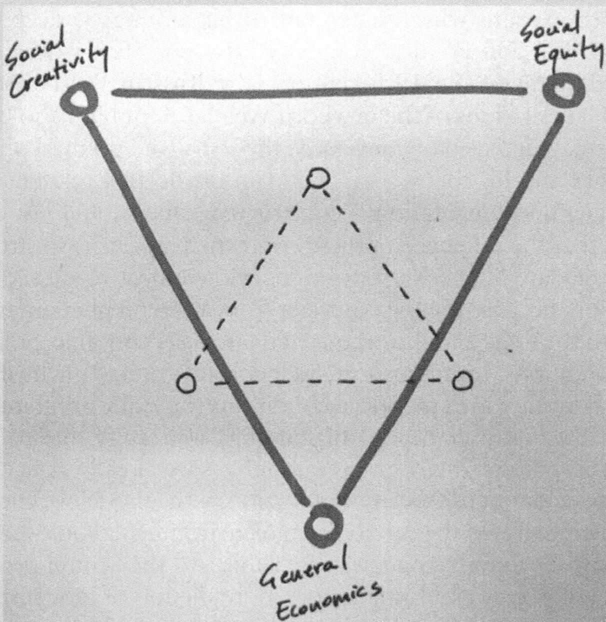

Figure 14.6 The new, transformed model g of the social into two poles.

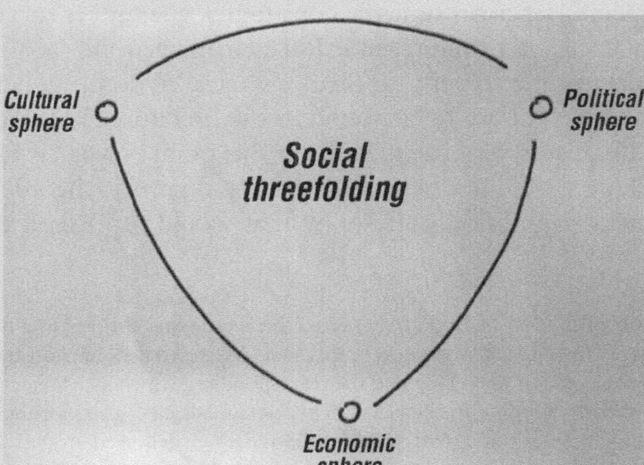

Figure 14.7 Rudolf Steiner's "social threefolding" model.

lecturing activity only really started after the war.[12] Steiner's very pragmatic stance explains why the content of his writings is deeply related to the historical and political context of the war. The workers' movements of the end of the 19th century (the Russian Bolshevist revolution and the hesitations of the newborn Weimar Republic) also belong to this contextual dimension. Conversely, the social and political philosophy of Karl Marx and his followers is the main intellectual reference against which Steiner's argumentation is constructed. One should not conclude therefrom that the influence of these contextual aspects is so strong as to limit or invalidate Steiner's propositions with respect to our contemporary situation, because the former result from a careful phenomenological inquiry into the logic and the origins of the social, economic, political and spiritual conditions. In this respect, Steiner shows quite convincingly how Marx's own inquiry into the social conditions was quite accurate but that the materialist interpretation of his findings was faulty due to an inadequate anthropology. Interestingly enough, as we already noticed earlier, the bipolar – economics/society – reduction of the model we end up with after having integrated the economic and environmental pole into a more comprehensive economic one, is quite similar to the central problematic of Marx's philosophy and his famous "contradiction of capitalism". Marx asks how it is possible to reconcile both poles, that is, to transform the dialectics into a superior synthesis in order to imagine an economic system respectful of the social aspirations of the workers. In terms of sustainable development, the question reads as follows: is it possible to reconcile the economic and the social pole, that is, to transform the duality into a dialectics whereby we could imagine an economic/technological system that is respectful of environmental constraints, guarantees social and generational justice and promotes individual emancipation?

It is not possible to sum up Steiner's social philosophy here. Several good syntheses of it are to be found in the literature.[13] Moreover – and Steiner himself was very categorical on this point – what is called "philosophy" here should not be considered as a dogmatic theoretical system of economics or political philosophy that would replace all the others.

12 The written contribution of Rudolf Steiner to the social issue consists in a book and a series of 13 articles. The original works are in German. My references are mostly in French, but English-speaking readers may easily find the corresponding works since Steiner's books and lectures are classified according to the GA system (Gesamtausgabe) used for the complete works (Steiner 1985). Besides these written basic works, Steiner delivered around 200 lectures and courses on social threefolding to which should be added his lectures on education, agriculture, and therapy, in which practical, economic, and organizational issues were raised and discussed. For more information, see the website of the Goetheanum (www.goetheanum.org), seat of the Freie Hochschule für Geisteswissenschaft, the Free University created by Steiner in 1923, which has a department named "Sektion für Sozialwissenschaft" (Section for Social Science).

13 Among many others, Nouyrit (2004) and Steiner (1996). See also the Talking Economics Bulletin available at www.talkingeconomics.com.

The threefold character of social life is an empirical account based on a phenomenological observation that we all could and should carry out by ourselves. What it reveals is that there are three fundamental qualities that characterize (or "color') our relationships with other human beings, depending on the specific field on which the relationship is focused. The three fields, according to Steiner, are the "economic" (Wirtschaft), the "political" (Recht) and the "cultural/spiritual" (Geist). These should be considered as independent, meaning that they each obey a distinct logic. He points to the fact that the three "logics" used to be integrated into a unique system in earlier history (theocracy), the cultural aspect being the determining ruling force (religion). The two other forces progressively departed on an independent path in successive times in history: the political in the Roman Empire, the economic during the Industrial Revolution.

In the economic sphere, says Steiner, the relationships between individuals are based on needs and mediated by goods, so that there is a strong mutual interdependence; its guiding logics is solidarity or *fraternity*, not in the moral but in the strict economic sense. The political sphere, that of public rights, is the world of essentially human relationships, in the sense that we consider the other not as a provider of some good or service, but as an equal; its guiding logics is therefore *equality*. In the cultural sphere, the relationships are based on the mutual recognition of one's own and idiosyncratic capacities, competences and potentialities, which can only be cultivated and developed in a non-oppressive intellectual, artistic or religious environment; its ruling logics is *liberty* (Figure 14.8). The three dimensions of fraternity, equality, and liberty are potentially present in every project. According to the threefolding principle, their respective singularity must be strictly respected when deliberating about economic, political, and cultural issues which arise in the course of the project so that, for instance, no political consideration should ever interfere with matters of cultural or economic import. Likewise, the cultural sphere should be carefully preserved from any economic or political interference, and so on. One must admit that this sounds almost nonsensical when checked against our contemporary habits, but isn't that exactly how Thomas Kuhn described what a new paradigm was to feel like at its beginning?

The analogy of the threefolding principles with the slogan of the French Revolution is not only a useful mnemonic artifice but is also historically grounded. The analogy helps to understand the pathologies of the contemporary social and economic philosophies, where the three distinct guiding logics are put at the wrong place. By putting, for instance, the economic life under the rule of liberty, one experiences the drifting of extreme liberalism and its perverted or wicked consequences. Likewise, by insisting on equality being the basic value of cultural life, for instance in education, one runs into the paralyzing effects of egalitarianism, that is, the levelling by the lower mean.

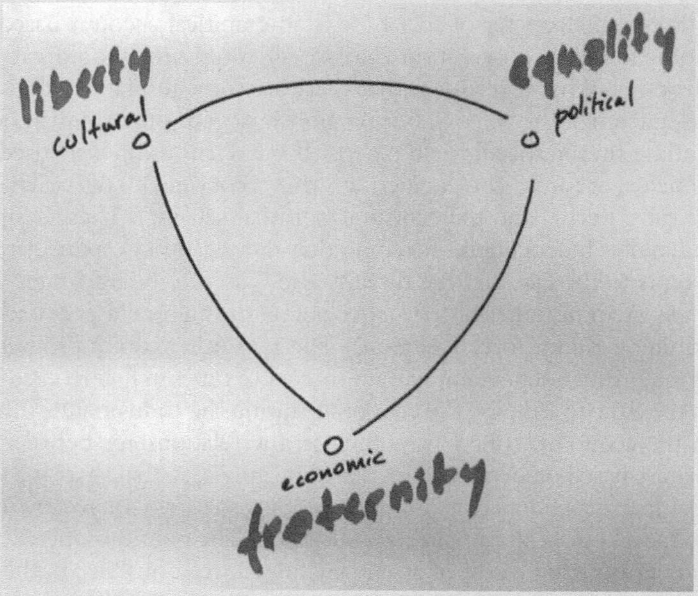

Figure 14.8 The threefolding model and its corresponding guiding "logics".

The very phenomenological and non-dogmatic nature of the threefolding model should discourage any discussion of it in a mere speculative, rhetorical, manner. As indicated above, it should be considered as a working hypothesis in the purest Jamesian pragmatic sense and as such taken out into the field so that its fruitfulness may be experienced (or not). There exist already a number of initiatives around the world where Steiner's threefolding model has been put into practice, in fields as diverse as agriculture, banking, business and management, educational and therapeutic institutions.[14] The next step of my own research will precisely consist in this. Its current state has not yet allowed me to gather a comprehensive review of experiences (if any) carried out in sustainable design projects with the threefolding perspective. An outstanding example would be the network of biodynamic farms in operation worldwide. They should be considered indeed, for their management and landscape design principles, as potentially interesting case studies of contributions to sustainable development.

In the specific field of design, it would be interesting to analyze the extremely original and convincing experiences carried out under the

14 The most ambitious contemporary example is Nicanor Perlas' Agenda 21 Program for the Philippines (www.cadi.ph/nicanor_perlas.htm). Other noteworthy cases are the Sekem initiative, founded 1977 in Egypt (www.sekem.org), whose initiator, Ibrahim Abouleish, was granted the so-called Alternative Nobel Prize together with Perlas in 2003 and the banking enterprises of GLS (G), Triodos (NL), Mercury (UK), Merkur (DK), Nef (F).

advisorship of Ezio Manzini in Milan[15] in the light of the threefolding conceptual framework. Although Manzini's central theoretical and pragmatic framework draws mainly on Nobel Prize winner Amartya Sen's economic theory, I can see no contradiction at first sight with Steiner's guidelines.[16]

Some remarks on metaphysical issues

The essay has already pointed to the relevance of a reflection on the metaphysical aspects of design.[17] The reason these aspects should be considered important is that they determine our way of dealing with design projects, both in theory and practice. They provide us with – or imprison us in – our intellectual habitus and influence our interpretive activity. It is therefore a useful exercise to bring them to light so that their deterministic effect can be recognised, evaluated, and eventually modified.

The one aspect I wish to draw attention to in the context of this paper is the dualistic, polarized scheme used in practical deliberation throughout Western intellectual culture. The archetypal figure of this polarizing pattern is the iconographic interpretation of the Last Judgement, especially visible in Catholic architecture, sculpture, painting and glasswork. The central figure of Christ is dividing the space into the two opposed realms of the good and the evil, of the saved and the damned. But one can find this moral topology also in non-Christian iconography, for instance in Greek-Roman mythology where some hero is bound to choose between vice and virtue, between hell and heaven, between beauty and ugliness, and so forth. What seems to have encouraged the emergence of such a strong polarity is the "ontologizing" of this central scheme into a static form. The origins of the scheme stem way back into the earliest cosmogonies of pre-Socratic philosophy but it was undoubtedly strongly reinforced by Platonic dialectics and Aristotelian ethics. What has been lost in the course of their respective interpretations is their dynamic character, leading to the

15 Manzini's basic principles for "strategic design" are synthesized in his "Design, Ethics and Sustainability. Guidelines for a Transition Phase" in Sotamaa et al. (2006). The same proceedings at www.cumulusassociation.org also include three case studies conducted at the Ph.D. program of Milan Polytechnics by S. Maffei, B. Villari, A. Meroni, and C. Cipolla.

16 Nouyrit (2004:84–85) criticizes Sen's 'methodic reflection upon the very complex and infinitely diverse play of the equality–liberty couple for not having thrown strong enough a bridge between both poles; the mediating role should, according to Nouyrit, be given to fraternity ("affectio societatis"), in order for his theory to become more practical and fruitful. Nouyrit's professional experience has been in the cooperative movement in agriculture and in the construction of the European Community. He was one of the founders of the Nef bank (Nouvelle économie fraternelle) in France in 1978 and was recently, until his death, mayor of a small village in the southwest of France.

17 See footnote 3 and Per Galle's contribution in this respect. In an earlier essay on the Bauhaus, I have already hinted at some metaphysical aspects of design theory and ideology (Finceli (1999/2000).

"ontologizing" phenomenon mentioned earlier. The two opposing entities have ended up being considered as the final outcome of a process, whereas they were originally to be thought of as a process in its own right, namely, as the very movement of human thinking and acting. When considered from such a dynamic viewpoint, there always appears a third member in the dyad. Thus, for instance, the iconology of the Last Judgement is truly incomprehensible without the central figure of Christ. Particularly sophisticated and interesting in this respect is Aristotle's definition of virtue. In his ethics, virtue is not the opposite of vice but virtue is a mean between two extremes, between two vices, one in excess and one in default.[18] A mean, however, which is not arithmetical or "Darwinian", but a mean that is the result of human creativity and agency and is to be renewed, re-invented, re-designed in each new situation. The topology of Aristotle's principle is therefore triadic and dynamic. A careful meditation of these observations may lead us to the following conclusion: whenever, by analyzing a phenomenon, we discover it is structured along a polar contradiction (as in Marxian philosophy for instance), we should be looking for the third member of the configuration in order to transform the dyadic into a triadic gestalt.[19] Very evocative in this respect is Rudolf Steiner's own artistic interpretation of this fundamental principle. The monumental wood sculpture he personally conceived and constructed at the end of his life with sculptor Edith Maryon, known as the "Representative of Humanity", shows the central figure of the human being who is torn apart between the two poles of spiritual and material temptations, Lucifer and Ahriman. Not between good and evil, but between two figures of evil, where intellectual arrogance on one hand and the intemperance of materialistic hedonism on the other may be recognized as typical incarnations of the Western mind.

The archetypical character of Steiner's threefolding model derives from such anthropological anchoring. This is precisely the main reason why its pragmatic and operational fruitfulness deserves to be experienced in design, more specifically in sustainable design. Such is both the conclusion of this essay and the starting point of the next phase of the corresponding research project.

Conclusion

The research originates in the empirical evidence that the use of the 3E or 3P model as the main reference for sustainable design projects

18 Aristotle, Nichomachean Ethics, for instance II, 104a 10–25 or 1106b 22–28.

19 French philosopher François Julliens's very thought-provoking comparative study of Chinese/Eastern and Greek/Western philosophical traditions is extremely helpful to uncover our own metaphysical guiding axioms. His conclusions are usually quite useful for design thinking and project theory. See, for example, Jullien (2004:109) where one can read that "China enlightens the tragic character of European thinking".

consistently faces insuperable contradictions and dissatisfactions among the project carriers (designers, clients, users and so forth). A critique of the central model appeared therefore to be necessary. The main inconsistencies of the model can be uncovered when the critique is carried out both from a systemic and an anthropological- philosophical perspective. In effect the former appeared only as a preliminary step toward a more fundamental reconsideration of the logical foundations of the model. It appears possible to surmount the so-called anthropological unsustainability of the model by an inversion of its triadic structure and a renaming of the corresponding poles. The model then becomes analogous to the threefolding model Rudolf Steiner proposed after the First World War in order to construct a more human social order. This model has since proved fruitful for the design and management of various economic and social initiatives. The strong analogy this research has made obvious between the threefolding and the modified 3E or 3P model used in sustainable development leads to the conclusion that it might be fruitful to work with the latter in further sustainable design projects. The three poles of the rectified model are the following: "comprehensive economic" (or ecological economics), "social equity" and "human creativity". The ruling logics of the respective poles are those identified by Steiner, namely, the logics of fraternity (or solidarity), equality and liberty (Figure 14.9).

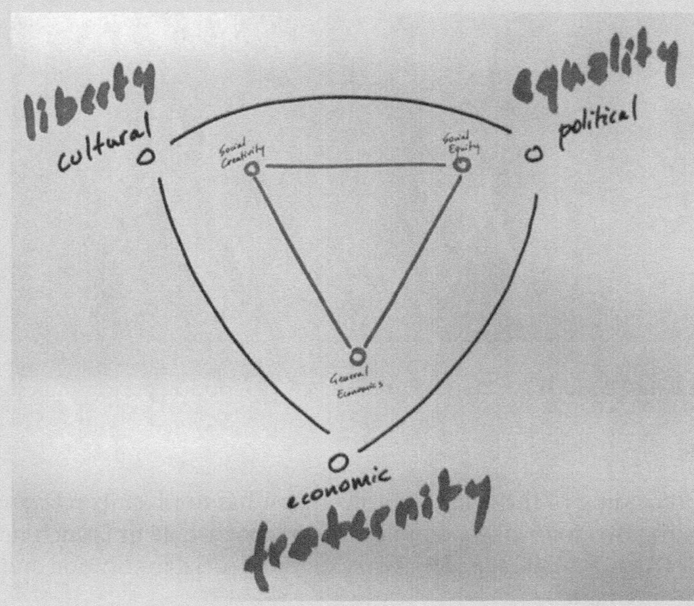

Figure 14.9 The analogy between both models.

As stated at the beginning of the essay, the methodology adopted in this first phase of the research project is mainly speculative but it was preceded by a careful phenomenological observation of projects in action, mainly but not exclusively in the field of design. The general argumentation is not as well supported on empirical and bibliographical evidence as one might wish or expect.

However, the pragmatic viewpoint adopted all over allows research to proceed along concurrent paths, according to the methodological foundations of project-grounded research in design. My conclusion is far from being . . . conclusive. This research in progress is now entering a new phase, in the form of a situated "reflection-in-action" experience, whereby the potential of our new threefolded model will be put to work and to test (Figure 14.10).

Figure 14.10 Models are heuristic, not merely descriptive, devices in design.

Epilogue

My own rereading of the essay confirms that it has not lost its relevance, as attested by the two following examples. In a recent article in French newspaper *Le Monde* (Aug. 5th 2022, p. 29), two experts in social innovation and sustainable finance remark:

> The absence of commonly agreed definition and nomenclature [of social impact] may lead to risks of confusion among clients and consumers, as well as for funders and companies themselves. . . . The loss

> of consensus on what societal impact actually is and how it can be measured may disorient public action and economic actors, disperse the commitment of companies and associations, discredit those involved in an ambitious impact study, while others satisfy themselves with mere corporate and government communication policies and the false promises of impact washing.

Later (*Le Monde* Aug. 29th 2022, pp. 20–21), an economist expert in world income and pollution inequities, writes that "there can be conjunctions between climate and social issues". The good news is that the problems raised in the original essay are now not only widely discussed, but also theoretically and sometimes practically adequately addressed. Clive Splash (2011), for instance, endeavors to develop a "social ecological economics" in order to "understand the past to see the future". Wironen and Erickson (2020), lamenting "the lack of a biophysical grounding of economic processes", calls for a "a critically modern ecological economics for the Anthropocene", while Gowdy and Erickson (2005) had already explained why their approach of an ecological economics was "the only heterodox school of economics consistently focusing on the human economy both as a social system and as one constrained by the biophysical world".

However, in order to facilitate a better understanding of the essay for a readership unfamiliar with design research and some critical issues it has raised throughout these years, I assume some additional contextual elements could be of some help.

The paper is presented as a first step in a research program supposed to be in progress. The discipline of design still being a newcomer in the university, design research can be considered to be barely in its third generation since a handful of pioneers made their first steps in the early 1960s. One of the central issues in fundamental research was, and still is, to better understand the epistemological underpinnings of design practice, that is, what sort of reasoning is taking place in the individual or collective minds of designers when involved in a project. As the stakes, the scope, and the complexity of their projects were increasing, design practitioners became more aware of not only the aesthetic, but also the social, environmental, and cultural consequences of their projects. Consequently, in the 1990s, design ethics and later design ontology were added to the agenda of design research groups and labs. The research program mentioned in the paper resulted from two projects I had undertaken previously: one dealt with the responsibility of designers in the framework of an ethics of technology ('Prometheus Enlightened' 1990–93), the other, based on the conviction that the theory of complex systems would provide a sound epistemological foundation for design, led to the creation of a master's program in "Design and Complexity" (2001) at the University of Montreal (Canada). The urgency for a more sustainable design practice and for methods of life cycle assessment emerged at the same time, but focused mainly on the ecological pole of the tripolar model.[20] Indeed, he latter seemed to me and my colleagues to be suitable to frame a more responsible practice of design but, as argued in the essay, its theoretical foundations and practical workability deserved to be questioned.

The opportunity to do so was provided by an invitation of University of Nîmes (France) to introduce design research in a country where design was still narrowly considered as applied art and essentially taught in art schools. In this perspective, the project to create both a specialized master's program and a research lab in what we provisionally called "Social Design" was launched and supported by the institution. The master in "Design-Innovation-Society" opened in 2011 and the research lab "Projekt/Social Innovation by Design" in 2013. There remained to ground such initiatives on sound theoretical foundations, guided by an appropriate axiology. This is precisely what the aforementioned "research program" consisted in. In such perspective, the idea that Rudolf Steiner's worldview could be of some help was a result of my own study of anthroposophy as well as some practical personal experience in biodynamics, Waldorf education, and anthroposophical healthcare.

Steiner's social threefolding is by far not the most difficult or apparently obscure part of his work. It remains however that it may sound somewhat arbitrary if considered at first sight, but becomes clearer and more coherent if one is familiar with his complex anthropology, that is, his model of the human being. The latter displays a triple partition in body, soul, and spirit, each also composed of three constituents, where in particular the cognitive, affective, and conative activities of the soul are related to the three main organic systems (nervous, rythmic, metabolic). What Steiner did is transpose his anthropology to the social field, constantly insisting on the fact that such models (or theories) are not to be considered valid unless experienced in action. It is fruitfulness *(Fruchtbarkeit*, a Goethean concept), and not truth that is the core criterion, the function of a model (or theory) being to serve as guideline for human endeavor. Keeping this central pragmatic maxim as a compass, my colleagues and I devoted the years that followed the essay to put it to work in our design practice and teaching, and by so doing to try to better understand the conceptual model of social threefolding.[21]

One of the main difficulties we struggled with is the problem of scale. Steiner's original attempts were on the level of European geopolitics, a macrolevel that does not correspond to the meso- or microlevels of social design projects like that of a hospital, an educational institution, a national, regional, or local public policy, an urban district or rural community, etc. Although also focusing on a macroscale, the noteworthy cases, mentioned in the essay, of Nicanor Perlas' Phillipine Agenda 21 or Ibrahim Abouleish's Sekem initiative were

20 One of the early books is Tischner (2000).

21 Such is also the basis of the methodology of "project-grounded research" mentioned in the chapter. For an update on this topic, see Findeli, A. (2010).

inspiring examples of rather successful interpretations and implementation of social threefolding. Sekem's guiding conceptual model, for example, is interesting in our perspective since it shows how to manage the economic, societal, and cultural fields of the source model with the requirements of sustainable development (Figure 14.11).

My efforts to design the management principles of Projekt, our own research lab, according to social threefolding, although welcomed with skepticism by some rationalistic colleagues, proved helpful, as well as Friedrich Glasl's conflictology (Glasl 1980)[22] to solve potential and actual conflicts between academic research and granting institutions, senior researchers, junior researchers, doctorate students, and support office and maintenance staff. The introduction

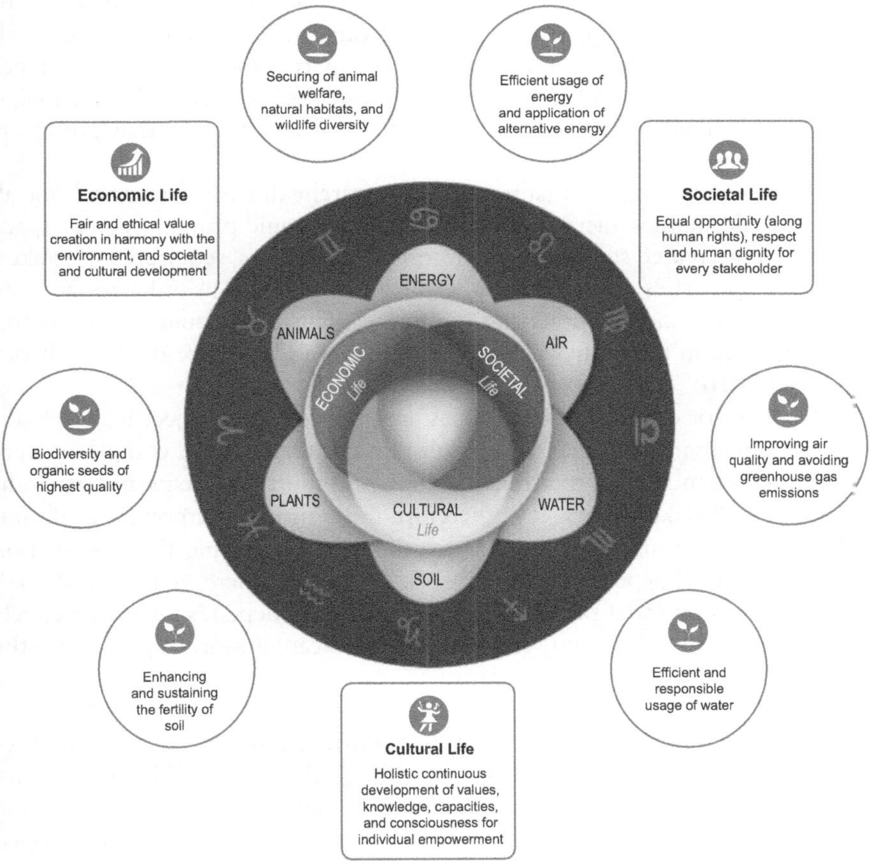

Figure 14.11 SEKEM's "Sustainability Flower", its basic model of sustainable development, and social threefolding (Sekem 2010:2).

22 Likewise, and more recently, Yves Citton (2021), following Baptiste Morizot (2020), proposes a "meta-polemic algorithm" to solve complex, potentially conflictual situations, by acting "with" instead of "against" in ecological systems.

into the master's curriculum of pedagogical activities inspired by Waldorf education and Waldorf teachers' training were sometimes also considered exotic by the students, but eventually stimulating. Finally, the explicit use of the threefolding principles in the projects carried out by the students with our partners is still in progress and would, in the future, need some extra time and collaboration with experts (my contract having come to its end in 2017, this task has unfortunately remained unachieved). Here is how we understood figure 16.6 of the original essay and undertook its implementation in our projects:

- "Social creativity" implies that a space be created where all stakeholders, including the end-users, would actively participate in the early conception phases (sometimes called "fuzzy front-end"). No priority is to be given to the designers, the supposed experts in creativity. But whereas in the much popularized "brainstorming" activities creativity is meant to be instantaneous, spontaneous, and not reflective, in our case we came closer to what Goethe named ***anschauende Urteilskraft***, that is, a process that grows organically and ripens thoughtfully.
- "Social equity" implies that no a priori hierarchy due to professional, social, economic, gender, racial, cultural, etc. capital should prevail. Again, a space should be created so that the intentions, expectations, requirements, stakes, etc. of all parties could be made explicit and justified in order to facilitate the necessary negotiations and consensus. This pole could also be better interpreted in the framework of Hartmut Rosa's "social resonance" theory (Rosa 2016).
- "General (or ecological) economics" implies that the project team, by adequate mapping methods, renders visible and readable the complex system of interdependencies, interrelationships, and mutual benefits and expenditures of all stakeholders. It is therefore not a moral interpretation of "fraternity" that must prevail here, but the actual and objective observation that we are all dependent on the work of others: a mere look around one's life space shows that the objects and services one depends on are effectively made by others. An interpretation in terms of care is more in phase with the second than with the third polarity.

At the same time the original essay was written, I undertook the study of Otto Scharmer's Theory U (Scharmer 2007),[23] the result of his long inquiry inspired, among others, by Steiner's anthropology, Glasl's conflictology, and Goethean phenomenology. I realized that his model, originally developed at MIT for a better and more responsible leadership in management, could

23 Scharmer later published a shortened version of his original book (Scharmer 2018). It would be presumptuous and not reasonable to try to summarize Scharmers Theory U in such a short essay. The U-School for Transformation he prototyped at the Presencing Institute offers all the necessary tools. It is, in their terms, a "school for transformation that democratizes access to the tools and methods for activating awareness-based collective action".

actually be considered as a – quite challenging – theory of the design project, a transfer which called to be tested in practice. Having had the privilege to attend a hands-on workshop with Scharmer on the occasion of the publication of the French translation of his book, and having registered to the pedagogically innovative massive open online course on Theory U he developed with his team at MIT, my first intuition switched into a strong conviction. The three steps of the left, descending branch of his U-shaped methodology ("Open Mind' or "Seeing", "Open Heart" or "Sensing", "Open Will" or "Presencing") reflect quite accurately what Steiner named the capacities of Imagination, Inspiration, and Intuition (with capital I in order to differentiate them from their usual meaning).[24] These capacities of a "higher knowledge" may well be what the complexity of our contemporary and future world and its various crises are indisputably calling forth: a new type of intelligence (*metanoïa*). The respective counter-forces of these capacities, named by Scharmer "Voice of judgment", "Voice of cynicism", and "Voice of fear", also find their equivalent in Steiner's lessons. The three steps of the right, ascending branch of the U ("Crystallizing", "Prototyping", "Performing") are more familiar to designers since they correspond to the development of the insights, or "primary generators" gained during the conception phases of a project (Figure 14.12).

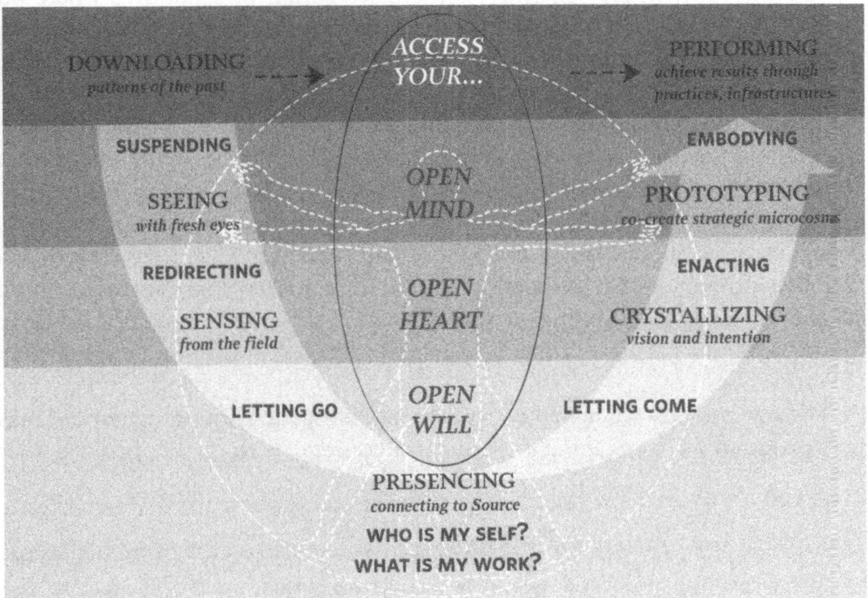

Figure 14.12 Otto Scharmer's Theory U model (2011:5).

24 See Steiner (2009) as well as Holland (2008) for artist Josef Beuys' interpretation of the triad.

My conviction that Scharmer's model was definitely the most insightful and relevant theory and methodology of the design project for the coming decades was further strengthened by his following book, written with Katrin Käufer (Scharmer and Käufer 2013), a book "that has the potential to transform civilization from one based on a rapacious, ego-driven economics to a viable, ecological, awareness-based model", writes Arthur Zajonc, former president of the Life and Mind Institute. Not only do what the authors call "the three divides", that is, the symptoms that characterize our contemporary world (the "ecological", "social", and "spiritual–cultural" divides) accurately reflect the three poles of the transformed tripolar model of Figure 14.6, but it is their concept of "blind spot" which, in my view, crystallizes the most radical and urgently required transformation needed today and tomorrow, that is, the necessity and capacity, in their words, "to shift the inner place from which we operate", or in anthropologist Philippe Descola's terms, to get rid of our "lazy certainties" and of "the descriptive categories [we] carry in our backpack" (Descola 2021:12, 9).[25] As landscape architects and educators know well, the beauty and richness of a landscape in our outer world can only be recognized and appreciated by someone who has developed a rich inner world. It is precisely this outer world/inner world polarity which is at the core of the U process and path, and which, in my view, should deserve the most insightful and critical examination in the future (Findeli 2018). "The blind spot, writes Scharmer, is the place within or around us where our attention and intention originate. It's the place from where we operate when we do something. The reason it's blind, is that it is an invisible dimension of our social field, of our everyday experience in social interactions" (Scharmer 2007), therefrom concluding, quoting his colleague Bill O'Brien, that "the success of an intervention [for instance a design project, A.F.] depends on the interior condition of the intervener" (Scharmer 2007:6–7). The U-approach is precisely a very effective methodology to uncover the individual and consequently the collective blind spots, their potentialities and weaknesses, and concurrently to help the complex system of the project to become so to speak conscious of itself, so that the adequate healing or transformation devices, policies, or processes can be imagined ("Crystallizing"), validated ("Prototyping"), and implemented ("Performing").[26]

The few projects we managed to carry out with the U-approach in our master's program were engaging enough to encourage us to further investigate its

25 In his seminal book, *Par-delà nature et culture* (2005), Descola showed that our Western "ontology", that is, the specific view we adopt since the Enlightenment to interpret the world he names "naturalism", is only one of four possible and existing ontologies besides "animism", "totemism", and "analogism".

26 The *Journal of Awareness-based Systems Change*, launched in 2021 by Scharmer and colleagues, provides "a platform to make visible and accessible the growing knowledge base that supports societal transformation across [the three divides] through awareness-based systems change" and "draws upon theoretical and practical foundations that connect awareness-based approaches to individual and collective transformation to effect systems change.

relevance and fruitfulness. Consequently, my conclusion is almost too simple and straightforward: the tripolar model tentatively – if not laboriously – developed in the original essay has actually – and auspiciously – found its incarnation in Otto Scharmer's "theory U" and "eco-system economy". To put such conceptual and praxic frames to work in various situations remains the most pressing, if arduous, task.

References

Boutaud, A., 2006. *Le développement durable: penser le changement ou changer le pansement?* Thesis (PhD). Saint-Etienne: École des Mines et Université Jean-Monnet.

Brunel, S., 2004. *Le développement durable.* Paris: PUF.

Citton, Y., 2021. *Faire avec. Conflits, coalitions, contagions.* Paris: Les Liens qui libèrent.

Descola P., 2021. *Les Formes du visible*, Paris: Seuil.

Descola, P., 2005. *Par-delà nature et culture.* Paris: Gallimard.

Findeli, A., 1999/2000. The Bauhaus project: An archetype for design education in the 21st century. *The Structurist*, 39/40, 36–43.

Findeli, A., 2004. Die projektgeleitete Forschung: eine Methode der Designforschung. In: Michel, R. (ed.). *Erstes Design Forschungssymposium.* Zurich: Swiss Design Network, 40–50.

Findeli, A., 2010. Searching for Design Research Questions. A Conceptual Clarification. In: Rosan Chow, Wolfgang Jonas and Gesche Joost (eds), *Questions, Hypotheses and Conjectures.* Berlin: iUniverse, 286–303.

Findeli, A., 2018. The metamorphosis of the designer. A prerequisite to social transformation by design. In: Förster, M., Hebert, S., Hofmann, M., and Jonas, W. (eds), *Un/Certain Futures.* Bielefeld: [transcript] Verlag, 104–114.

Glasl, F., 1980 (10th ed. 2011). *Konfliktmanagement. Diagnose und Behandlung von Konflikten in Organisationen (= Organisationsentwicklung in der Praxis, Band 2).* Bern and Stuttgart: Haupt Verlag.

Gowdy, J. and Erickson, J., 2005. The approach of ecological economics. *Cambridge Journal of Economics*, 29, 207–222.

Holland, A., 2008. *Joseph Beuys & Rudolf Steiner. Imagination, Inspiration, Intuition.* Melbourne: National Gallery of Victoria (exhib. Catalog).

Jullien, F., 2004. *Du mal/Du négatif.* Paris: Seuil.

Lalonde, P., 2007. *Structuration de l'espace agissant de la pratique de la gestion de projet.* Thesis (PhD). Montréal: Ecole Polytechnique.

Morizot, B., 2020. *Manières d'être vivant: Enquêtes sur la vie à travers nous.* Paris: Babel.

Nouyrit, H., 2004. *Fraternité, Egalité et Liberté. Actualité de la pensée sociale de Rudolf Steiner.* Laboissière: Triades.

Rosa, H., 2016. *Resonanz: Eine Soziologie der Weltbeziehung.* Berlin: Suhrkamp.

Scharmer, O. 2007. *Theory U: Leading from the Emerging Future.* San Francisco: Berrett-Koehler.

Scharmer, O., 2018. *The Essentials of Theory U: Core Principles and Applications.* San Francisco: Berrett-Koehler Publishers.

Scharmer, O. and Käufer, K. 2013. *Leading from the Emerging Future: From Ego-System to Eco-System Economies.* San Francisco, Berrett-Koehler Publishers.

SEKEM, 2010. *Report on Sustainable Development.* Cairo, Egypt: Cairo-Belbeis, 5.

Sotamaa, Y., Salmi, E., and Anusionwu, L., 2006. *Nantes Cumulus Working Papers.* Helsinki: UIAH University of Art and Design, 9–15.

Splash, C., 2011. Social Ecological Economics: Understanding the Past to See the Future. *American Journal of Economics and Sociology*, 70(2), 340–375.

Steiner, R., 1985. *Fondements de l'organisme social (GA 23–24).* Genève: Ed. Anthrop. Romandes.

Steiner, R., 2009. (German original ed. 1904). *Imagination, Inspiration, and Intuition.* Hudson (NY): Steiner Books. Also published as *The Gates of Knowledge* (2012). London/New York: Putnam & Sons.

Tischner, U., 2000. *How to Do Ecodesign? A Guide for Environmentally and Economically Sound Design.* Frankfurt: Verlag form.

Wironen, M. and Erickson, J., 2020. A critically modern ecological economics for the Anthropocene. *The Anthropocene Review*, 7(1), 62–76.

15 Outlook and key topics for the construction of critical sustainability sciences

Stephan Rist, Beat Dietschy, Patrick Bottazzi, and Johanna Jacobi

Where it all began

This book was created under the auspices of the volcanoes of Ruka Pillán and the more distant Llaima, nearby the lake of Villarrica and the adjacent village of Pucón. Many of the co-authors met in this part of the Araucaria region, in southern Chile, between June 17 and 23, 2019, where they began the reflections that eventually flowed into this volume. According to the Mapuche Nation's language and cosmovision, *Ruka Pillán* is considered the "home of the great spirit" *Pillán*, who dwells in the bowels of the volcano and is closely linked to the "master spirit of nature" *Ngen*.

Ngen is seen as the guardian and owner of the volcano. The other volcano in this "living geography", *Llaima,* is known as the "home of bad omens for sleepers". Through the lake of Villarrica, the spirits of the volcanos are linked with the river Toltén that flows into the Pacific Ocean. The rich landscape around Pucón represents the veins through which the lives of the heavens, the volcanos, and the lowlands are connected with the life of the Pacific Ocean and hence, with the rest of our planet, also known as Gaia. Pucón, representing the waters at the entrance to the Cordillera, is hence the entrance to world in which humans (*mapu*) and all other living beings in, on, and above the physical body of Mother Earth are interconnected and are continually interacting.

Drawing on Escobar (2016), we refer to this dense network of interlinkages as a "relational ontology", in which all things and beings are expressed based on their relations, and are viewed as *not* existing prior to or separately from these relations. Something similar might be said about the present volume. It did not emerge from any preexisting conceptual agenda of a particular institution prescribing what it should or should not contain. It emerged from a free and open dialogue of researchers from different countries in Latin America and Europe.

It was born in critical discussion with a preparatory paper by the book editors and in a Mapuche area – the Wallmapu – where Mother Earth communicating with us in her own language as well. Much like objects of nature flowing through the watercourses from the volcanos to the Pacific Ocean, and from there to the rest of Gaia/Mother Earth, certain book contributions were

DOI: 10.4324/9781003043577-15

stranded along the way while others continued with the current, staying in motion and evolving further. In this way, the catchment area of this dialogue program has finally expanded to other continents and now includes contributions from 18 authors from nine countries. They all shared their lived experiences, in view of questioning and further reflecting on the past, present, and future of sustainable development and its relations to the whole spectrum of sciences, including various forms of more comprehensive sciences, as emerging from relational ontologies, existing all over the world.

Why "sustainability" is not sustainable

Despite the authors' wide diversity of lived experiences and related conceptual expressions, there is an identifiable red thread running throughout the book: It is our concern about people's betrayed hopes for true sustainable development – development that enables the emancipation of humans and all other living beings and parts of the cosmos from the unbearable exploitation perpetrated by a small elite of powerful people and their organizations (Rothkopf and Medish 2008). Our starting point was hence the lack of an *emancipatory sustainable development* in the corporate- and state – government-driven 2030 Agenda and corresponding sustainable development goals (SDGs).

With this deception in mind, we began looking for the possible reasons behind the failures of ongoing mainstream sustainability policies as embodied by the 2030 Agenda and SDGs. While there are clearly many reasons for today's "sustainability malaise", our analysis made clear that one major obstacle involves the current ways of understanding and practicing *sustainability science.*

The first two chapters of this book show that current sustainability science was not able to provide enough critical insights into actual socio-ecological dynamics. Its shortcomings made it incapable of properly identifying the main root causes of unsustainable development trends. As sustainability science was heavily involved in the processes shaping the 2030 Agenda and corresponding SDGs, we conclude that it contributed to establishment of a form of sustainable development that does not mobilize real change (see Chapter 1). Wittingly or not, sustainability science helped to establish concepts and policies of development based on never-ending economic growth and accumulation, rooted in techno-scientific and reductionist materialist visons of societal progress. Today, the rules of the development game are increasingly defined by means of top-down – even authoritarian – policies, making effective democratic and social control of current and future policies by the people more and more difficult.

One key weakness we identified is sustainability science's lack of critical reflection on its own underlying dualist ontology and epistemology (see Chapter 2). Dualist ontologies lead to binary thinking, assuming that sustainability problems – and solutions – must be understood and addressed based on fundamental divisions, for example, between culture and nature, mind and matter, the natural and social sciences/humanities, objects and subjects, etc. The unreflected consequences of this dualist ontology contribute to today's prevailing forms of sustainable development, which ultimately perpetuate highly problematic, unsustainable modes of civilization and human organization.

This Western-driven, dualist mode of reasoning is, as Vandana Shiva (2023) points out in her Preface to this book, ultimately based on colonial expansion, assumed superiority, and division – that is, the construction cognitive of apartheid and oppression. The unreflective use of a dualist ontology in sustainability science means ignoring that many problems of unsustainable development – such as oppressive private- or state-based forms of capitalism as well unilaterally materialist notions of scientific, technological, and institutional development – are direct expressions of dualist notions of development as well. In the words of Vandana Shiva, they are the basis of a "monoculture of mind" that conceives of humans as outside and above nature, with the latter viewed as a limitlessly available and controllable resource. This separation is exemplified by the reasoning of Descartes, who regarded his dog as a soulless thing and only considered himself connected with the divine perfection through his "I think".

Toward critical sustainability sciences

Against this backdrop, we propose to overcome the dualist ontology of sustainability science, by radically rethinking it in view of relational ontologies, emancipation, interculturality, critical theory, post-development, decoloniality, and political ecology. As we show in Chapters 3 to 14, this completely transforms the current social, cognitive, and cultural features of scientific inquiries with regard to present and future policies of societal evolution.

As we show in this book, the ontological, conceptual, and methodological transformation of sustainability science can enable a new field of scientific examination of socio-ecological problems and solutions we refer to as *critical sustainability sciences*. The starting point of *critical sustainability sciences* is engagement with existing and new forms of relational thinking that make it possible to overcome the prevailing dualist ontology in science.

This book shows that relational thinking is found all over the world. It is at the heart of diverse human groups engaged in the search for existential, emancipatory, and cognitive self-determination. As shown in the different Chapters of this book, these groups can be found in indigenous communities and people engaged in the revitalization of health traditions (Chapters 3–5), new philosophies of the world and societal change (Chapters 5–7), as well in groups living relational ontologies dealing with living labor and commons (Chapters 8 and 9), alternative forms of agriculture, food, and nutrition (Chapters 10 and 11); and comprehensive sciences (Chapters 12–14).

Often the problem of dualism is identified, yet we continue to reproduce it unintentionally by speaking of inner–outer world, society–nature, and mind–matter. As Findeli (2023) notes in Chapter 14, mind-body dualism, for example, shapes our language so that even to express how something could be different involves using the language that has developed to embody those cultural assumptions".

Relational ontologies

Relational ontologies offer ways of overcoming dualist thinking and related concepts. It means renouncing substantialist binaries and replacing them with concepts that are not just expressions of abstract representations of lived experience; they are part of an "ecosophy" or body of wisdom "of which we are not the patrons" (Panikkar 2008), which Sepúlveda (2023) describes as a cosmosophy. It is Earth's wisdom, not ours, and one that can be perceived through mindful observation and introspection. This can take the form of rituals (see Chapters 3 and 5), the inclusion of the human brain as a sixth sense for perceiving the spiritual world (see Chapter 4), or mindful introspection (Chapters 12–14). Common to all these relational practices is the rooting of conceptual activity in lived experiences, resulting in life in a living cosmos. The concepts and ideas that emerge from mindful experience are borne from a form of observation of phenomena "freed from the grim torture chamber of empiricism, mechanism and dogmatism" (Goethe). This paves the way for creating ideas as "formative forces" inherent to the multiple expressions of life in its material, biological, psychological, social, economic, cultural manifestations within the Pluriverse. Examples of how this occurs in the practice of critical sustainability sciences can be found in Chapters 3–14.

Notably, prevailing understandings of sustainable development rooted in dualism tend to view relational ontologies as obstacles to progress and development. Decolonization offers an avenue to reclaim people's right to worldviews based on relational ontologies, as part of efforts toward decolonized, depatriarchalized lifeworlds, institutions, and societies. This goes beyond using other concepts or language. It is not a matter of introducing language regulations, as occidental patterns of thought can no longer withstand the pressing socio-ecological multi-crises.

Much more is at stake. We must change the ontological grounds, and change, through this, the rules of the game, not just rename it. With respect to nature, this means nothing less than abandoning today's reigning anthropocentrism and instead striving for a bio- or cosmocentric world relationship. As shown in Chapters 3 to 7, cosmosophic knowledge refers to the wisdom of Earth, of which we are an integral part. This means overcoming the split between material and spiritual worldviews, past and present or "pre-modern" indigenous and modern Western ways of thinking.

Emancipation, autonomy, and interculturality

In order to build into the foundations of *critical sustainability sciences*, not only the manifold relational ontologies, but also to do this in an emancipatory perspective, the book offers a lot of food for thought on how emancipation is shaped by the different cultures that live according to relational ontologies. Moreover, in order to avoid relying exclusively on Western notions of emancipation, we propose constructing ideas of emancipation in a context of plurality, autonomy and exchange through intercultural dialogues.

The authors of this book come from diverse backgrounds and apply different existential and theoretical frameworks. The process of exchange and collaboration on this volume has confirmed for us the importance of a plurality of approaches. At the same time, we are also aware that simple acceptance of the diversity of existential and cognitive cultures is far from sufficient to articulate the richness of the pluriverse to which we belong. In order to go beyond domination-blind liberal multiculturalism, we believe that relationality and interdependence must be upheld as core guiding principles and processes of intra- and intercultural collaboration. As proposed in the Chapter 5 (Millar et al., 2023) and Chapter 10 (Maluf, 2023), dealing with pluriepistemical paradigms means we have to ask ourselves: How can bridges be built between ontologies, cosmovisions, cultures, ideologies, identities, histories, space, and time to strengthen sciences that promote emancipatory forms of sustainable development?

Last but not least, more attention must be paid to the art of translating between different ontologies and related scientific paradigms. This is an important part of intercultural collaboration, yet it reveals not only similarities and complementarities, but also contradictions and areas of tension. All three aspects are important elements for mutual learning and definition of possible or impossible areas of intercultural collaboration.

Key topics for advancing critical sustainability sciences

Our book concludes that three topic areas are crucial to the construction of critical sustainability sciences. These refer to realms of societal reorganization, living labor and commons, and holistic paradigms of transdisciplinary knowledge production.

Societal reorganization and relational thinking

Putting relational thinking at the heart of critical sustainability sciences enables us to overcome simplistic notions of societal organization based on substantialist dualisms, such as the individual versus the collective, economy versus politics, human resources versus natural resources, agency versus structure, etc.

How critical sustainably sciences can benefit from relational thinking on societal organization is vividly illustrated by Fraser's (2013) recommendation to transition from Polanyi's dualist notion of a "double movement" toward a more relational understanding of societal structuring, which she refers to as a "triple movement". Fraser (2013) critiques Polanyi's idea of a double movement between marketization and social protection from the perspective of feminism and social justice. For example, she describes how progressive feminist laws gradually afforded women a certain liberation from and within households. However, women's increasing participation in the labor market – previously restricted mainly to men – captured them in new forms of oppression, being characteristic of women's participation in capitalist economies. In the sense of a double movement, Fraser describes how one form of subjugation (of patriarchic households) was exchanged for another – that is, the equally patriarchal structure of private or public enterprises in capitalist markets as well as the patriarchal practices through which states protect the rights of women. She concludes that societal transformation should be based on the integration of emancipation, not as an implication of social protection, but rather as a normative orientation in its own right.

As a result, she suggests enhancing the idea of a double movement with a triple movement. This means reorienting societal organization so as to balance market forces and social protection in such a way that they enhance emancipation. In this way, the idea of a triple movement is highly relevant in critical sustainability sciences. Incorporating the factor of emancipation vis-à-vis market forces – in addition to issues of socio-ecological protection – would make visible the overlooked mechanisms of subjugation of humans and other beings (nature). Restoring emancipation in the orientation of sustainable development would reconnect sustainability discourses with their origins in environmental movements. Additionally, taking the notion of emancipation seriously would help to counter growing tendencies toward top-down, technocratic, and sometimes even eco-authoritarian modes of governance in the East and West (Shahar 2015; Kloepfer 1994). The question is not only that of which institutional mechanisms and technologies would be most effective in protecting humans and nature from exploitation, but rather that of which policies would be most effective in protecting the rights of humans and nature consistent with the *highest possible level* of emancipation and autonomy.

In a similar vein are relational approaches to understanding societal organization based on ideas of "Earth Democracy" (Shiva 2005). The concept of Earth Democracy strives to overcome "dead democracy and killing economies" (Shiva 2010). More concretely:

> Earth Democracy is both an ancient worldview and an emergent political movement for peace, justice and sustainability. Earth Democracy connects the particular to the universal, the diverse to the common, and local to the global. It incorporates what in India we refer to as *vasudhaiva kutumbkam* (the earth family) – the community of all beings supported by the earth. Native American and indigenous cultures worldwide have understood and experienced life as a continuum between human and nonhuman species and between present, past, and future generations.
> (Shiva 2005:1)

The ideas of Boff (2011) echo this perspective. Emphasizing the need to reconceptualize and broaden existing anthropocentric political constitutions in order to integrate the cosmic dimension, he proposes new "societal contracts" based on a relational ecology:

> Ecology does not deal only with questions related to green or endangered species. Ecology presupposes a new paradigm, that is, a way of organizing the set of relationships of human beings among themselves, with nature and with their meaning in this universe. It inaugurates a new alliance with creation, an alliance of veneration and fraternity. We have not been created to place ourselves above nature as those who dominate, but to be at its side as those who live together as brothers and sisters. We thus discover our cosmic roots and our earthly citizenship. Today it is not only the poor who must be freed from the captivity of a development model that denies them their dignity, squanders their resources and breaks the equilibrium elaborated over millions of years of cosmic work. The cry of the poor thus joins the cry of the Earth. The ecological experience allows a new recovery of the sacred in creation, a new image of God, a broader and more cosmic conception of the Christian mystery and a new spirituality.
>
> (Boff 2011: back of book cover, translation by authors of this chapter)

Similar points are raised by Dietschy (2023) in Chapter 7. Drawing on German philosopher Ernst Bloch's idea of a "nature alliance", Dietschy recommends enhancing sustainability sciences by means of an intercultural, transdisciplinary, and emancipatory approach. The concept of a "nature alliance" calls for re-embedding human practice in the web of nature. Of course, this is not enough to achieve the radical changes called for by critical sustainability sciences. But in conjunction with other approaches such as intercultural philosophy and decolonial thought and practices, it can help to formulate new building blocks for critical sustainability sciences rooted in relationality. Forging a non-exploitive relationship with nature also requires freeing up human capacities to transform their destructive societal relations.

Findeli (2008) demonstrates how it is possible to transform the current model of sustainability into a new, essentially relational understanding of societal organization, known as "social threefolding". It is based on three independent – but interacting – realms, namely, comprehensive economics, social equity, and human creativity. Each realm of life expresses one fundamental value of the "threefold" societal organization: fraternity is at the core of comprehensive economics; equality is the key value in democratically defined principles of a constitutional state; and freedom is the fundamental value in the cultural/creative realm of life. Here, the overall dynamic of society emerges from the *relations* between people's autonomous forms of self-organization and self-management in the three realms of life.

In Chapter 14, Findeli (2024) presents an extended version of these ideas, also demonstrating how "social threefolding" can express a cosmosophical understanding of the coevolution of human beings and other beings.

Living labor and commons

In previous sections, we highlighted the importance of transferring the principles of emancipation, liberation, and self-management into the economic sphere, not only the political sphere. From the perspective of existential and cognitive self-determination, emancipation from oppressive forces of private, state-based, or mixed forms of capitalism is fundamentally important for critical sustainability science. This task requires overcoming two negative tendencies: on the one hand, capitalist economic relations are the source of alienation of humans from themselves, from their communities, from nature, and from the cosmos; on the other, the relations imposed by capitalism are key sources of widening dualistic divisions between the human, natural, and cosmic communities.

Overcoming the oppressive forces of capitalist forms of labor appropriation starts by taking account of the fictitious character of labor, land, and money as pointed out in the "Great Transformation" by Polanyi (2001 [1944]). In this work, Polanyi carefully reviews how British feudalism was transformed into a "modern" capitalist society. He observes that *labor* (representing human activities), *land* (representing nature), and money as a means of exchange (representing social relations) were transformed into a central commodity to fuel capitalist stabilization and modernization. He concludes that these three fundamental resources (labor, land, money) of "modern" capitalist regimes of accumulation are nothing more than *fictitious* commodities. At the same time, Polanyi observes (2001 [1944]:75–76):

> Labor is only another name for a human activity which goes with life itself, which in its turn is not produced for sale but for entirely different reasons, nor can that activity be detached from the rest of life, be stored or mobilized; land is only another name for nature, which is not produced by man; actual money, finally, is merely a token of purchasing power which, as a rule, is not produced at all, but comes into being through the mechanism of banking or state finance.

For critical sustainability sciences, Polanyi's notion of fictitious commodities demands finding ways to decommodify human labor, land, and money as a central condition for truly sustainable, emancipatory societal organization. Critical sustainability sciences must therefore explore processes of labor organization beyond the current forms of capitalist appropriation. Notably, labor in its commodity form is always a derivative of "living labor", which is identical with the source of creative and concrete human activity. Living labor exists before, within, and after capital (Dussel 2018). Though Marx formulated the concept of living labor, his adherents did not link living labor with resistance to capitalist labor appropriation. It was often assumed that living labor only emerges after socialist or communist revolution.

Cabaluz and Torres López (2020) show that living labor is present in community formations in rural, peri-urban, and urban spaces in Bolivia and other

Latin American countries. They conceive of living labor as a non-capitalist form of social action and human creativity, forming the foundation of the community, though its members may also participate in capitalist wage labor. The experiences and aspirations of living labor may expand into social movements, involvement in political struggles, and collaboration with other political movements.

The concept of living labor implies applying a relational approach to work. It sees work as a series of multiple labor practices inseparable from personal relationships of kinship, friendship, and community. It emphasizes the interconnectedness of work performed in socioeconomic spheres that transcend the poles of market/non-market, formal/informal, public/private. A relational approach to work reveals that living labor is often embedded in informal spheres consisting of networks of reciprocity, kinship commitments, mutual obligations and care, moral considerations, and solidarity, in addition to formal, market, or workplace spheres (Alacovska 2018).

In Chapter 8, Räthzel (2024) demonstrates that labor – more than just being a commodity – serves as a mediator between human and non-human lives. Living labor is therefore a decisive element in shaping society–nature relationships that favor either the reproduction of capitalism or the emergence of resistance to the reduction of living labor to a dead commodity. Existence of a relational ontology that promotes resistance to dualist notions of life versus work can also foster struggles against the commodification of living labor.

Negri (2018) further points out that the concept of living labor does not contradict modern forms of production based on high degrees of division of labor:

> The general productivity of labor made a huge leap forward, but above all it intensified the social cooperation of productive subjects, because cognitive work thrives thanks to linguistic cooperation, the knowledge that makes it what it is, and of the singular innovation it produces. Thus, *the independence* of living labor grows in the face of the dead labor that wishes to organize it. Thus is imposed *the common* of cooperation.
>
> (Negri 2018:7)

In Chapter 9, Bottazzi (2024) points out that labor consists in a set of practices, institutions, and power relations, playing a major role in coevolutionary processes between societies and ecosystems. Drawing on political ecology, he proposes the concept of "regenerative work". According to the concept, human work should be considered as both "material and immaterial commons" to become regenerative. Here, human work is viewed as a physical and cognitive resource that is the product of collective effort. At the same time, human work should benefit human and non-human beings through the creation of common values of use to both, based on agreed rules.

Going further, the cosmosophical dimension of critical sustainability sciences demands broadening our analysis on behalf of a non-anthropocentric concept of labor. Analyzing living labor as an expression of all human creativity – connected to that of the cosmos – remains an unexplored, but highly relevant avenue for the development of critical sustainability sciences.

Holistic scientific paradigms

Putting relational ontologies at the center of critical sustainability sciences means advancing old and new, more holistic paradigms of scientific knowledge creation. This enables critical sustainability sciences to engage with many other rich scientific traditions laying outside or at the margins of mainstream scientific thinking, institutionalized in public and private academia.

This type of paradigm shift presents a double challenge: first, it implies finding common ground regarding what is knowable by means of the natural and social sciences, and humanities. Second, it requires freeing the spiritual dimension of the coevolution of human beings and other beings from its secular confinement in the realm of subjectivity. It means liberating the spiritual dimension of life and the world from its supposedly "illusionary, accidental, and arbitrary character". Without this "spiritualization" of scientific work on behalf of critical sustainability sciences, it will be impossible to engage with relational ontologies in a meaningful way.

A growing community of scientists and practitioners working in the field of "subtle agroecology" – dealing with the "hidden half of nature" (Wright 2021) – calls for adoption of two new ontological perspectives that reach beyond the prevailing secular-materialistic worldview:

> One is the basic existence of an invisible dimension to nature. The second concerns the nature of this dimension, as potentially involving not only light- and sound wave-based and magnetic field-based phenomena – some of which we now have the technology to capture and measure – but also more ethereal aspects of consciousness and/or spirit. The term spirit is variously interpreted, and here it is defined as a vital principle or animating force within all living things which is inextricably linked to consciousness . . . this definition of spirit holds resonance across indigenous cultures. Returning to Brooke's historical review of the relationship between science and spirituality, any conflicts between the two have, he reminds us, been overplayed and are simply a relatively recent development of Western Europe, instigated by political, social and economic factors (Brooke 1991).
>
> (Wright 2021:xxix)

This perspective on the demands of holistic scientific paradigms highlights two other aspects that should be considered by critical sustainability sciences: First, the integration of knowledge from the natural and social sciences and humanities can only be achieved based on concepts and ideas. The "natural laws" established by natural science are necessarily expressed in languages and words that refer back to concepts and ideas. This means that integration of the natural and social sciences together with the humanities requires traveling through the "hidden half nature". Indeed, even natural laws must be expressed in terms of concepts and ideas. Putting concepts and ideas at the center of new holistic science paradigms paves the way for integration of the spiritual dimension of human beings and other beings, all belonging to the Pluriverse. Even if the origins of spiritual experiences are of immaterial nature – for example, based in emotions, intuition, dreams, or visions – they become part of a shared objective world to the degree we can understand language, words, concepts, and ideas in terms of cosmosophy, as proposed in Chapter 6 by Sepulveda Pizarro (2023).

Second, integration of the spiritual dimension of human and non-human beings means accepting that human cognition unfolds as an interplay between an inner and an outer world, encompassing unique forms of reality. This means that integration of the material, social, and spiritual dimensions of ourselves only becomes possible when it is based on mindful observation of the interrelations between our inner world – consciousness, soul, and spirit – and our outer world, comprising our physical body and its physical senses as well as the rest of the material world beyond our own materiality. In Chapter 13 Brook (2023) gives a wonderful example for this, referring to Goethe's scientific method.

Woiwode et al. (2021) emphasize that reconnection of the neglected inner world with the external world is key to a more holistic perspective on sustainability. The authors emphasize how the inner world needs to be transformed for real change to happen in the outer world. The "inner transformation relates to various aspects of human existence and interactions such as consciousness, mindsets, values, worldviews, beliefs, spirituality, and human–nature connectedness" (Woiwode et al. 2021:841).

Referring to the Andean relational ontology, Choquehuanca Céspedes (2017) makes clear that the inner and outer world cannot be viewed as separate:

> We are the cosmos that thinks, we are cosmos made flesh and spirit. In the community no one has to feel free for destroying the other, whether animal or plant, because we all complement each other, because we are all sisters and brothers, raised by Mother Earth, because we all drink from the milk of Mother Earth, which is water. Human beings are similar to other living beings, because human beings are part of the natural and cosmic totality, and because in cosmic life, we are energy and life.
> (Choquehuanca Céspedes 2017:38)

In Chapter 6, Sepúlveda Pizarro (2024) expands on conceptualization of the inner and outer spaces through which the spiritual part of the world flows and interconnects with all dimensions of life in the Pluriverse. She illustrates how cosmosophy is a highly interesting point of departure for creation of the cognitive foundations of critical sustainability sciences. She demonstrates how cosmosophy lays the groundwork for approaching the presence of a wisdom in a living cosmos; cosmogenesis, understood as the wisdom of liberation in a coevolutionary process of becoming; a sacred and reverential cosmos that implies the cosmic responsibility of the human being; and, finally, a relational cosmos that is the source of the paradigm of inter-independence.

In Chapter 14, expanding on critiques of the classical tripolar concept of sustainability, Findeli (2024) suggests engaging with Steiner's ideas of social threefolding (1985 [1919]) as well as Scharmer's "new type of intelligence (metanoïa)" (2018). The latter can be understood as the result of transforming dualist, materialist, or constructivist ontology into relational ontology. This transformation endeavors to overcome merely intellectual reasoning, expressed as "the voices of judgment, cynicism, and fear". These three "negative voices" are exposed – through a mindful inner process that aims at interrelating and integrating the capacities of thinking, feeling, and willful acting – to three related "positive counterforces", called "crystallizing, prototyping, and performing" (Scharmer 2018).

Of particular importance for critical sustainability science is becoming aware that this process of transforming merely intellectual reasoning into an empirical experience and insight into the reality of cosmosophic knowledge means engaging with a fundamental "blind spot". In this inner space, encompassing the blind spot

> crystallizes the most radical and urgently required transformation needed today and tomorrow i.e. the necessity and capacity . . . to shift the inner place from which we operate, or in anthropologist Philippe Descola's terms, to get rid of our lazy certainties and of the descriptive categories [we] carry in our backpack.
>
> (Findeli 2024:282)

We posit that the "blind spot" in our inner space – where inner and outer spirituality meet – might be a crucial aspect to the construction of an area in which ontologies can begin interacting on behalf of critical sustainability sciences. The contribution of Shankar (2024) in Chapter 4 points in this direction. Based on his deep involvement in the revitalization of Ayurvedic healing and health traditions, he lays out the Ayurvedic relational ontology. It is based on the yoga knowledge system's understanding of the holistic and relational reality of nature and the universe (ontology). He concludes that Ayurvedic knowledge is similar in principle to the source of knowledge of any other knowledge tradition. Knowledge comes from the conscious use of the six natural human senses: the five known senses and the sense of mind, generally ignored by dualist Western sciences. Shankar (2024) assumes that all cultures use the same senses to discover the world. The cultural–ontological differences in how they perceive and experience themselves in the holistic world come from using these senses differently and in different social and natural settings; or also by leaving aside the mind as a true sense that allows humans to connect with the existing intangible spiritual world, similar to how the eye enables connecting with the tangible existence of light.

Facilitating exchange between the diversity of relational and other ontologies is thus not an impossible task. As suggested, a key starting point for dialogue and possible cooperation among different ontologies may emerge from examination of how different ontologies relate to the "blind spot".

The perspective of African relational ontologies also sheds light on aspects of incorporating spirituality or the mind as a sixth sense of humans. In Chapter 5, Millar et al. (2024) describe how the Dagara cosmovision and corresponding relational ontology holds that

> life is cyclical and does not end at death, a comprehensive and cyclical understanding of sustainability is at the center of their ontology. This is reflected in a popular African adage to "treat the world well; it was not given to you by your Ancestors; it was lent to you by your children". *This manifestation is the perfect state, which strives for absolute sustainability.*
> (Chapter 5, Millar et al. 2024:89).

For critical sustainability research, the culturally different ways of recognizing human reincarnation do not present an obstacle to exchange between relational and dualistic, materialist, or constructivist ontologies. Joint inquiry into the relevance of reincarnation as a possible foundation for sustainability can enrich all involved.

In Chapter 12, Zunino (2024) contributes to critical sustainability sciences with a novel social research perspective and comprehensive methodology, compatible with the principles of relational ontologies. According to the author, our prevailing knowledge system captures only one side of reality. It is limited to making rational statements about the visible material world, which ultimately remains alien to our consciousness. Based on a holistic science inspired by Goethe, Deleuze, and Guattari, the author understands the world as an organism that is in constant change. The knowledge of this world arises from a creative act performed by individuals who are at the same time part of the whole world. The author shows how critical sustainability research can combine the concepts of plane of immanence, creative forces, and social field in such a way that they all point to an understanding of the material world as a manifestation of complex forces that can be grasped through intuitive inner experiences.

In Chapter 13, Brook (2024) presents a new holistic paradigm that can be used for critical sustainability sciences. It aims at overcoming mainstream dualistic–materialistic natural science approaches. The author examines insights about the natural world from the perspective of a relational ontology. The approach was originally developed by the German poet and scientist Johann Wolfgang von Goethe in the 19th century. It was his response to the mechanistic and reductionist scientific approaches that were beginning to take hold at the time. Goethe's relational and collaborative slant on knowledge creation was based on a method he called "delicate empiricism". This relational approach to generation of knowledge about "Nature" sees it as emerging from our six senses (thus including the mind-sense) and the "blind spot", which forms part of the inner sphere of life, entailing soul and mind.

Brook (2024) maintains that the "blind spot" is not an empty space. Goethe's method of "delicate empiricism" activates the inner realm of life in which humans can connect with imagination, inspiration, and intuition. The knowledge that emerges from this goes beyond abstract and intellectual reasoning; it is based on both a personal commitment and a transformation of our very nature of thinking and being. It holds that, in our humanity, we cannot consider nature simply as a resource for ourselves. Reflection might even lead us to turn away from nature as if we were the problem. However, through a holistic science, we can become aware of how nature herself invites to teach us, together with her, about the relationships that define human and non-human beings in the Pluriverse.

In Chapters 4 and 5, Shankar (2024) and Millar et al. (2024), respectively, describe how people who believe in life after death relate differently to the Earth than those who do not. In the context of critical sustainability sciences, this means that it is not enough to know the features of a relational ontology – for example, that of the Dagara in Africa. Attention must also be paid to the practical consequences of such cosmovisions. Until recently, Dagara smallholders used compost and farmyard manure to fertilize their crops. Their worldview implies that they treat planet Earth as a living organism that can only produce healthy plants when nourished with "natural" fertilizers. This contrasts starkly with chemical-based agricultural practices that focus on nourishing plants at the expense of the living quality of the soil and are informed by a materialistic worldview promoting human domination of the earth.

From this perspective, critical sustainability sciences must draw a line from engagement with relational ontologies to practical renewal in all realms of life. As shown in Chapters 10 and 11, there are social movements whose starting points are that of defending the practical aspects of relational ontologies – for example, movements related to agroecology (Chapter 11 of Spirito et al. 2024) or struggles for "real food" in Brazil (Chapter 10, Maluf 2024). These movements have an important role to play in the creation of critical sustainability sciences. Besides obtaining recognition and support for their practical alternatives, they could benefit from intra- and interontological dialogue with other relational ontologies, thereby making the spiritual dimension of their claims more explicit. This could mean linking up with Ayurvedic, Andean, African, or Goethean ways of creating cosmosophic insights and knowledge about the interrelations of humans and non-human beings inhabiting the Pluriverse.

Broadening the ontological scope of existing, integrative fields of knowledge and disciplines belonging to academia represents another important task of critical sustainability sciences. For instance, facilitating dialogue between agroecology and relational ontologies could enhance the former via engagement with the spiritual dimension, for example, by examining the possible links of agroecology to reincarnation and cosmosophy. This, in turn, could show how agroecology might be transformed into a more expansive field of agroecosophy. This could pave the way for ontological renewal of other disciplines, for example, transformation of geography into geosophy, ecology into ecosophy and cosmosophy (see Chapter 6) – similar to the transformation of classical anthropology into anthroposophy (see Chapter 14).

References

Alacovska, A., 2018. Informal creative labour practices: A relational work perspective. *Human Relations*, 71(12), 1563–1589.

Boff, L., 2011. *Ecología: grito de la tierra, grito de los pobres*. Trotta Madrid, Traducido: Herranz, Juan Carlos Rodríguez.

Bottazzi, P., 2024. Regenerative work: From commodity to commons. In: Rist, S., Bottazzi, P., and Jacobi, J., eds. *Critical Sustainability Sciences: Intercultural and Emancipatory Perspectives*. London: Routledge.

Brook, I., 2024. Goethe's scientific method: The road not taken. In: Rist, S., Bottazzi, P., and Jacobi, J., eds. *Critical Sustainability Sciences: Intercultural and Emancipatory Perspectives*. London: Routledge.

Brooke, J. H., 1991. *Science and Religion: Some Historical Perspectives*. Cambridge: Cambridge University Press.

Cabaluz, F. D., and Torres López, T. T., 2020. El concepto de trabajo vivo desde el marxismo latinoamericano. Notas a partir de la obra de Enrique Dussel y Álvaro García Linera. *Izquierdas*, 49, 1397–1423.

Choquehuanca Céspedes, D., 2017. *Manifiesto del Vivir Bien. "Nuestro mundo es posible"*. La Paz: Canilleria del Estado Plurinacional de Bolivia.

Dietschy, B., 2024. Towards a "nature alliance" – Why sustainability must be rethought in terms of relationality. In: Rist, S., Bottazzi, P., and Jacobi, J., eds. *Critical Sustainability Sciences: Intercultural and Emancipatory Perspectives*. London: Routledge.

Dussel, E., 2018. Lebendige Arbeit. In: Haug, W. F., et al., eds. *Historisch-kritisches Wörterbuch des Marxismus 8/I*. Bonn: Rosa Luxembourg Stiftung, 746–747.

Escobar, A., 2016. Thinking-feeling with the Earth: Territorial struggles and the ontological dimension of the epistemologies of the South. *AIBR. Revista de Antropología Iberoamericana*, 11(1).

Findeli, A., 2008. Sustainable design: A critique of the current tripolar model. *The Design Journal*, 11(3), 301–322.

Findeli, A., 2024. Sustainable design: A critique of the tripolar sustainability model 15 years later. In: Rist, S., Bottazzi, P., and Jacobi, J., eds. *Critical Sustainability Sciences: Intercultural and Emancipatory Perspectives*. London: Routledge.

Fraser, N., 2013. A triple movement. *New Left Review*, 81.

Kloepfer, M., 1994. Is there the threat of an authoritarian ecological state. *European Energy and Environmental Law Review*, 3, 112.

Maluf, R., 2024. Food, food systems and sustainability: Elements of the "real food" debate in Brazil. In: Rist, S., Bottazzi, P., and Jacobi, J., eds. *Critical Sustainability Sciences: Intercultural and Emancipatory Perspectives*. London: Routledge.

Millar, D., Santuah, N., and Ba-an, M., 2024. Cosmovisions and critical sustainability sciences – An African ontology of "Vurr" (an energy) amongst the Dagara of Southwest Burkina Faso and Northwest Ghana. In: Rist, S., Bottazzi, P., and Jacobi, J., eds. *Critical Sustainability Sciences: Intercultural and Emancipatory Perspectives*. London: Routledge.

Negri, A., 2018. Starting again from marx. *Radical Philosophy*, 2(3).

Panikkar, R., 2008. *Ecosofía*. www.grupotortuga.com/Ecosofia (Accessed 18.1.2023).

Polanyi, K., 2001 [1944]. *The Great Transformation.* Boston, MA: Beacon Press.
Räthzel, N., 2024. Society – Labor – Nature: The potential of conflict. In: Rist, S., Bottazzi, P., and Jacobi, J., eds. *Critical Sustainability Sciences: Intercultural and Emancipatory Perspectives.* London: Routledge.
Rothkopf, D., and Medish, M., 2008. *Superclass: The Global Power Elite and the World They are Making.* Carnegie Endowment for International Peace. Transcript by Federal News Service, Washington, DC.
Scharmer, O., 2018. *The Essentials of Theory U: Core Principles and Applications.* Berrett-Koehler Publishers.
Sepúlveda Pizarro, J., 2024. Contributions of the notion of cosmosophy to the formulation of critical sustainability sciences. In: Rist, S., Bottazzi, P., and Jacobi, J., eds. *Critical Sustainability Sciences: Intercultural and Emancipatory Perspectives.* London: Routledge.
Shahar, D. C., 2015. Rejecting eco-authoritarianism, again. *Environmental Values*, 24(3), 345–366.
Shankar, D., 2024. Relational Ontologies in health sciences and practices in India. In: Rist, S., Bottazzi, P., and Jacobi, J., eds. *Critical Sustainability Sciences: Intercultural and Emancipatory Perspectives.* London: Routledge.
Shiva, V., 2005. *Earth Democracy: Justice, Sustainability and Peace.* London: Zed Books.
Shiva, V., 2010. Earth democracy: Beyond dead democracy and killing economies. *Capitalism Nature Socialism*, 21(1), 83–95.
Shiva, V., 2024. Foreword. In: Rist, S., Bottazzi, P., and Jacobi, J., eds. *Critical Sustainability Sciences: Intercultural and Emancipatory Perspectives.* London: Routledge.
Spirito, F., Blanco, V., and Montalba, R., 2024. Agroecology as a transformative approach to sustainable food systems. In: Rist, S., Bottazzi, P., and Jacobi, J., eds. *Critical Sustainability Sciences: Intercultural and Emancipatory Perspectives.* London: Routledge.
Steiner, R., 1985 [1919]. *Renewal of the Social Organism.* Interlochen, MI: SteinerBooks.
Woiwode, C., et al., 2021. Inner transformation to sustainability as a deep leverage point: Fostering new avenues for change through dialogue and reflection. *Sustainability Science*, 16(3), 841–858.
Wright, J., 2021. *Subtle Agroecologies: Farming with the Hidden Half of Nature.* London, New York: CRC Press.
Zunino, H., 2024. Through the veil: A relational and participatory perspective to knowledge production and sustainability. In: Rist, S., Bottazzi, P., and Jacobi, J., eds. *Critical Sustainability Sciences: Intercultural and Emancipatory Perspectives.* London: Routledge.

Index

For Product Safety Concerns and Information please contact our EU representative GPSR@taylorandfrancis.com Taylor & Francis Verlag GmbH, Kaufingerstraße 24, 80331 München, Germany